IMMUNOLOGY

IMMUNOLOGY

I KANNAN
(General Secretary,
Indian Association of Applied Microbiologists)
Lecturer and Head In-charge
Department of Microbiology
Madras Christian College
Chennai-600 059

MJP PUBLISHERS
Chennai 600 005

ISBN 978-81-8094-290-7 (HB)

MJP Publishers
No. 44, Nallathambi Street,
Triplicane,
Chennai 600 005

MJP 023

Publisher : C. Janarthanan

To my son

K Santhosh Kumar

PREFACE

The field of immunology emerged only at the beginning of the 20th century, yet it has developed profoundly than any other field. Immunology is an interdisciplinary subject encompassing a number of other fields like microbiology, biochemistry, molecular biology, biotechnology, zoology, clinical medicine, pathology, etc. Hence an attempt has been made in this book to cover almost all the important areas of immunology.

The approach towards the subject is changing gradually according to growing needs in the various other fields. Currently more emphasis is given to the molecular aspects of immunology. This has opened the floodgates towards the new approach for the treatment and diagnosis of many diseases including cancer. Bearing all these aspects in mind, apart from the conventional chapters, various new chapters like immunological synapse, molecular immunology, immunohistochemistry, tumour immunology, etc. have been included. Thus this book comprehensively covers all basic and advanced aspects of immunology in a simple manner in order to keep in pace with the modern developments in this field. It provides up-to-date information with necessary illustrations that will help students to understand the recent concepts in every area of immunology.

As the subject of immunology occupies an important position in the entire field of life sciences, I am extremely confident that this book will be helpful to all the students and teachers in the field of life sciences.

I. Kannan

ACKNOWLEDGEMENT

First and foremost, it is my duty to thank my mentor and microbiology guru, Prof. Dr. P. Rajendran, Department of Microbiology, AL Mudaliar Postgraduate Institute of Basic Medical Sciences, who was not only responsible for bringing me to this fascinating field of immunology but also for providing his constant encouragement and support to sustain in this field.

I will be failing in my duties if I do not thank my family members, especially my wife Dr. K. Cecilia, who herself is a microbiologist and thus was able to support me a lot in this endeavour. I should thank my son Master K. Santhosh Kumar for sparing his valuable time with me to write this book.

I should not forget the support of Dr. Alexander Mantramurthi, former principal, and Prof. Winfred Chelliah, former co-ordinator, SFS of Madras Christian College, in all my endeavours.

I should thank profusely Dr. C. Livingstone, Professor and Head, Department of Botany, Madras Christian College, a man of great wisdom from whom I learnt the skill of hardwork.

I would like to place in record my gratitude to Dr. Major M. Jailani, Principal, Mohamed Sathak College of Arts and Science who moulded my career and gave me the confidence that nothing is impossible in life.

I am grateful to Dr. V.J. Philip, Principal, and Prof. J. Chandradas, Co-ordinator, SFS, Madras Christian College for their constant support and encouragement.

I cherish my association with my colleagues Mrs. V. Mahalakshmi, Dr. S. Niren Andrew, Dr. Preethi,

Mrs. K. Rohini, Mr. K.V. Premkumar and Ms K. Kavitha for their constant support in the day-to-day affairs of the department thereby enabling me to find time to concentrate on preparing this manuscript.

I must avow my gratitude to MJP Publishers, for their patience and constant guidance in the preparation of this manuscript.

I owe my gratitude to all the scientists in the field of immunology without whose contribution, this book would not have been possible.

Last but not least, I thank God the Almighty for making my maiden attempt a reality.

I. Kannan

CONTENTS

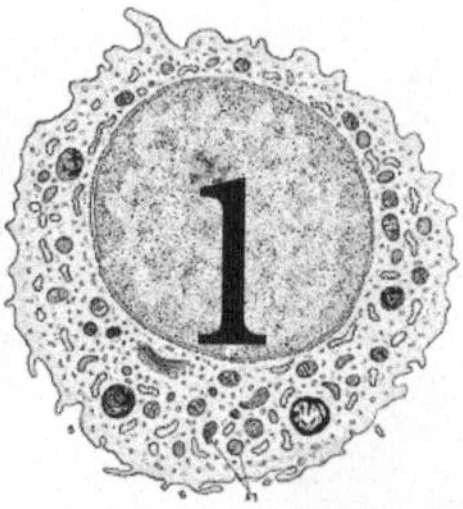

HISTORY OF IMMUNOLOGY

The history of immunology can be traced back to the end of the 18th century when **Edward Jenner** in 1796, discovered that cowpox can induce protection against human smallpox, a fatal disease of those days. However, Jenner knew nothing about the germ theory of disease or about the immunity developed against it.

Edward Jenner

It was in the late 19th century that the gates of immunology opened with the discovery of **Robert Koch** who proved that the infectious diseases are caused by microorganisms. The discovery of Koch and strategy of Jenner's vaccination made **Louis Pasteur** to device a vaccine against chicken cholera in the 1880s. Pasteur discovered the methodology to prepare rabies vaccine (the first human vaccine to be prepared). When his first trial of rabies vaccination in a boy bitten by a rabid dog became successful, the search for the mechanism of immunity started all over the world.

Robert Koch

Louis Pasteur

In the same period **Elie Metchnikoff** proposed the hypothesis of phagocytosis. According to him many microorganisms could be engulfed and destroyed by a group of specialized cells in the body. He further proposed that these phagocytic cells move to the site of infection and effectively engulf the infectious microorganisms. Metchnikoff continued his work on phagocytosis at Pasteur Institute Paris. In 1908, he received the Nobel Prize for his contribution to the study of the mechanism of inflammation along with **Paul Ehrlich**.

Elie Metchnikoff

Paul Ehrlich

Even though Metchnikoff's cellular based immunity was well developed, some scientists were able to find some soluble factors which are also involved in immunity. The soluble factors were named humoral factors. It was in 1903 that **Almroth Wright** and **Stewart Douglas** were able to find out that a humoral component could render bacteria susceptible to phagocytosis. They named the factor opsonin.

In 1890, **Emil Von Behring** and **Shibasaburo Kitasato** did experiments on toxins of diphtheria and tetanus. They inoculated these toxins into animals to produce a neutralizing antitoxin serum. This marked the birth of passive immunization, a vaccination technique still practised in modern medicine.

In 1894, **Pfeiffer** worked on *in vivo* cytolysis of *Vibrio cholerae* and found that cytolysis was mediated by a serum factor namely complement. Thus Pfeiffer elucidated the complement-mediated cytolysis and this was further extended by **Butchner** and **Jules Bordet.**

Another important breakthrough in immunology came in the early 1900s with the discovery of blood groups by **Karl Landsteiner**. Thus the transfusion of blood in man became possible. This finding proved to be important not only in immunology but also in field of medicine.

Karl Landsteiner

The 20th century marked an important era in the field of immunology. Many discoveries were made in the different facets of immunology (Table 1.1). The field still seems to be wide open for many more findings in the years to come.

Table 1.1 Milestones in the history of immunology

Year	Scientist	Discovery
1798	Edward Jenner	Smallpox vaccination
1877	Paul Ehrlich	Mast cell recognition
1879	Louis Pasteur	Attenuated chicken cholera vaccine development
1883	Elie Metchnikoff	Phagocytosis
1885	Louis Pasteur	Rabies vaccination development
1888	Pierre Roux and Alexandre Yersin	Bacterial toxins
1888	George Nuttall	Bactericidal action of blood
1891	Robert Koch	Delayed type hypersensitivity
1894	Richard Pfeiffer	Bacteriolysis
1895	Jules Bordet	Complement and antibody activity in bacteriolysis
1900	Paul Ehrlich	Antibody formation theory
1901	Karl Landsteiner	A, B and O blood groupings
1901–08	Carl Jensen and Leo Loeb	Transplantable tumours
1902	Paul Portier and Charles Richet	Anaphylaxis
1903	Almroth Wright and Stewart Douglas	Opsonization reactions
1906	Clemens von Pirquet	Coined the word "allergy"
1907	Svante Arrhenius	Coined the term "immunochemistry"
1910	Emil von Dungern and Ludwik Hirszfeld	Inheritance of ABO blood groups
1910	Peyton Rous	Viral immunology theory
1914	Clarence Little	Genetic theory of tumour transplantation
1915–20	Leonell Strong and Clarence Little	Inbred mouse strains
1917	Karl Landsteiner	Haptens

(Contd.)

Table 1.1 (Continued)

Year	Scientist	Discovery
1921	Carl Prausnitz and Heinz Kustner	Cutaneous reactions
1924	L Aschoff	Reticuloendothelial system
1926	Lloyd Felton and GH Bailey	Isolation of pure antibody preparation
1934–38	John Marrack	Antigen–antibody binding hypothesis
1936	Peter Gorer	Identification of the H-2 antigen in mice
1940	Karl Landsteiner and Alexander Weiner	Identification of the Rh antigens
1941	Albert Coons	Immunofluorescence technique
1942	Jules Freund and Katherine McDermott	Adjuvants
1942	Karl Landsteiner and Merill Chase	Cellular transfer of sensitivity in guinea pigs (anaphylaxis)
1944	Peter Medwar	Immunological hypothesis of allograft rejection
1948	Astrid Fagraeus	Demonstration of antibody production in plasma B cells
1948	George Snell	Congenic mouse lines
1949	Macfarlane Burnet and Frank Fenner	Immunological tolerance hypothesis
1950	Richard Gershon and K Kondo	Discovery of suppressor T cells
1952	Ogden and Bruton	Discovery of agammaglobulinaemia (antibody immunodeficiency)
1953	Morton Simonsen and WJ Dempster	Graft-versus-host reaction
1953	James Riley and Geoffrey West	Discovery of histamine in mast cells

(Contd.)

Table 1.1 (Continued)

Year	Scientist	Discovery
1953	Rupert Billingham, Leslie Brent, Peter Medawar, and Milan Hasek	Immunological tolerance hypothesis
1955–59	Niels Jerne, David Talmage and Macfarlane Burnet	Clonal selection theory
1957	Ernest Witebsky *et al.*	Induction of autoimmunity in animals
1957	Alick Isaacs and Jean Lindemann	Discovery of interferon (cytokine)
1958–62	Jean Dausset *et al.*	Human leucocyte antigens
1959	James Gowans	Lymphocyte circulation
1959–62	Rodney Porter *et al.*	Discovery of antibody structure
1961–62	Jaques Miller *et al.*	Discovery of thymus involvement in cellular immunity
1961–62	Noel Warner *et al.*	Distinction of cellular and humoral immune responses
1963	Jaques Oudin *et al.*	Antibody idiotypes
1964–68	Anthony Davis *et al.*	T and B cell cooperation in immune response
1965	Thomas Tomasi *et al.*	Secretory immunoglobulin antibodies
1967	Kimishige Ishizaka *et al.*	Identification of IgE as the reaginic antibody
1971	Donald Bailey	Recombinant inbred mouse strains
1974	Rolf Zinkernagel and Peter Doherty	MHC restriction
1975	Kohler and Milstein	Monoclonal antibodies
1984	Robert Good	Failed treatment of severe combined immunodeficiency (SCID, David, the bubble boy) by bone marrow grafting

(Contd.)

Table 1.1 (Continued)

Year	Scientist	Discovery
1985	Tonegawa, Hood *et al.*	Identification of immunoglobulin genes
1985 Onwards		Rapid growth in identification of genes for immune cells, antibodies, cytokines and other immunological structures
1985–87	Leroy Hood *et al.*	Identification of genes for the T-cell receptor
1990	Yamamoto *et al.*	Molecular differences between the genes of O and A blood groups between those of A and B blood groups
1990	NIH team	Gene therapy for SCID using cultured T cells
1993	NIH team	Treatment of SCID using genetically altered umbilical cord cells

NOBEL LAUREATES IN THE FIELD OF IMMUNOLOGY

- 1901—E.A. Von Behring (Germany), for the work on serum therapy especially its application against diphtheria.
- 1905—R. Koch (Germany), for the investigations concerning tuberculosis.
- 1908—E. Metchnikoff (Russia) and P. Ehrlich (Germany), for their work on immunity (phagocytosis/cellular theory and humoral theory respectively).
- 1913—C.R. Richet (France), for the work on anaphylaxis.
- 1919—J. Bordet (Belgium), for the discoveries relating to immunity (complement).
- 1930—K. Landsteiner (Austria/USA), for the discovery of human blood groups.

- 1951—M. Theiler (South Africa), for the discoveries and developments concerning yellow fever.
- 1957—D. Bovet (Italy/Switzerland), for the discoveries related to histamine and compounds which inhibit action of histamine and other substances on the vascular system and the skeletal muscles.
- 1960—Sir F. McFarlane Burnet (Australia) and Sir P.B. Medawar (Great Britain), for the discovery of acquired immunological tolerance.
- 1972—G.M. Edelman (USA) and R.R. Porter (Great Britain), for their discovery concerning the chemical structure of antibodies.
- 1977—R. Yalow (USA), for the development of radioimmunoassays of peptide hormones.
- 1980—B. Benacerraf (USA), J. Dausset (France) and G.D. Snell (USA), for their discoveries concerning genetically determined structures on the cell surface (major histocompatibility complex) that regulate immunological reactions.
- 1982—S. K. Bergstrom (Sweden), B. I. Samuelsson (Sweden) and J. R. Vane (UK), for their discoveries concerning prostaglandins and related biologically active substances.
- 1984—N.K. Jerne (Denmark/Switzerland) for theories concerning the specificity in development (lymphocyte clonality) and control of the immune system; G.J.F. Köhler (Germany/Switzerland) and C. Milstein (Argentina/Great Britain), for the discovery of the principle for production of monoclonal antibodies.
- 1987—S. Tonegawa (Japan/USA), for the discovery of the genetic principle for generation of antibody diversity.
- 1990—J.E. Murray and E.D. Thomas (USA), for their discovery concerning organ and cell transplantation in the treatment of human diseases.

- 1996—P.C. Doherty (Australia/USA) and R.M. Zinkernagel (Switzerland), for their discoveries concerning the specificity of the cell-mediated immune defence ("dual recognition").
- 1997—S.B. Prusiner (USA), for the discovery of prions as a new biological principle of infection.
- 1999—G. Blobel (USA), for discoveries concerning signal transduction.

POINTS TO REMEMBER

- Developments in the field of immunology started in the late 19th century with Robert Koch's discovery that infectious diseases are caused by microorganisms.
- The discovery of Koch and strategy of Jenner's vaccination made Louis Pasteur to device a vaccine against chicken cholera in the 1880s.
- Elie Metchnikoff proposed a hypothesis of phagocytosis. According to him many microorganisms could be engulfed and destroyed by a group of specialized cells in the body.
- It was in 1903 that Almroth Wright and Stewart Douglas discovered a humoral component that could render bacteria susceptible to phagocytosis. They named the factor opsonin.
- Pfeiffer elucidated the complement-mediated cytolysis and this was further extended by Butchner and Jules Bordet.
- In the early 1900s Karl Landsteiner discovered blood groups.
- The 20th century marked an important era for the field of immunology.

REVIEW QUESTIONS

1. Write short notes on:
   i. Louis Pasteur
   ii. Edward Jenner
   iii. Robert Koch
   iv. Elie Metchnikoff
   v. Karl Landsteiner
2. Describe the developments in the field of immunology during the nineteenth century.
3. Why is the 20th century considered important in the history of immunology?

CELLS OF THE IMMUNE SYSTEM

INTRODUCTION

The cells of the immune system are so adapted that some of them mediate specific immunity and some can get involved in non-specific immunity. The cells of the immune system are derived from the pluripotent stem cells in the bone marrow (Figures 2.1 and 2.2). They follow either of the following two cell lineages:

- Myeloid cell line
- Lymphoid cell line

The myeloid cells mediate the non-specific immune response which is mostly phagocytic in nature and they include:

- Monocytes
- Macrophages
- Granulocytes

On the other hand the lymphoid cells are mostly involved in specific immune response. They are:

- T lymphocytes

- B lymphocytes
- Large granular lymphocytes

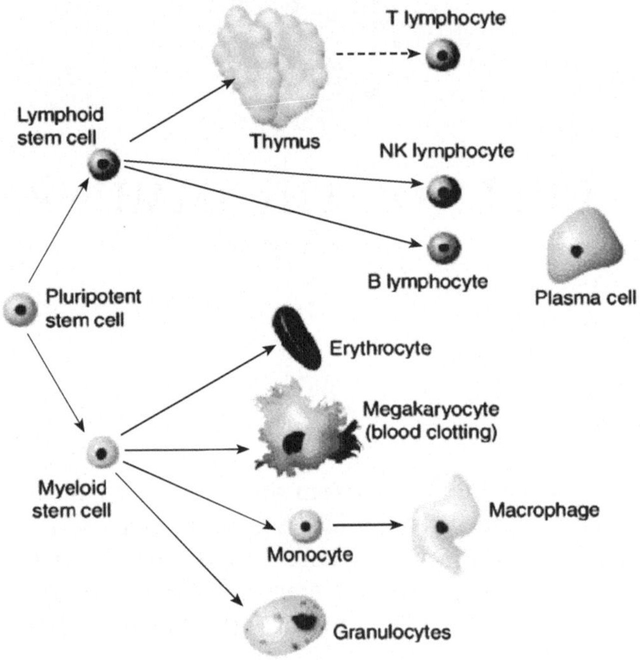

Figure 2.1 Origin of various immunocompetent cells

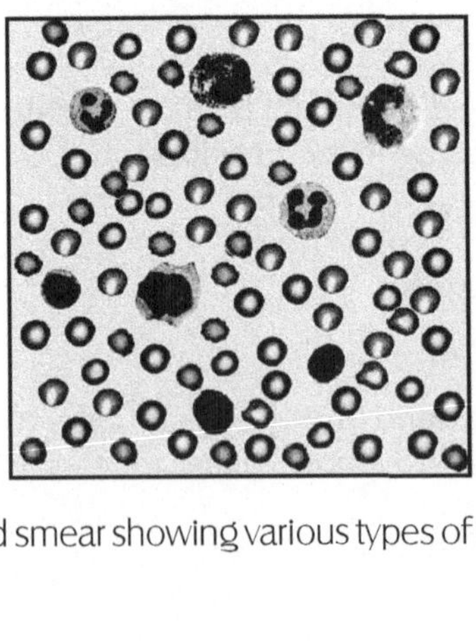

Figure 2.2 Blood smear showing various types of immunocompetent cells

MYELOID CELLS

The myeloid cells are involved in non-specific immunity, i.e., they can act on any foreign substance. There are different types of myeloid cells.

Monocytes

They are normally referred to as the **mononuclear phagocytic cells**, found as free circulating cells in the bloodstream (Figure 2.3). They constitute approximately 4–10% of the nucleated cells in blood. The monocyte has a diameter of 12 to 17 μm with a characteristic horseshoe-shaped nucleus and cytoplasmic azurophilic granules. The monocytes are the precursors of the tissue-bound macrophages. They migrate to the tissues and become macrophages.

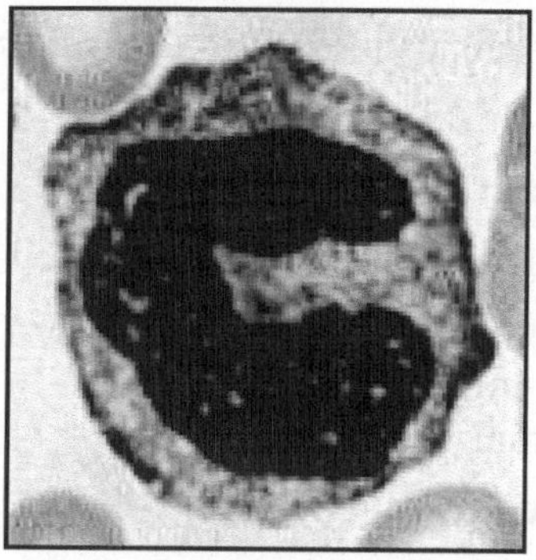

Figure 2.3 Monocyte

Macrophages

Normally the monocytes circulate in the blood for about 8–10 hours, get enlarged and migrate into the tissues to become the tissue-specific macrophages (Figure 2.4). After migration, they undergo a lot of changes in their structure. The cell gets enlarged and the cell organelles show increase in their size and number. Most importantly, the phagocytic property of the cell is increased.

The macrophages and dendritic cells do not possess specific antigenic receptors. However they participate in the immune response together with lymphocytes. These non-lymphoid cells are thus called as "accessory cells".

Macrophages contain azurophilic lysosomal granules that contain myeloperoxidase, lysozyme, acid hydrolases such as β-glucuronidase, phosphatase, etc.

Macrophages can actively participate in phagocytosis of various pathogens and tumour cells. The tissue macrophages will either get fixed to it or may be wandering phagocytes that can travel by amoeboid movement throughout the tissues. They are called by different names in different tissues. Some examples, are given below.

- **Histiocytes** in loose connective tissue
- **Kupffer cells** in the liver
- **Osteoclasts** in bone
- **Microglial cells** in nervous tissue like the brain
- **Langerhans' cells** in the epidermis (they recognize antigens, ingest them and present them to lymphocytes for eventual destruction)
- **Glomerular mesangial cells** in the kidney
- **Pulmonary alveolar macrophages** in the lungs
- **Macrophages** in the spleen and lymph nodes
- **Monocytes** in the blood

The macrophages can be activated by various stimuli. The phagocytosis itself forms an important stimulus. Further, the macrophages can be activated by **interferon gamma (IFN-γ)**, an interleukin secreted by a type of lymphoid cell called T helper cells. Further the macrophages can also be activated by the components of the bacterial cell wall.

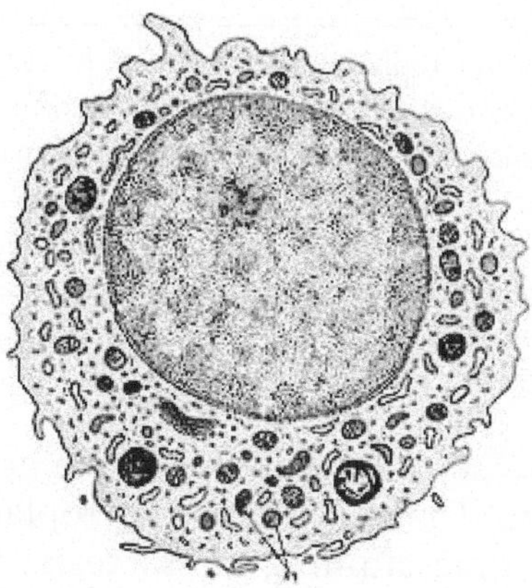

Figure 2.4 Macrophage

The activated macrophages are very effective in eliminating the engulfed pathogen and secrete the inflammatory mediators required for further immune response.

The activated macrophages express high level of MHC class II molecules, the ones that are required to present the antigen to the T helper cells, which is a prerequisite for immune response. Due to this activity the macrophages are referred to as **"antigen-presenting cells"**.

The activated macrophages can also secrete soluble factors such as **tumour necrosis factor alpha (TNF-α)** that can kill a variety of cells including the tumour cells.

Macrophages express certain characteristic molecules on their cell surface, which can be identified by specific monoclonal antibodies. A systematic nomenclature has been attributed to those molecules, in which CD stands for **cluster of differentiation**. They are assigned with different numbers according to the monoclonal antibodies to which they respond. The different **CD markers** of the macrophages are **CD64, CD32** and **CD16** which acts as Fc receptor of the antibody. The macrophages also express complement receptor molecule **CD35.**

Inflammatory macrophages are present in various exudates. They may be characterized by various specific markers, e.g. peroxidase activity, and since they are derived exclusively from monocytes they share similar properties. The term exudate macrophages designate the developmental stage and not the functional state.

Granulocytes

Granulocytes are so called because of the granules present in their cytoplasm. These cells have cytoplasmic differences resulting in unique staining characteristics. This staining characteristic differentiates these cells into three types. They are:

1. **Neutrophils** They have a multilobed nucleus, granulated cytoplasm and stain both with acid and basic dyes.
2. **Eosinophils** They have a bilobed nucleus and a granulated cytoplasm, and stain with acidic dyes such as eosin and hence its name.
3. **Basophils** They have a lobed nucleus and are heavily granulated. They stain with basic dyes like methylene blue.

Neutrophils The neutrophils are the predominant leucocytes in blood contributing to about 70% of the nucleated cells and 90% of the granulocytes (Figure 2.5). As they possess a multilobed nucleus, they are normally referred to as **polymorphonuclear leucocytes (PMN)**. The granules of the PMN contain various enzymes responsible for its phagocytic activity. They are the important cells seen predominantly at the site of inflammation. The substances that are secreted during the inflammatory reaction serve as chemotactic factors that leads to the accumulation of neutrophils at the site of inflammation.

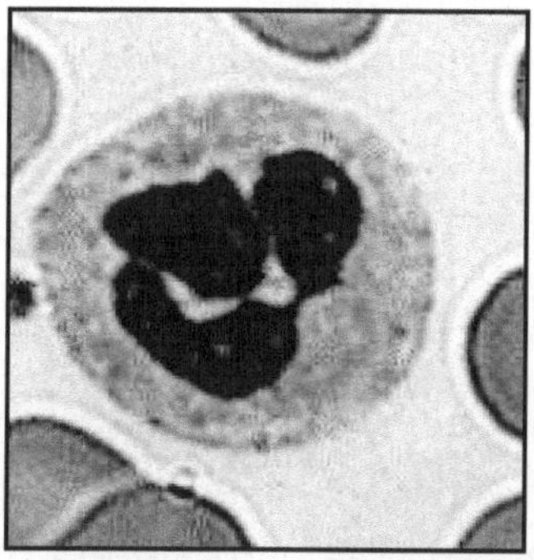

Figure 2.5 Neutrophil

The process by which the circulating neutrophils enter into the tissue space is called **extravasation**. This mainly involves penetration of cell through the vascular endothelium (blood vessel wall). The first step in extravasation is the adhesion of cells to the endothelium at the site of inflammation. Then it penetrates into the gap in between the two endothelial cells to reach the vascular basement membrane. Finally it penetrates and moves into the tissue space.

The neutrophils are active phagocytic cells. They have two types of granules, viz. primary granules and secondary granules. The primary granules are like lysosomes, larger in size and contain peroxidase, lysozyme and various other hydrolytic enzymes. The secondary granules are smaller in size and contain enzymes like collagenase, lactoferrin and lysozyme. When the foreign substance is engulfed by the neutrophils, it is enclosed in a sac-like structure called phagosomes. Then the primary and secondary granules fuse with this phagosome and eliminate the foreign substance.

The neutrophils also express certain molecules on their cell membranes. They are CD35, a complement receptor, and CD16, the receptor for Fc portion of immunoglobulin.

Eosinophils A normal, non-allergic individual will have about 2 to 5% of eosinophils in the leucocyte population (Figure 2.6).

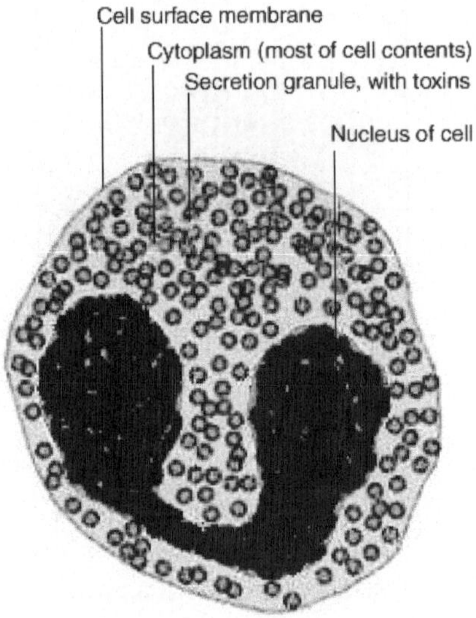

Figure 2.6 Eosinophil

The percentage may increase in allergic individuals. These cells can be stained with an acid stain like eosin. Even though eosinophils are phagocytic in nature their role in phagocytosis is less significant. However, it has been understood that they are involved in the immunity against parasitic infections. They kill helminthes by releasing chemical mediators stored in the granules as they cannot phagocytose them since they are larger organisms.

Basophils The basophils are less in number when compared to other granulocytes (Figure 2.7). They contribute just 0.2% to the total leucocyte population. They are heavily granulated and the granules contain mainly histamine and other vasoactive substances. They possess receptors for a unique class of immunoglobulins, viz. IgE which mediate allergic reactions. Thus the basophils are important cells involved in allergic reactions. The stimulus caused by the allergic substances to IgE-attached basophils will release the chemical mediators which are responsible for the allergic reactions.

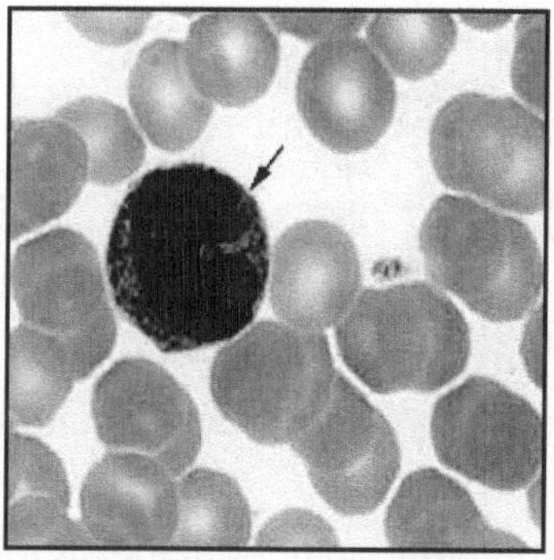

Figure 2.7 Basophil

LYMPHOID CELLS

About 20% of leucocytes are lymphocytes. Two morphologically distinct lymphocytes can be observed in the blood. The first type is the small agranular lymphocyte having high nuclear-to-cytoplasm ratio (N/C). These small lymphocytes are further divided into two types based on their function. They are T lymphocytes and B lymphocytes. The second type is a large cell, which is granulated, and possesses low nuclear-to-cytoplasmic (N/C) ratio. They are called as large granulated lymphocytes (LGLs). Lymphocytes normally possess specific receptors for antigens and thus mediate specific immunity (Figure 2.8).

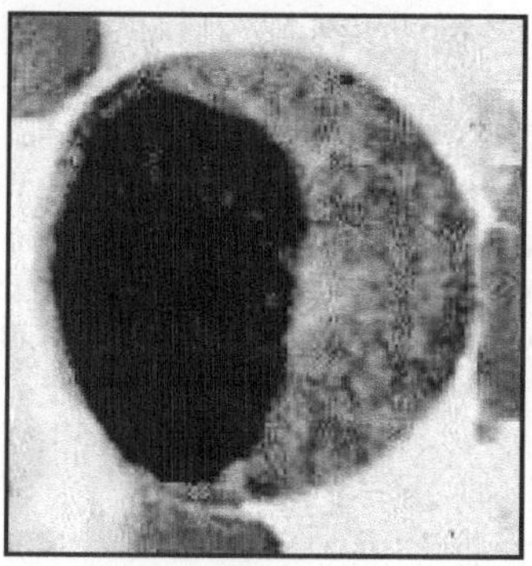

Figure 2.8 Lymphocyte

T Lymphocytes

The T lymphocytes are found in the bone marrow as **pre T lymphocytes**. Then they enter the **thymus** to become matured T lymphocytes. Most of the circulating lymphocytes are T lymphocytes and contribute to about 80% of the lymphocyte population.

The T cells play two important functions—**effector** and **regulatory**.

The effector function includes cytolysis of cells infected with microbes and tumour cells and lymphokine production. The regulatory functions are either to increase or to suppress other lymphocytes and accessory cells.

T cells are of two types (Table 2.1):

T helper lymphocytes (T_H cells) that help B lymphocytes to produce antibodies and help phagocytes to destroy ingested microbes.

Cytolytic or cytotoxic lymphocytes (CTLs or T_C cells) that kill cells harbouring intracellular microbes.

These two types of T cells show differences in their surface markers. The T_H cells express **CD4 marker** on their cell surface, whereas the T_C cells express **CD8 marker** on their cell surface. Thus the T_H cells will be referred to as **CD4 cells** and T_C cells as **CD8 cells**.

> Monoclonal antibodies clustered as CD4 detect most thymocytes and a subpopulation of peripheral blood T cells, called T helper cells. In addition, CD4 is expressed on monocytes and macrophages. The CD4 antigen is a 55-kDa glycoprotein which plays a role in the recognition of foreign antigens presented to T cells by MHC class II molecules. Furthermore, CD4 acts as a receptor for HIV-1 by binding the viral protein gp120.

Apart from the above markers, all the T cells possess two important markers namely **CD2** and **CD3**. Further all the T cells have antigen receptor on their cell surface. This is called as **T-cell receptor** or **TCR**.

The **CD2 molecule** acts as the **sheep RBC receptor**. The CD3 molecule will be seen associated with the TCR. The CD4 molecule acts as the receptor for MHC class I molecule.

Thus from the above discussion it is evident that the two T cell subset expresses two different CD markers especially CD4 and CD8 and are functionally distinct.

Helper T cells are capable of influencing the action of a variety of immune cells, and the response generated (including the extracellular signals such as cytokines) is essential for a successful outcome from infection.

Proliferating helper T cells that develop into effector T cells differentiate into two major subtypes of cells known as T_H1 and T_H2 cells (also known as type 1 and type 2 helper T cells respectively). These subtypes are defined on the basis of the specific cytokines they produce. T_H1 cells produce interferon-gamma (or IFN-gamma) and lymphotoxin (also known as tumour necrosis factor-beta or TNF-beta), while T_H2 cells produce interleukin-4 (IL-4), interleukin-5 (IL-5) and interleukin-13 (IL-13), among numerous other cytokines. The interleukin-12 (IL-12) plays an essential role during T_H1 development, but IL-12 is not produced by helper T cells, but rather by certain professional APCs, such as activated macrophages and dendritic cells. Interleukin-2 is associated with T_H1 cells, and its production by helper T cells is necessary for the proliferation of cytotoxic $CD8^+$ T cells, but this association with T_H1 may be misleading; IL-2 is produced by all helper T cells early in their activation.

It has been suggested that both T_H groups play separate roles during an immune response. **T_H1 cells** are necessary in maximizing the killing efficacy of the macrophages and in the proliferation of cytotoxic $CD8^+$ T cells, therefore their primary role during an immune response is to activate and proliferate these cells. T_H2 cells express many cytokines, many of which are essential in stimulating B cells with antibody class switching and increased antibody production; T_H2 cells are therefore considered necessary for the full maturation of the humoral immune system.

> Monoclonal antibodies clustered as CD8 detect most thymocytes and a subpopulation of both NK cells and peripheral blood T cells. CD8 acts as a co-receptor with the TCR in recognizing antigens presented by MHC class I and plays a role in the T-cell mediated immune response. CD8 monoclonal antibodies are commonly used in routine immunomonitoring and in the determination of CD4/CD8 ratios in HIV/AIDS patients.

A cytotoxic T cell (or T_C) has antigen receptors on its surface, which can bind to fragments of antigens displayed by the class I MHC molecules, somatic cells and tumour cells infected by virus or other intracellular pathogen.

Once activated by an MHC-antigen complex, T_C cells release the cytotoxins perforin and granulysin, which form pores in the target cell's plasma membrane. This causes ions and water to flow into the target cell, making it expand and eventually lyse. T_C also releases granzyme, a serine protease, that can enter target cells via the perforin-formed pore and induce apoptosis (cell death).

A second way to induce apoptosis is through an interaction between cell-surface molecules on the T_C and the infected cell. When a T_C is activated, it starts to express the surface cytokine–Fas ligand, which can bind to Fas molecules on the target cell. This Fas–Fas-ligand interaction is the main route to dispose unwanted T lymphocytes during their development.

Most T_C cells have the protein CD8 present on the cell surface, which is attracted to portions of the class I MHC molecule. This affinity keeps the T_C cell and the target cell bound closely together during antigen-specific activation. T_C cells with CD8 surface protein are called $CD8^+$ T cells.

B Lymphocytes

They comprise only 5% to 15% of the total lymphocyte population. They are very important in **antibody-mediated immunity** as they secrete specific immunoglobulins in response to antigenic stimulus. The B cells form and mature in the bone marrow. Apart from secreting immunoglobulins these cells express surface immunoglobulin (sIg), a unique feature of the B cells. The surface immunoglobulin on the majority of the cells is monomeric IgM and IgD. Some B cells may express other classes of immunoglobulins. The B cells associated with mucosal-associated lymphoid cells express IgA class immunoglobulin. These surface immunoglobulins act as the antigen receptor sites of the B cells.

The other markers that are expressed on mature B cells are CD16, an Fc receptor of immunoglobulin and CD11b and CD35, the complement receptors. Further B cells also can express MHC class II molecules.

The B cells are of two subsets. They are:

1. **T-cell-independent cells** which do not require the help of T_H cells for the production of immunoglobulins. They generally recognize multimeric sugar or lipid antigens of microbes.
2. **T-cell-dependent cells** which require the help of T_H cells for the production of immunoglobulins. They are considered to be the conventional B cells and respond to a wide variety of antigens.

The differentiation of B cells into "plasma cells" (Figure 2.9) occurs as the cell continues to divide in the presence of cytokines. Different cytokines are known to stimulate B cells to become plasma cells secreting different classes of antibodies such as IgG, IgA, IgM, etc. (Figure 2.10). This change in antibody production from surface IgM to other classes is called **"class switch"**.

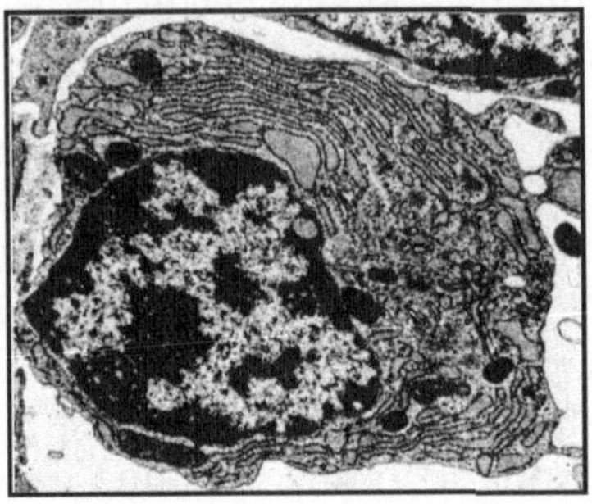

Figure 2.9 Plasma cell

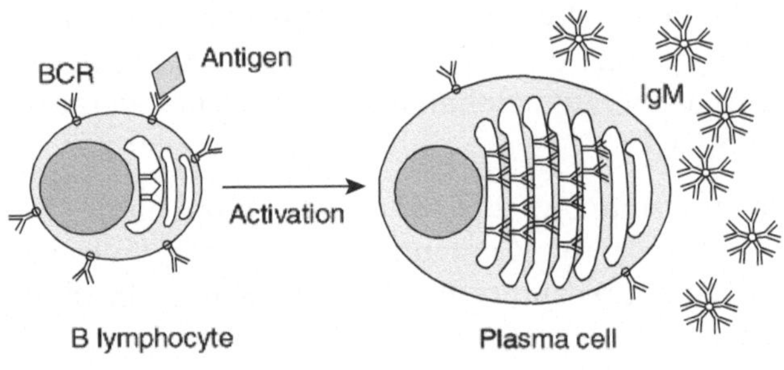

Figure 2.10 Development of plasma cell from the B cell

Plasma cells are the final stage of development of B cells which have recognized antigen and been stimulated by T cell-derived cytokines. These plasma cells reside in the spleen and lymph nodes and secrete the antibodies found in the circulation. Therefore, antibody is secreted in response to antigen. Plasma cells lose their capacity to divide. Some plasma cells live for 2 to 3 days while others continue to produce antibodies for several weeks.

Waldever introduced the term "plasma cell" in 1875. Ramon Y. Cajal accurately characterized plasma cells in 1890 in a study of syphilitic condylomas. He stated that the unstained perinuclear area contained the Golgi apparatus. In 1895, Marschalko described the essential characteristics of plasma cells, including the blocked chromatin, an eccentric position of the nucleus, a perinuclear pale area and a spherical or irregular cytoplasm.

Plasma cells are large lymphocytes with a large cytoplasm-to-nucleus ratio and a characteristic appearance on light microscopy. They have basophilic cytoplasm and an eccentric nucleus with heterochromatin in a characteristic **cartwheel** arrangement. Their cytoplasm also contains a pale zone which on electron microscopy shows an extensive Golgi apparatus and centrioles. Abundant rough endoplasmic reticulum combined with a well-developed Golgi apparatus makes plasma cells well-suited for secreting immunoglobulins.

Memory B cell Memory B cells are functionally and physically distinguishable from naive B cells. They often have membrane IgG, IgA or IgE and higher levels of ICAM-1 and CR than naive B cells, and are thought to live longer. Low levels of antigen may remain on FDC for years, so that B cells may be continually activated at low levels to replenish memory cell populations.

Class Switch

Class switch is a process whereby the B cell, as it develops into a plasma cell, can switch the class (also called isotype) of antibody it produces while retaining the same antigen specificity.

The class of antibody (IgG, IgM, etc.) is defined by the Fc portion of the heavy chain. Class switch involves rearrangement of this area of the immunoglobulin gene. Other rearrangements that take place in the Fab regions of the genes of native B cells define antibody diversity (breadth of antigen recognition). Class switch does not occur until after B-cell activation and proliferation. It is under the control of cytokines such as IL-4 and IL-5.

Since B cells initially express IgM they initially use the constant (C) heavy chain gene for IgM (the *mu* gene). The immunoglobulin heavy chain gene locus has the constant region genes for mu, delta, gamma, epsilon and alpha in tandem. In class switch to IgG, for example, the *mu* gene is spliced out such that the portion of the gene that defines the variable region of the heavy chain is brought into apposition with the gamma gene. This results in IgG being produced.

$CD5^+$ B cells $CD5^+$ B cells (sometimes referred to as B-1 cells) form a population that is distinct from conventional B cells (sometimes referred to as B-2 cells). They have the following characteristics:

- They are the first B cells to appear in ontogeny.
- They express surface IgM, but little or no IgD.
- They produce immunoglobulins, mainly IgM, from unmutated or minimally mutated germ-line genes.
- They produce antibodies of low avidity that are polyreactive (i.e., bind multiple different antigens, mainly bacterial polysaccharides and double-stranded DNA).
- They contribute most of the IgM found in adult serum.
- They do not develop into memory cells.
- They are self-renewing in adults (i.e., do not continue to arise from a stem cell in the bone marrow as the conventional B cells do).
- They reside in peripheral tissues and are the predominant lymphocytes in the peritoneal cavity.

Large Granular Lymphocytes

Apart from T cell and B cell certain groups of lymphocytes do not express the membrane-bound molecules and receptors that the T cell and B cell possess. They also have less N/C ratio and are large and granulated. Very importantly they do not express any antigen-binding receptors. These cells are normally called as large granular lymphocytes (LGLs) or simply as null cells.

Most of the membranes of the null cells belong to a special category called natural killer cells (NK cells). These NK cells contribute to about 5% to 10% of the lymphocytes displaying cytotoxic activity against tumour cells and certain viral infected cells. Thus these cells react only against the "self cells" which become abnormal.

Their activity depends on certain important cell-surface molecules called killer activation receptors (KARs) and killer inhibitory receptors (KIRs). The NK cells kill the self cells which contain virus and also tumour cells. These self cells exhibit certain abnormal surface molecules that can be recognized by the NK cells and the self cells are killed. When NK cells bind to uninfected self cells, NK receptors, KIRs provide a negative signal to the NK cell, preventing it from killing the self cell. The NK cells kill the cells through the release of its granules which are mainly proteins like perforins and granzyme. The perforins form holes in the cells and the granzyme will enter the cell and destroy the cell content.

The NK cells also exhibit a unique feature against virus-infected cells. When NK cells are activated by virus-infected cells, they secrete a substance called interferon-gamma. This helps to protect surrounding cells from virus infection.

Another interesting feature of these cells is the expression of CD16 molecules on their cell surface. These act as receptors for the Fc portion of IgG molecule. Thus they attach to these antibodies and can destroy those target cells to which the antibodies can bind. This is an example of a process known as antibody-dependent cell-mediated cytotoxicity (ADCC).

MAST CELLS

Mast cells are sessile and are found in various tissues like skin, connective tissues of various organs and mucosal epithelial tissues of the respiratory, genito-urinary and digestive tracts (Figure 2.11).

The mast cell precursors are formed in bone marrow and remain in circulation as undifferentiated cells. Once they leave the blood and enter the tissues they become mast cells. These cells resemble the basophils in having large numbers of cytoplasmic granules that contain histamines and other pharmacologically active substances.

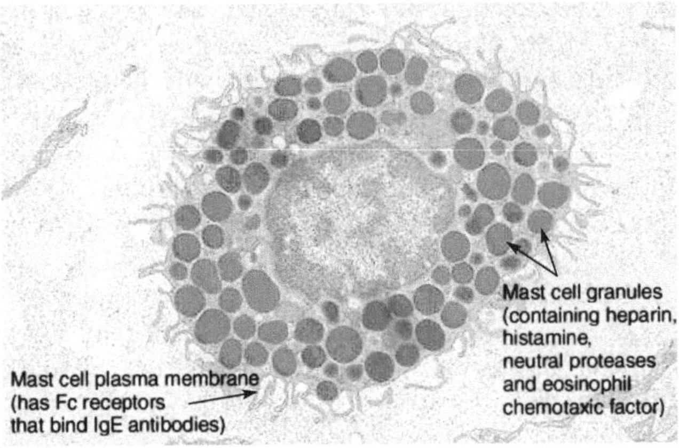

Figure 2.11 Mast cell

Mast cell degranulation (by binding to C3a and C5a, the complement proteins) regulates the traffic of inflammatory cells and molecules through endothelial tight junction during a local inflammatory response. However, the IgE-mediated mast cell degranulation triggers various allergic systems related to type I hypersensitivity.

DENDRITIC CELLS

The dendritic cells are so named as they resemble the dendrites of a neuron on their cell-surface extensions (Figure 2.12). The dendritic cells descend from the myeloid cell lineage. They circulate in blood as immature cells and mature as complete dendritic cell in the tissue.

The dendritic cells express high levels of both class I and class II MHC molecules. They are potent antigen-presenting cells (APCs). Most dendritic cells process the antigen and present it to T_H cells.

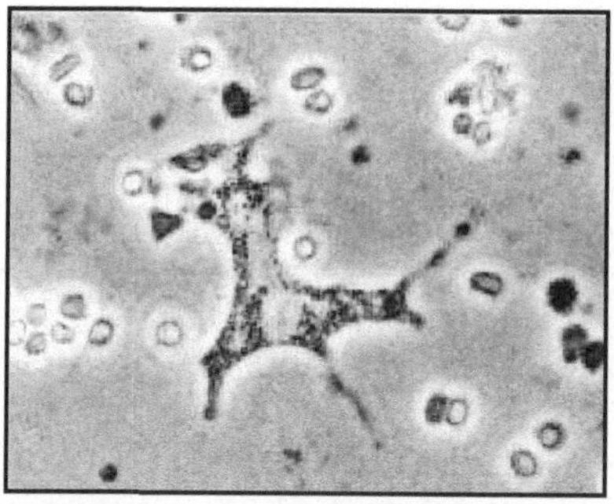

Figure 2.12 Dendritic cell

There are different types of dendritic cells. They are:

1. **Langerhans' cells** They are the APCs found in epidermis of the skin and also in the mucous membranes.
2. **Interdigitating dendritic cells** These cells are present in the T cell areas of the secondary lymphoid tissue.
3. **Interstitial dendritic cells** They are found in many organs like heart, lungs, liver, kidney, gastrointestinal tract, etc.
4. **Veiled cells** The circulating dendritic cells are called as veiled cells and are found in blood and lymph.

Apart from the above four types of dendritic cells there is another type of dendritic cell called the **follicular dendritic cell**. It appears to have a different origin and function. They are neither APCs nor express high level of MHC class II molecules. They are predominantly found in organized structures of lymph follicles and hence the name. The lymph follicles are rich in B cells. The follicular dendritic cells have receptors for antigen–antibody complex. Thus they bind to the complex and express it to the B cells of lymph node and B cells will be activated. They also play a role in developing memory B cells within the follicle.

Myeloid dendritic cells (MDCs) are most similar to monocytes. MDCs are made up of at least two subsets: the more common MDC-1 and the extremely rare MDC-2, which may have a function in fighting wound infection.

Dendritic cells start as immature dendritic cells. These cells are characterized by high endocytic activity and low T cell activation potential. Dendritic cells constantly sample the surroundings for pathogens such as viruses and bacteria. This is done through pattern recognition receptors (PRRs) such as the toll-like receptors (TLRs). TLRs recognize specific chemical signatures found on subsets of pathogens. Once they come into contact with such a pathogen, they become activated into mature dendritic cells. Immature dendritic cells phagocytose pathogens and degrade their proteins into small pieces and upon maturation present those fragments at their cell surface using MHC molecules. Simultaneously, they up-regulate cell-surface receptors that act as co-receptors in T-cell activation such as CD80 and CD86, greatly enhancing their ability to activate T cells. They also upregulate CCR7, a chemotactic receptor that induces the dendritic cell to travel through the bloodstream to the spleen or through the lymphatic system to the lymph node. Here they act as antigen-presenting cells. They activate helper T cells and killer T cells as well as B cells by presenting them with antigens derived from the pathogen, alongside non-antigen-specific co-stimulatory signals.

Plasmacytoid dendritic cells (PDC) look like plasma cells, but have certain characteristics similar to myeloid dendritic cells. They can produce high amounts of interferon-alpha and have thus become known as IPC (interferon-producing cells) before their dendritic cell nature was revealed.

Every helper T cell is specific to one particular antigen. Only professional antigen-presenting cells (macrophages,

B lymphocytes and dendritic cells) are able to activate a helper T cell which has never encountered its antigen before. Dendritic cells are the most potent of all the antigen-presenting cells.

In 1883 a Russian zoologist, Eli Metchnikoff first demonstrated the role of phagocytic cells in the immune process and proposed the theory of cellular immunity. It occurred while he was on a holiday near the sea. He decided to examine some starfish larvae that he found. He pushed a splinter of wood into one of the animals and was surprised to see that many phagocytic cells surrounded the "foreign object". His subsequent investigations elucidated much of the basic role of phagocytic cells in dealing with infections. Much of Metchnikoff's work was carried out at the Pasteur Institute.

Thus, each and every cell present in the immune system is involved in the immune response in one way or the other. This has been summarized in Table 2.1.

Table 2.1 Types of cells of the immune system and their functions

Cell group	**Surface components**	**Function**
B lymphocytes	Surface immunoglobulin (antigen recognition)	Direct antigen recognition
	Immunoglobulin Fc receptor	Differentiation into antibody-producing plasma cells
	Class II Major Histocompatability Complex (MHC) molecule (antigen presentation)	Antigen presentation within class II MHC
T lymphocytes	CD3 molecule T-cell receptor (TCR, antigen recognition)	Involved in both humoral and cell-mediated responses

(Contd.)

Table 2.1 (Continued)

Cell group	**Surface components**	**Function**
Helper T cells (T_H)	CD4 molecule	Recognizes antigen presented within class II MHC
		Promotes differentiation of B cells and cytotoxic T cells
		Activates macrophages
Suppressor T cells (T_S)	CD8 molecule	Downregulates the activities of other cells
Cytotoxic T cells (CTL)	CD8 molecule	Recognizes antigen presented within class I MHC
		Kills cells expressing appropriate antigen
Accessory cells	Variable	Phagocytosis and cell killing
Macrophages	Immunoglobulin Fc receptor	Bind Fc portion of immunoglobulin (enhances phagocytosis)
	Complement component C3b receptor	Bind complement component C3b (enhances phagocytosis)
	Class II MHC molecule	Antigen presentation within class II MHC
		Secrete IL-1 promoting T-cell differentiation and proliferation
		Can be “activated” by T-cell lymphokines

(Contd.)

Table 2.2 (Continued)

Cell group	Surface components	Function
Dendritic cells	Class II MHC molecule	Antigen presentation within class II MHC
Polymorphonuclear cells (PMNs)	Immunoglobulin Fc receptor	Bind Fc portion of immunoglobulin (enhances phagocytosis)
	Complement component C3b receptor	Bind complement component C3b (enhances phagocytosis)
Killer cells	Variable	Direct cell killing
NK cells	Unknown	Kills variety of target cells (e.g. tumour cells, virus-infected cells, transplanted cells)
K cells	Immunoglobulin Fc receptor	Bind Fc portion of immunoglobulin
		Kills antibody-coated target cells (antibody-dependent cell-mediated cytotoxicity, ADCC)
Mast cells	High-affinity IgE Fc receptors	Bind IgE and initiate allergic responses by release of histamine

POINTS TO REMEMBER

- There are two types of cell lineage that are present in the immune system. They are myeloid cell line and lymphoid cell line.
- Myeloid cell line includes monocytes, macrophages, and granulocytes.
- Lymphoid cell line includes T lymphocytes, B lymphocytes and large granular lymphocytes.
- Macrophages, dendritic cells and polymorphonuclear cells are involved in antigen processing and phagocytosis and are referred to as accessory cells.
- T cells and B cells are involved in specific immunity.
- T cells are further divided into helper T cells and cytotoxic T cells.
- B cells, on stimulation, become plasma cells which produce the antibodies.

REVIEW QUESTIONS

1. What short notes on:
   i. Histiocytes
   ii. Kupffer's cells
   iii. Osteoclasts
   iv. Microglial cells
   v. Langerhans' cells
   vi. Follicular cells
   vii. T helper cells
   viii. NK cells
   ix. Plasma cells
   x. Neutrophils

xi. Basophils

xii. Eosinophils

xiii. Mast cells

2. Give the structure and function of macrophages.
3. Give a detailed account on lymphocytes.
4. What are dendritic cells? Explain in detail its types and functions.

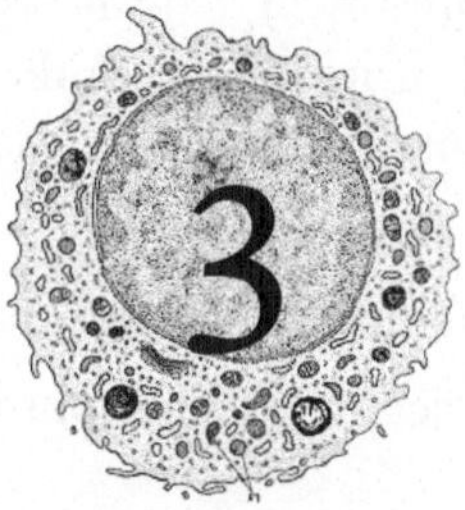

ORGANS OF THE IMMUNE SYSTEM

INTRODUCTION

The organs of the immune system or lymphoid organs are not represented as distinct organs. They are present in the form of

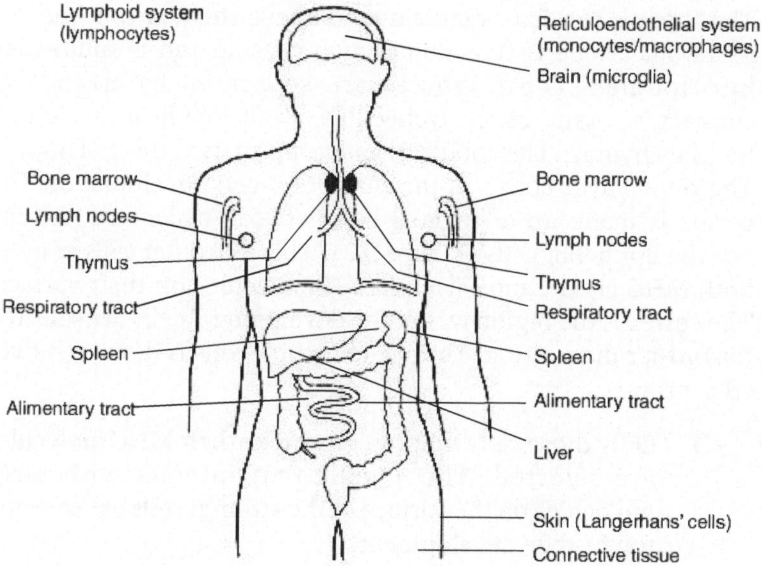

Figure 3.1 Lymphoid organs in human

organized tissues (Figure 3.1). These organized tissues contain large number of lymphocytes in the midst of non-lymphoid cells. The lymphoid organs are broadly classified into two types based on their function:

1. Central or primary lymphoid organs
2. Peripheral or secondary lymphoid organs

PRIMARY LYMPHOID ORGANS

In the primary lymphoid organs, the immature lymphocytes generated in haemopoietic organs undergo the following changes:

- get proliferated
- become differentiated
- become immunocompetent cells (ICC)

Thymus

The thymus is a flat organ situated above the heart. It has two lobes. Each lobe is surrounded by a capsule and is subdivided into lobules. These lobules are separated by strands of connective tissue called trabeculae. Each lobule forms a unit for the thymus. The lobule is made up of two distinct layers. The outer layer cortex is the place of T-cell development. The cortex is made up of stromal cell network which are nothing but the epithelial cells (Figure 3.2). These stromal cells express both MHC class I and MHC class II molecules on their surface. The cortex is the region where the developing T cells are selected for further maturation. The selection of T cells is done as a two-step process:

1. Only those cells that can recognize their MHC molecules are selected. The T cells that interact with MHC molecules on the surface of the stromal cells are selected for further development.

2. The T cells which bind strongly to self antigens are strongly rejected.

Schematic diagram of a normal thymus

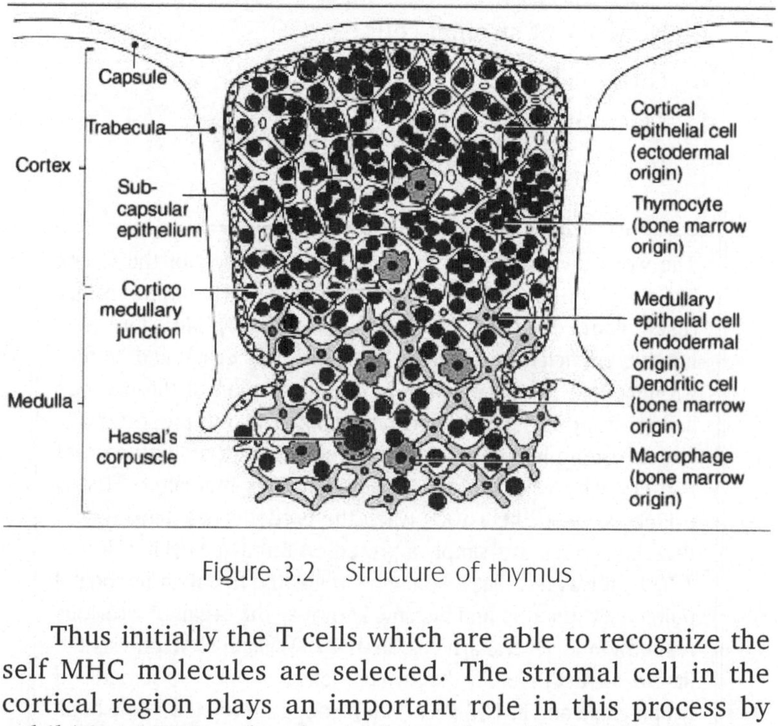

Figure 3.2 Structure of thymus

Thus initially the T cells which are able to recognize the self MHC molecules are selected. The stromal cell in the cortical region plays an important role in this process by exhibiting MHC molecules on its surface. In the second stage the T cells which bind strongly to self antigens are eliminated. Thus those T cells which bind self-MHC molecules and which bind weakly to self antigens and strongly to foreign antigens are allowed to mature. The remaining thymocytes (90–97%) are subjected to apoptosis (programmed cell death). The dead cells are phagocytosed by the macrophages that are present along with the other cells of the cortex region. The cortex also contains another type specialized epithelial cells called nurse cells which are seen surrounded by immature thymocytes. Along with these, the cortex also contains interdigitating dendritic cells. Thus the cortex contains five important types of cells:

1. Thymocytes at different stages of maturation
2. Network of stromal cells
3. Nurse cells
4. Macrophages
5. Interdigitating dendritic cells

The word thymus comes from a Latin derivation of the Greek *thymos*, which means warty excrescent, because of its resemblance to the flowers of the thyme plant. The first notable physician to mention the thymus was Galen and he felt, wrongly, that it played a role in the purification of the nervous system; however, he was correct in noting that it was proportionally largest during infancy. He actually referenced it as the organ of mystery, which held up for several centuries thereafter. That is evidenced by in the 1600s when the predominant thinking was that the thymus was simply a protective thoracic cushion. In the 1700s, it was felt that it somehow regulated foetal and neonatal pulmonary function and became known as the organ of vicarious respiration in several articles. In 1777, William Hewson was the first to correctly identify the thymus gland as some sort of modified lymph gland and in 1832 Sir Astley Cooper devoted a whole book to the anatomy of the thymus gland, which contained detailed cadaveric dissections. In 1846, a major breakthrough came in the study of the thymus when Hassall and Vanarsdale used improvements in compound microscopy to study the thymus more thoroughly and described differences between the thymus and other lymphoid tissues, specifically the characteristic histological feature that became known as Hassal's corpuscles.

The medulla of the thymus consists of mature thymocytes, medullary epithelial cells, macrophages and dendritic cells. Along with this they also possess special type of cells called Hassal's corpuscles. However, it is believed that these cells are involved in cell destruction. Figures 3.3 and 3.4 depict the histology and anatomical position of thymus respectively.

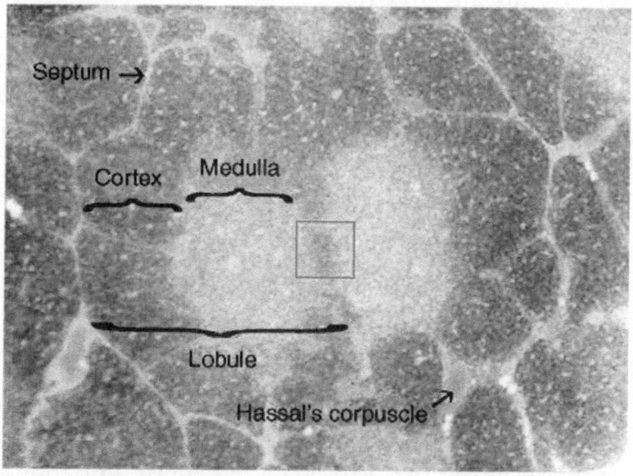

Figure 3.3 Histology of thymus

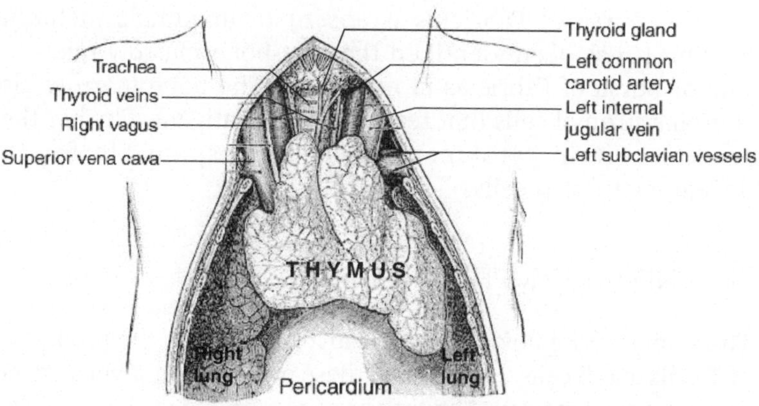

Figure 3.4 Anatomical position of thymus

The importance of thymus in immune function can be studied by performing neonatal **thymectomy** (surgical removal of thymus) in mice. The mice failed to show the production of T cells and cell-mediated immunity. The importance of thymus in immune response can also be studied in congenital defect in humans **(DiGeorge's syndrome)** and in certain mice (nude mouse) in which the thymus fails to develop. In both the cases the T cells are absent and fail to show cell-mediated immunity.

> DiGeorge's Syndrome is a congenital lack of the thymus, and leads to the increase in infections and an impaired cell-mediated immune response.

Bone Marrow and Bursa of Fabricius

In birds it has been identified that **Bursa of Fabricius** is the central lymphoid organ for the development of B cells. This bursa is present as the outpushing in the cloaca of the bird. The bursa is derived from the gut epithelium in the embryonic stage. The stem cells enter the Bursa of Fabricius and develop into mature B cells. On entering into the peripheral lymphoid organs they become capable of producing antibodies.

The Bursa of Fabricius is absent in mammals including human. It has been identified that the bone marrow takes the role of Bursa of Fabricius in mammals. The bone marrow also contains stromal cells that react with self antigens. Further they secrete certain cytokines which are responsible for the development of B cells.

SECONDARY LYMPHOID ORGANS

The primary lymphoid organs are involved only in the production of T cells and B cells. However, in these organs the antigens are not normally encountered. Antigens are encountered in the secondary lymphoid organs. The T cells and B cells leave the primary lymphoid organs and get organized into lymphoid follicles. These form the secondary lymphoid organs and can be found either as well organized organs or just as a diffused lymphoid tissue. The following are the secondary lymphoid organs:

1. Lymph nodes

2. Spleen
3. Mucosal-associated lymphoid tissues (MALT)

Lymph Nodes

The lymph nodes are capsulated bean-shaped structures present at the junction of lymphatic vessels (Figure 3.5). Morphologically the lymph node has 3 distinct regions. They are outer cortex followed by paracortex and medulla. The capsule penetrates into the lymph node to form trabeculae. The cortex of the lymph node contains lymphocytes (predominantly B cells), macrophages and follicular dendritic cells. They form an aggregate in the cortex called primary follicle. On antigenic challenge, the primary follicle enlarges to form secondary follicle containing the germinal centre. Beneath the cortex is the paracortex which is highly populated by T cells along with interdigitating dendritic cells. The interdigitating dendritic cells are derived from tissues through lymph. The cells express high levels of MHC class II molecules which are involved in antigen presentation to T helper cells. The paracortex region is called as T-cell-dependent area. It is the region where the antigens are processed. The processed antigen is presented to T helper cells which in turn stimulate the B cells. These B cells develop finally to antibody-producing

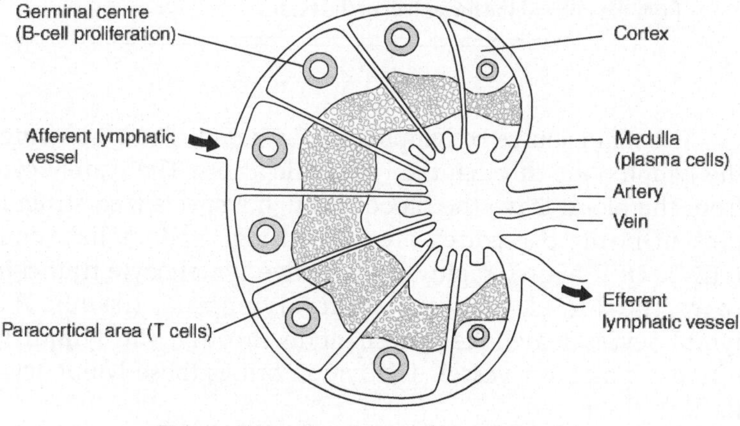

Figure 3.5 Structure of lymph node

plasma cells. The paracortex is followed by the medulla. The medulla contains mainly plasma cells actively secreting antibodies. These plasma cells and other lymphoid cells are seen in the form of medullary cords.

The afferent lymphoid vessels empty the lymph into the lymph node. The lymph penetrates the different layers of lymph node and leaves through the efferent lymphatic vessel. Any particulate antigens including pathogens will be trapped inside the lymph node and will be processed by the interdigitating dendritic cells.

Lymph is a clear body fluid which clots like blood. Lymph forms when dissolved proteins and solutes filter out of venules and capillaries because of local differences in luminal hydrostatic and osmotic pressure. Whenever the epithelial barrier is broached, the neuropeptide "substance P" is released by axons and increases lymph production through hyperaemic effects. Hyperaemia increases fluid transudate. Lymph flows unidirectionally towards lymph nodes because valves prevent backflow under normal physiological conditions. Lymph capillaries merge into larger lymphatics which drain into lymph nodes. Efferent lymph from regional lymph nodes may drain into one or more additional nodes before flowing into major efferent lymphatics. The thoracic duct carries lymph draining from the gut and the lower half of the body.

The main source of lymphocytes is blood. The blood enters the lymph node through the lymphatic artery. The lymphocytes from the blood leave the blood through a specialized structure present in the paracortex area called high endothelial venule (HEV). HEV allows rapid and selective lymphocyte trafficking from the blood into secondary lymphoid tissues. The lymphocytes leave the lymph node through the lymph via efferent lymphatic vessel. The lymph brings these lymphocytes back to the blood.

Spleen

The spleen is a capsulated large, ovoid organ situated in the left abdominal cavity (Figure 3.6). The capsule of the spleen extends into the organ in the form of projections called trabeculae to form many compartments. Each compartment contains two distinct regions namely red pulp and white pulp.

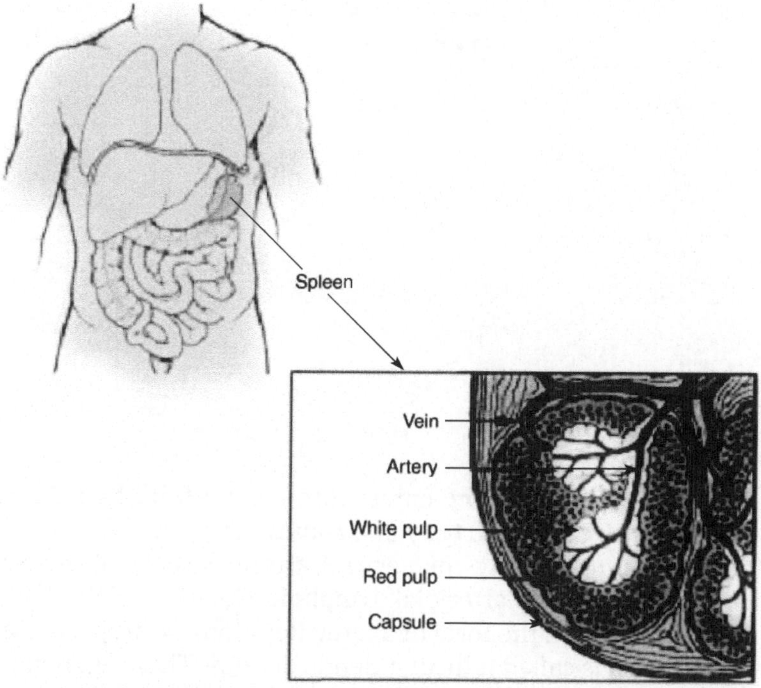

Figure 3.6 Location of spleen in human body

The red pulp contains a network of sinusoids in which macrophages and RBCs are present. The red pulp is the place where the worn out RBCs are destroyed and removed. The destroyed RBCs are engulfed by macrophages. The red pulp is separated from the white pulp by a diffused marginal zone. The white pulp is the area which surrounds the spleen artery (Figures 3.7, 3.8 and 3.9).

The white pulp consists of three important areas:

- Periarteriolar lymphoid sheath (PALS)
- B cell follicle
- Marginal zone

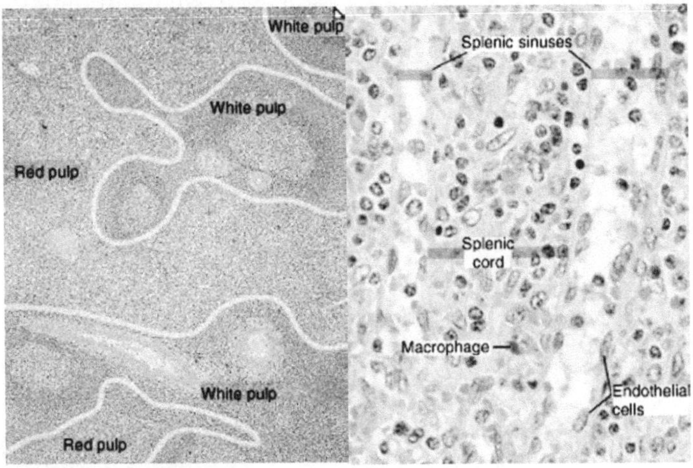

Figure 3.7 Histology of spleen

The **splenic artery** enters through **trabeculum** in the form of **trabecular artery** and comes out to form the central arteriole. Surrounding this central arteriole is the T cell-rich area called the periarteriolar lymphoid sheath (PALS). The B cells are found in the form of aggregates along with specialized dendritic cells called follicular dendritic cells. These aggregates are called primary follicles. The formation of these aggregates is mediated by the follicular dendritic cells. These dendritic cells secrete a cytokine called B lymphocytochemokine (BLC) for which B cells have receptors. Thus the B cells will be attracted towards the follicular dendritic cells to form aggregates. Once the B cells get antigenic stimulus, the selected B cells will start to proliferate to a distinct area called germinal centre surrounded by the resting, unstimulated B cells to form the mantle zone or corona. The germinal centre along with the mantle is the secondary follicle. These B-cell areas are surrounded by marginal sinus, which are branched from central arterioles.

The B-cell follicles are surrounded by a diffused region called **marginal zone**. The marginal zone shows the presence of B cell of non-migratory type. Further these B cells are different from the B cells present in the follicles in the presence of different surface molecules. The lymphocytes enter into the spleen from blood via splenic artery. The PALS region mainly receives the lymphocytes from the central arterioles whereas the B cell follicles receive most of the lymphocytes from the marginal sinus. Thus the splenic lymphocytes encounter the blood-borne pathogens and antigens in contrast to lymphocytes in the lymph nodes which encounter the pathogens and antigens from the tissues through the lymph.

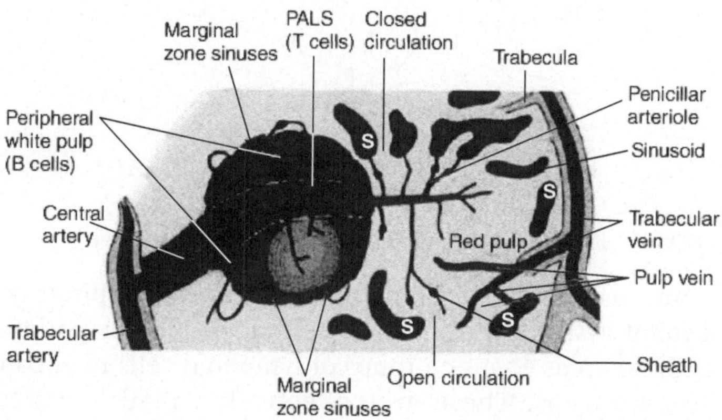

Figure 3.8 Structure of spleen

In spleen, the immune response mostly starts in the PALS (T-cell area). The interdigitating dendritic cells engulf the antigens and present it along with MHC class II molecules. These in turn activate the T helper cells. The T helper cells migrate to B-cell area to stimulate the B cells. The activated cells form the germinal centre.

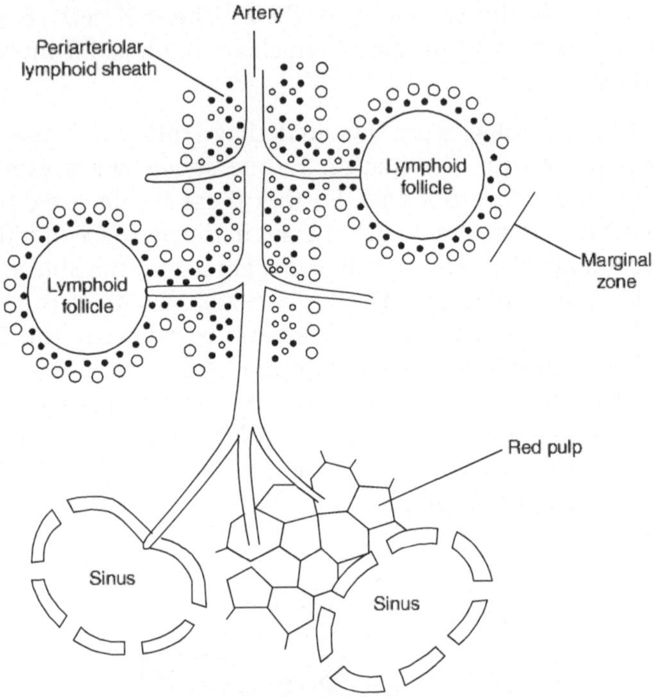

Figure 3.9 Area of white pulp in spleen

Mucosal-associated Lymphoid Tissue (MALT)

The mucous membrane lining of the digestive, respiratory and urogenital systems are major sites that encounter pathogens. Thus these areas contain groups of lymphoid cells to encounter these pathogens. These are collectively called as mucosal-associated lymphoid tissue (MALT). The MALT is called by different names depending upon the region where it is present.

- Gut-associated lymphoid tissues (GALT)
- Bronchus-associated lymphoid tissue (BALT)
- Cutaneous-associated lymphoid tissue (CALT)

The important MALTs are tonsils, Peyer's patches and appendix.

Bronchus-associated lymphoid tissues (BALT) The tonsils are of three types depending upon the region where they are present. They are **palatine tonsils**, **lingual tonsils** and **pharyngeal tonsils**. The tonsil is a nodular structure consisting of a network of reticular cells and fibres encompassing lymphocytes, macrophages, granulocytes and mast cells. They have well organized B cell follicles with a germinal centre. The follicles are surrounded by T cell area. Tonsils are very helpful in defending the entry of pathogens through nasal and oral epithelial routes. Among all the mucous membranes studied, the best studied is the mucous membrane lining the gastrointestinal tract.

Gut-associated lymphoid tissues (GALT) The mucosal Peyer's patches, appendix and isolated follicles in intestinal mucosa

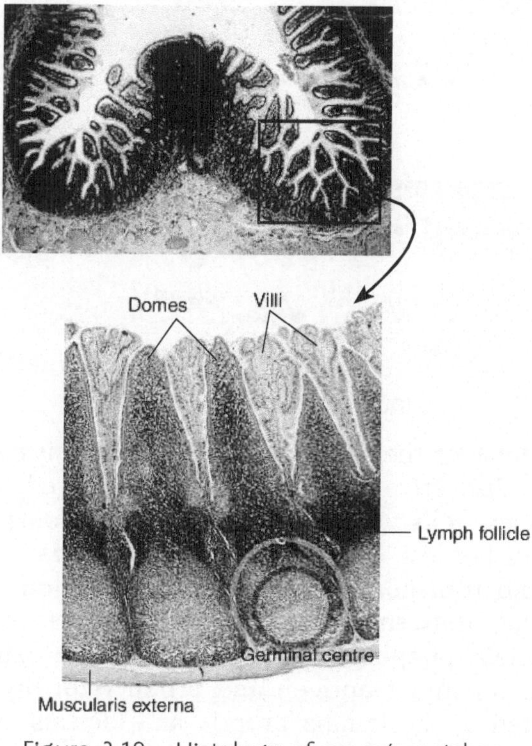

Figure 3.10 Histology of peyer's patches

vary with regard to type of surface epithelium (stratified squamous, ciliated columnar, or absorptive columnar) and relative proportions of T and B cells (Tonsils have 60% T cells compared to 25–40% T cells in respiratory or intestinal patches) but the similarity of these tissues to Peyer's patches is greater than the difference, especially since all have **"M" cells** in their follicle-associated epithelium (Figure 3.10).

Peyer's patch epithelium is specialized to simple antigens ingested along with food. Peyer's patches contain lymphoid compartments that are analogous to the deep cortex and follicles of lymph nodes, but there are no afferent lymphatics and no medullary cords for local accumulation of plasma cells. Each Peyer's patch contains multiple individual B cell follicles separated by diffuse lymphoid tissue in interfollicular areas (Figure 3.11).

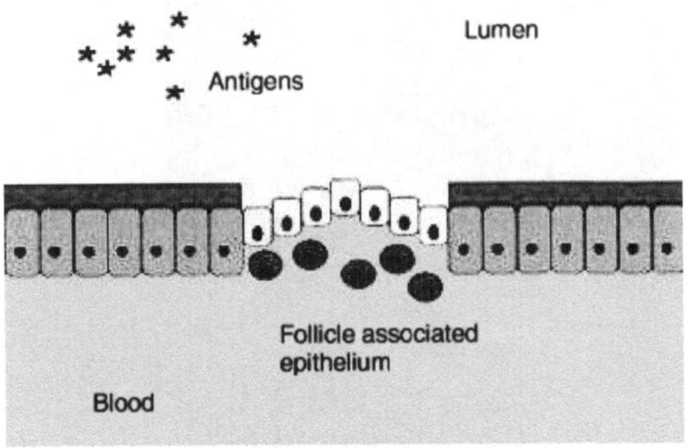

Figure 3.11 Peyer's patches

The dome epithelium covering each follicle is composed of cuboidal absorptive epithelial cells interrupted by delicate membranous cells which have luminal microfolds instead of microvillus borders. They are called as **M cells**. The M cells engulf and transport various materials without lysosomal degradation. Antigen is deposited into small lymphocytes and mononuclear phagocytes and also dendritic cells. Minute quantities of intact antigen and products of digestion are transported to the lamina propria and lacteals by ordinary absorptive epithelial cells anywhere in the small bowel. It is important to point out that these products enter interfollicular areas and do not have access to the dome area. Between the dome epithelium and the follicles, there is a thin region of reticulum, containing a delicate plexus of blood vessels and plasma cells.

Cutaneous-associated lymphoid tissues (CALT) Skin and epithelial surfaces are the first line of defence against antigens in the environment which threaten an organism's integrity. Epithelium with underlying loose connective tissue, blood vessels and lymphatics may be considered part of the diffuse lymphatic system, especially for contact-sensitizing antigens that bind to epithelial cells. The primary function of epithelium was used to be regarded as a physical barrier but recently it has been shown to permit the passage of drugs and antigens at a slow rate. Keratinocytes, mononuclear cells and nerve endings in the skin secrete numerous factors that may attract, activate and effect differentiation of T cells locally or in regional lymph nodes. Lymphocytes and dendritic epidermal cells are infrequently seen in the dermis or in intra-epidermal locations of normal skin. Because of the vast surface area, the total number of lymphocytes in the skin may be great despite their small number in any single location. Although there is considerable speculation about homing of lymphocytes to the skin, few are present in afferent lymph. Langerhans' cells and keratinocytes populate the epithelial layer above the basal cells.

Precursors for Langerhans' cells and monocytes enter the skin from the blood to replace cells that regularly exit into the afferent lymph of regional lymph nodes. Secondary lymphoid-tissue chemokine (SLC) and CC chemokine receptor 7 (CCR7) appear to participate in the emigration pathway of mature dendritic cells from the skin to regional lymph nodes. The mean turnover time of Langerhans' cells in mouse skin is about three weeks. Langerhans' cells in afferent lymph draining the skin are regarded as the sentinels of skin-associated lymph tissue. The so-called passenger leucocyte that initiates allograft rejection via antigen/class I MHC expression is also a Langerhans' cell, and removal of the cell or ablation of the afferent lymphatics prevents sensitization. The underlying connective tissue of skin is transformed into a "lymph node-like" microenvironment by alterations in the microvasculature and reticulum to accommodate local recirculation, lodging and proliferation of lymphocytes in situations where deposited antigen persists. Therefore, the skin has a barrier and sentinel function which depends upon a constant traffic of lymphocytes, Langerhans' cells and mononuclear cells through the skin and into regional lymph nodes via afferent lymph.

Nasal-associated lymphoid tissue (NALT) The NALT is the first organized lymphoid tissue that encounters inhaled antigen. Much recent evidence suggests that the NALT may play an important role in the development of the immune response towards invading pathogens through this site. However, little is known about this region. In humans, the NALT comprises the **Waldeyer's Ring** and in rodents it can be identified as an organized paired lymphoid organ at the entrance of the nasopharynx (O-NALT) and as more diffuse (D-NALT) cells lining the nasal passages. Almost all of the antibodies secreted by the D-NALT population are of the IgA isotype. The IgA isotype of antibody is particularly suitable for providing protection within mucosal surfaces as it is generally non-inflammatory, and at the same time efficient at neutralizing the antigen.

POINTS TO REMEMBER

- The lymphoid organs are of two types: primary and secondary lymphoid organs.
- Thymus and bone marrow are the primary lymphoid organs of human.
- T lymphocytes are produced in the thymus and B lymphocytes are produced in the bone marrow.
- The lymph node, spleen and mucosal-associated lymphoid tissues (MALT) are the secondary lymphoid organs of human.
- The B cells mature in germinal follicles of secondary follicles.

REVIEW QUESTIONS

1. Write short notes on:
   i. Peyer's patches
   ii. Germinal centre
   iii. Tonsils
   iv. PALS
   v. NALT
   vi. Bursa of Fabricius
   vii. Marginal zone
2. What are primary lymphoid organs? Explain their structure and function.
3. What are secondary lymphoid organs? Explain their structure and function.
4. Write a brief note on MALT.
5. Explain the structure and events that take place in the lymph node.
6. Give a detailed account of the structure and function of the thymus.

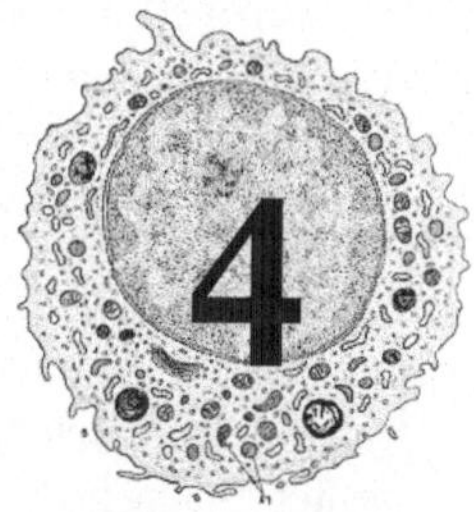

IMMUNOGLOBULIN

INTRODUCTION

In 1937, Tiselius electrophoretically separated the serum proteins into different fragments namely albumin, alpha-, beta- and gammaglobulin. Subsequently Tiselius and Kabat analysed all these fragments for their antibody properties and found that gamma globulins possessed antibody properties. Thus the term gammaglobulin has been used in the place of antibody until it was named as immunoglobulin, the term which had been internationally accepted for those proteins of animal origin that have the properties of antibodies.

Arne Wilhelm Kaurin Tiselius was born on August 10, 1902, in Stockholm. After the early loss of his father, the family moved to Gothenburg where he did his schooling and after graduation at the local "Realgymnasium" in 1921, he studied at the University of Uppsala, specializing in chemistry. He became a research assistant in The Svedberg's laboratory in 1925 and obtained his doctor's degree in 1930 on a thesis, "The moving-boundary method of studying the electrophoresis of protein."

Immunoglobulins are secreted by specialized cells called plasma cells that are derived from the B cells.

Elvin A. Kabat, was the founder of modern quantitative immunochemistry together with Michael Heidelberger, his doctoral mentor. During his long career the structural and genetic basis for specificity of antibodies was elucidated. It was he who first demonstrated that antibodies are gammaglobulins. Although his name is most associated with characterizations of the size and heterogeneity of antibody-combining sites, his contributions to modern biomedicine go well beyond this subject. His work advanced our understanding of fundamentals of developmental biology, inflammation, autoimmunity and blood transfusion medicine. Elucidation of structures of the major blood group antigens, embryonic stage-specific carbohydrate antigens, and functional carbohydrate markers of leucocyte subsets were either achieved by him and his associates, or made possible through meticulously characterized, invaluable compounds he generously made available to other investigators. Working with Tiselius (1939), he conducted the groundbreaking electrophoresis experiments that first demonstrated that anti-ovalbumin antibodies in sera of hyperimmunized rabbits were gammaglobulins (IgG).

STRUCTURE OF IMMUNOGLOBULIN

The immunoglobulins are glycoproteins. A typical immunoglobulin is composed of four polypeptide chains that are linked by disulphide bridges (Figures 4.1 and 4.2). Among the four larger chains, two larger ones are called as heavy chains and two smaller ones are called as light chains. Within a given immunoglobulin, the two heavy and the two light chains are identical to each other. Chemically and structurally there are five different heavy chains. Based on these, there are five different classes of immunoglobulins, viz. IgG (γ heavy chain), IgD (δ heavy chain), IgM (μ heavy chain), IgA (α heavy chain) and IgE (ε heavy chain). There are only two types of light chains. They are kappa (κ) and lambda (λ) light chains. Either of these may be present in any class of immunoglobulin and both the types will not be seen together in the same immunoglobulin molecule.

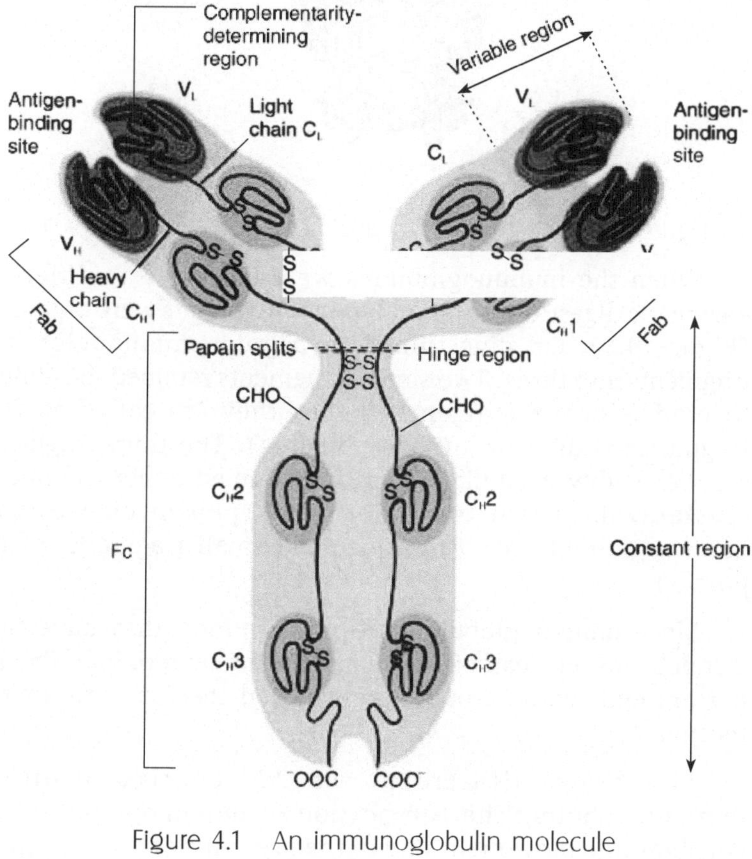

Figure 4.1 An immunoglobulin molecule

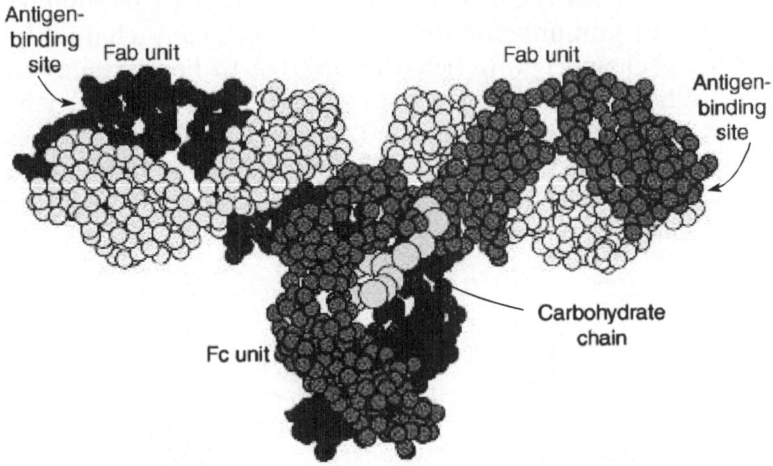

Figure 4.2 Three-dimensional structure of immunoglobulin

When the immunoglobulins were subjected to different enzymatic digestion, different types of fragments were obtained (Figure 4.3). On digestion with papain, immunoglobulin fragments into three. Two similar fragments retained the ability to bind with the antigen and thus they are called as Fab fragments (fragment antibody binding). The third fragment can be readily crystallized and was termed as Fc (fragment crystallizable). On the other hand, pepsin cleaves the immunoglobulin into Fab dimer and small fragments of Fc portion.

Since immunoglobulins are polypeptides, they have two distinct ends, viz. carboxy terminal and amino terminal. The Fc portion ends with carboxy terminal and the Fab with amino terminal.

The figure illustrates the "Y" configuration of immunoglobulin. The Fab portion of immunoglobulin has considerable movement due to the presence of hinge region.

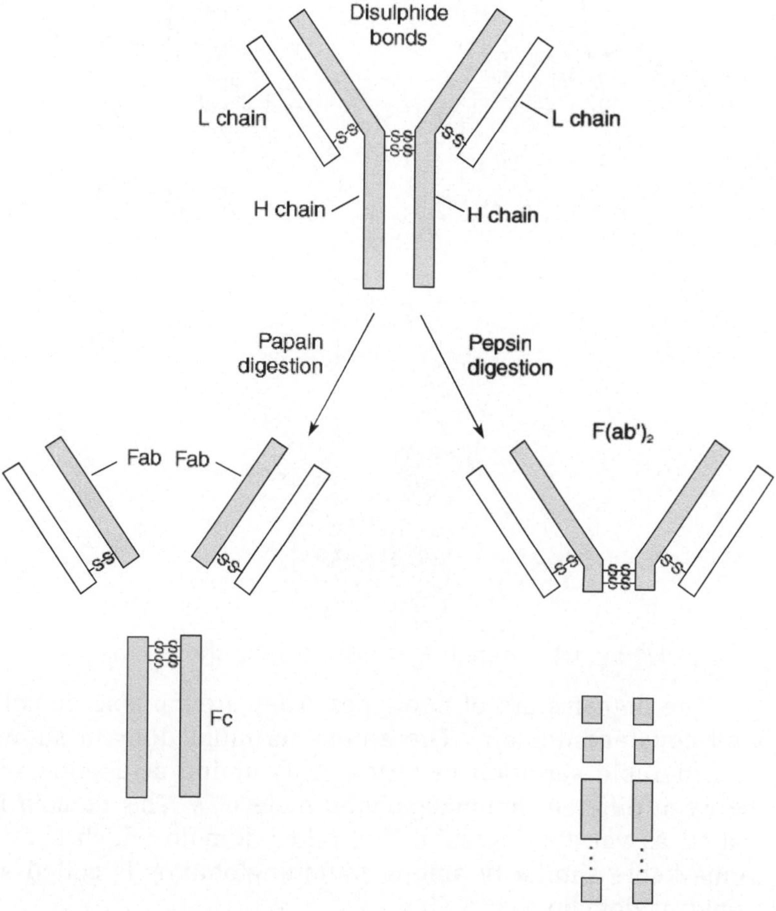

Figure 4.3 Digestion of immunoglobulin by papain and pepsin

Domains

Both the light chain and heavy chain consist of a series of similar globular subunits called domains (Figure 4.4). This domain has roughly a cylindrical configuration and is made up of up to 110 amino acids folded into two layers of β-pleated sheets held together by disulphide bridge.

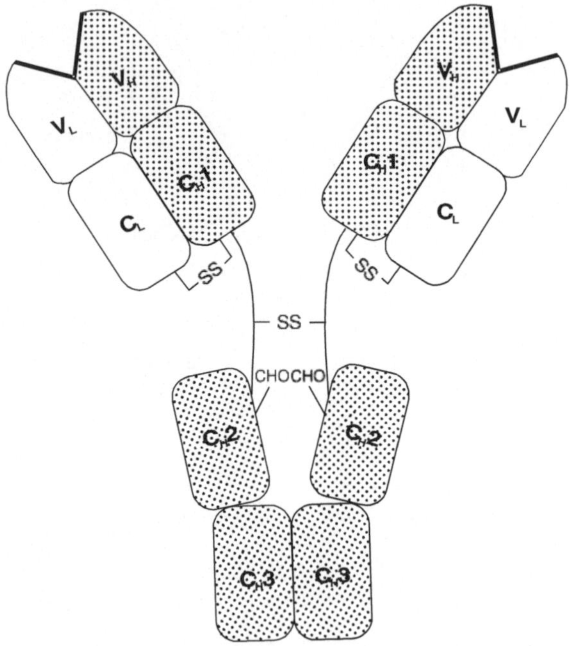

Figure 4.4 Immunoglobulin showing the domains

The domains are of two types. They are variable domain and constant domain. The amino terminal domain shows considerable variation in terms of its amino acid sequence between different immunoglobulin molecules. This domain is called as variable domain. The other domain which shows remarkable similarity among immunoglobulins is called as constant domain.

The light chain contains two domains, i.e., one variable domain (VL) and one constant domain (CL). The heavy chain contains 4 or 5 domains depending upon the class of immunoglobulin. They contain one variable domain (VH) and the others are constant domains (CH1, CH2, CH3 and CH4).

Antigen-binding Site

As seen earlier the Fab portion contributes to antigen binding. This region shows amino acid sequence variability between different antigen-specific immunoglobulins. Hence this region is called as variable region and the remaining portion is called as the constant region. One half of the light chain and one fifth of the heavy chain contribute to the variable region (Figure 4.5).

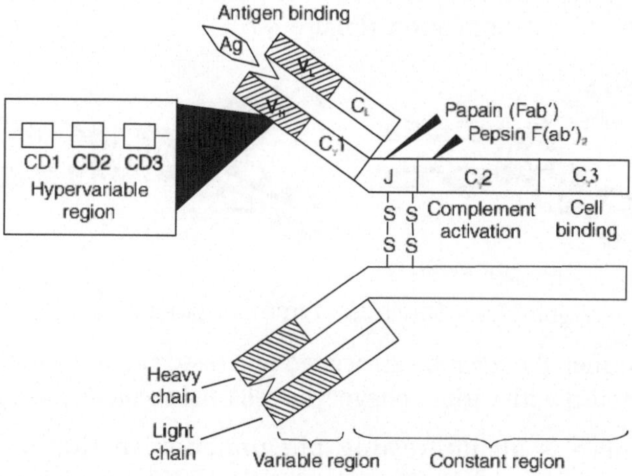

Figure 4.5 Immunoglobulin showing various binding sites

The variation is again seen specifically within some locations called as "hot spot" or hypervariable region. Thus the light chain has three hypervariable regions and the heavy chain has four hypervariable regions. In the variable domain, these hypervariable regions contribute to the formation of antigen-binding site. The intervening peptide segments are called as framework regions.

IMMUNOGLOBULIN CLASSES

Immunoglobulin G (IgG)

This is the major class of immunoglobulin which constitutes about 80% of the total immunoglobulins. They are produced particularly during secondary immune responses. It has a molecular weight of 150,000 daltons. They have the structure of a conventional immunoglobulin. The γ heavy chain contains three constant domains (Cγ1, Cγ2 and Cγ3) each of them involved in different effector functions. They have 4 subclasses, viz. IgG1, IgG2, IgG3 and IgG4. The subclasses IgG1 and IgG3 interact with Fc receptors expressed on various cell types. The Cγ2 and Cγ3 domains are involved in the Fc receptor-binding. All the four subclasses of IgG interact with Fc receptor in the placenta and are transported into the foetal circulation (Figure 4.6).

Figure 4.6 Structure of immunoglobulin G (IgG)

Further, the microbe-associated IgG (association by Fab region) can interact with various phagocytic cells and facilitate phagocytosis.

There is an interesting phenomenon in the bacterium *Staphylococcus aureus*. They express Fc binding proteins on their cell surface (protein A and protein G). Thus they bind the IgG which may aid in the survival of the bacterium.

Immunoglobulin A (IgA)

It is second most abundant immunoglobulin (10 to 13%) and the major immunoglobulin of external secretions. It is the major immunoglobulin in colostrums, saliva and tears.

IgA occurs in two forms. In serum it occurs as a monomeric molecule (MW = 160,000) resembling a conventional immunoglobulin molecule combined together. However in secretions it occurs as a dimer molecule, i.e., two units of immunoglobulin molecule combined together. The two immunoglobulin molecules are joined by a small piece of polypeptide chain called as J chain (joining chain). Further the secretory IgA contains T piece (transport piece) or S piece (secretory piece). The IgA, along with J chain is produced by plasma cells. After their production they gain access to the lumen of the mucosal site by attaching to the epithelial cells. The epithelial cells of this region will have receptor (poly Ig) for the J chain. The dimeric IgA binds to these receptors and gets transported across the cell along with the receptor. On reaching the other side of the cell, it gets detached from the receptor to get secreted into lumen. During this process, a piece of receptor attaches to the dimeric IgA molecules. This is nothing but the T piece or S piece. The secretory IgA confers immunity to the mucous surface. It is believed that IgA plays an important role in local immunity against respiratory and intestinal pathogens. The secretory IgA binds to these pathogens, forms a covering over them and prevents the adherence of the microorganism to the surface of the mucous membrane (Figures 4.7 and 4.8).

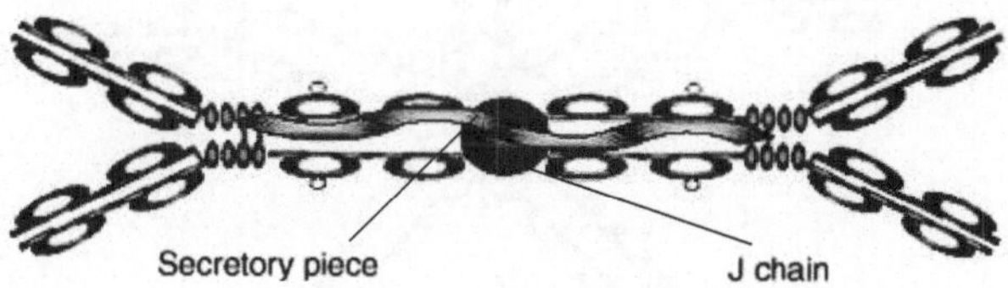

Figure 4.7 Structure of secretory immunoglobulin A (IgA)

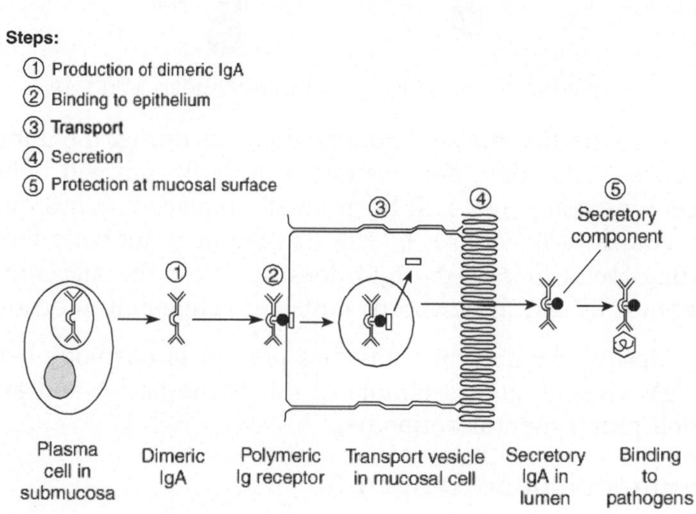

Figure 4.8 Formation of secretory immunoglobulin A (sIgA)

The IgA does not fix complements like IgG. However, it can activate the alternative complement pathway.

Immunoglobulin M (IgM)

The IgM has a heavy molecular weight of 1,000,000 daltons. They are pentameric molecules, made up of five basic immunoglobulin molecules joined together by J chain (Figure 4.9).

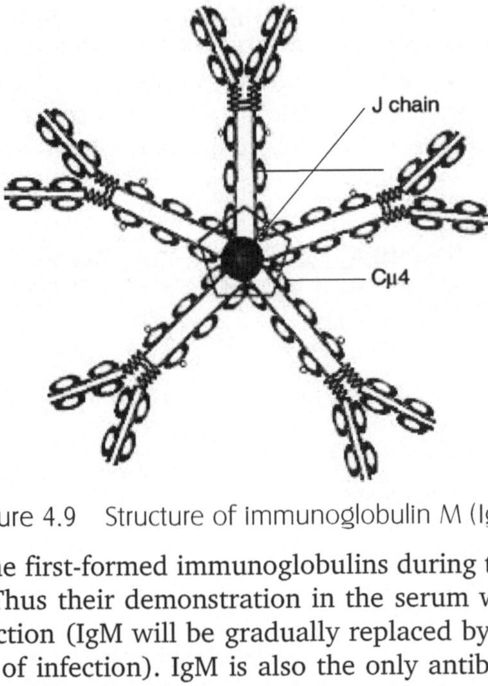

Figure 4.9 Structure of immunoglobulin M (IgM)

They are the first-formed immunoglobulins during the immune response. Thus their demonstration in the serum will indicate recent infection (IgM will be gradually replaced by IgG during the course of infection). IgM is also the only antibody formed during foetal life. As the IgM does not cross the placenta the presence of IgM in newborns indicates congenital infections.

Most of the natural antibodies present in our body belong to IgM class. A good example of this is the anti-A and anti-B blood group natural antibodies.

Immunoglobulin D (IgD)

Normally the IgD is present in minute quantities in blood and resembles IgG in its structure. These molecules are detected as surface immunoglobulins on any of the early B cells in conjugation with IgM. So far, the IgD has not been attributed to any effector function but however it is found to have a role as antigen receptor on early B cells (Figure 4.10).

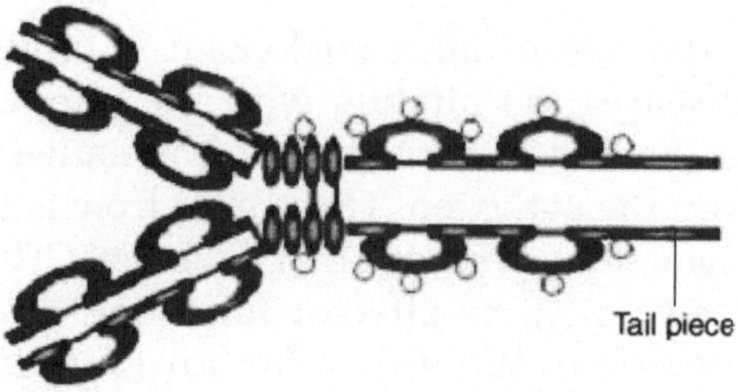

Figure 4.10 Structure of Immunoglobulin D (IgD)

Immunoglobulin E (IgE)

This is the least found immunoglobulin in the blood. They are monomeric immunoglobulin molecules and have the ability to bind to the Fc receptors present in basophils and mast cells (Figure 4.11). When it is further complexed with the antigen, it triggers the cells to release chemical mediators that are responsible for various clinical symptoms under type I hypersensitivity.

An elevated level of IgE is seen during the parasitic infestations. IgE binding to the Fc receptors of eosinophils are believed to be responsible for eliminating parasitic infestation.

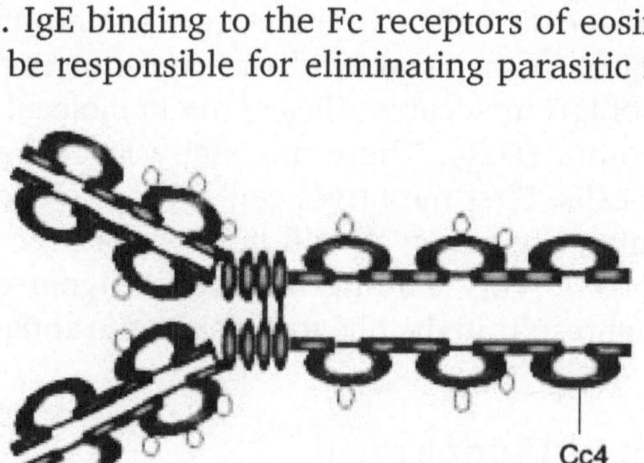

Figure 4.11 Structure of immunoglobulin E (IgE)

The properties of various classes of immunoglobulins are given in Table 4.1.

FUNCTIONS OF IMMUNOGLOBULIN

Considering the role of immunoglobulin in protecting against infectious diseases, its binding with the antigen is just not enough. The binding of antibodies to the antigen will neither kill nor remove the pathogen. Thus, apart from antigen binding, the immunoglobulins possess other functions. These functions are collectively called as effector functions. There are three important effector functions for antibodies. They are: opsonization, activation of complement and antibody-dependent cell-mediated cytotoxicity (ADCC). While the variable region is involved in antigen binding, the heavy chain constant regions are responsible for binding with various effector proteins to execute the other effector functions.

Opsonization

Opsonization is the process of promotion of phagocytosis of antigens by the phagocytic cells like macrophages and neutrophils. The substances that are involved in the process of opsonization are called as Opsonins. The immunoglobulins are opsonins.

The phagocytic cells express certain protein molecules on their cell surface that can bind the constant region of most subclasses of IgG molecules. These protein molecules are called as Fc receptors (FcR). There are many Fc receptors on the phagocytic cells. First the target cell will be coated with many IgG molecules. Then they all will bind with Fc receptors of the phagocytic cell. This binding initiates a signal transduction pathway that results in the phagocytosis of the antigen–antibody complex.

Activation of Complement

IgM and IgG can activate complement. Complement system is a collection of serum glycoproteins that can perforate cell

Table 4.1 Properties of immunoglobulin

Isotype	Structure	Placental transfer	Binds mast cell surfaces	Binds phagocytic cell surfaces	Activates complement	Additional features
IgM		–	–	–	+	First antibody in development and response
IgD	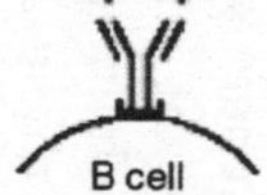	–	–	–	–	B-cell receptor
IgG		+	–	+	+	Involved in opsonization and ADCC; four subclasses; IgG1, IgG2, IgG3, IgG4
IgE	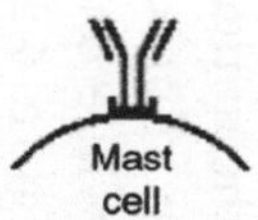	–	+	–	–	Involved in allergic responses
IgA	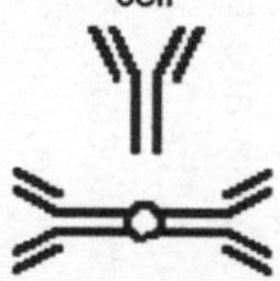	–	–	–	–	Two subclasses; IgA1, IgA2; also found as dimer (sIgA) in secretions

membranes on activation. An important by-product of the complement activation is C3b. This C3b fragment binds non-specifically to antigen and antibody complexes. The macrophages have receptors for C3b. Thus the macrophages can adhere to antigen–antibody complex and lead to phagocytosis of the cells or molecular complexes attached to C3b receptor.

Antibody-dependent Cell-mediated Cytotoxicity (ADCC)

The natural killer cells that have Fc receptor can bind to antibody bound to target cells and will kill the target cells.

ANTIGENIC DETERMINANTS OF IMMUNOGLOBULIN

Even antibodies can act as antigens. The immunoglobulins do have different antigenic determinant sites to which specific antibodies are produced. There are three different types of antigenic determinant sites in immunoglobulin.

They are:

1. Isotypes
2. Allotypes
3. Idiotypes

Isotypes

These are the antigenic determinant sites that are represented by the Fc portion of the constant region of the heavy chain molecule, while on the light chain they are on the constant region (Figure 4.12). Thus these antigenic determinants characterize classes and subclasses of heavy chains and types and subclasses of light chain. The antibodies that are produced against the antigenic determinants are called anti-isotypic antibodies. These antibodies are used for quantitation of Ig classes and subclasses in various diseases such as in the characterization of B-cell leukemia and in the diagnosis of various immunodeficiency diseases.

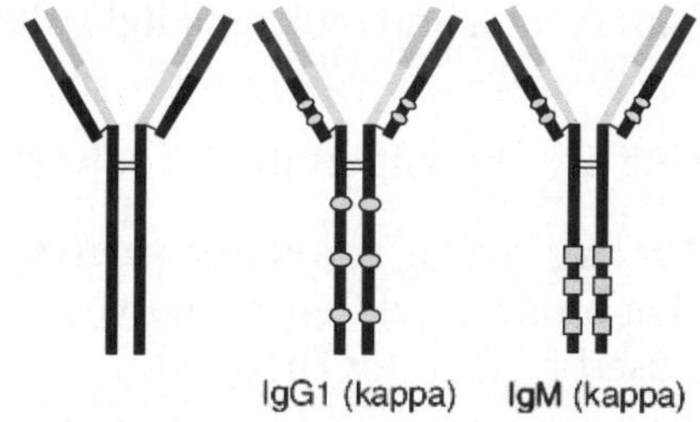

Figure 4.12 Immunoglobulins showing the isotype differences

Allotypes

The allotypic antigenic determinants are localized in the constant region of the heavy chain and light chain of the same class of immunoglobulin (Figure 4.13). Thus if we take IgG class immunoglobulin its isotype is same in all human beings but its allotype may differ between two individuals. They are antigenic determinants specified by allelic forms of the Ig genes. Allotypes represent slight differences in the amino acid sequences of heavy or light chain of different individuals. Even a single amino acid difference can give rise to an allotypic determinant, although in many cases several amino acid substitutions occur.

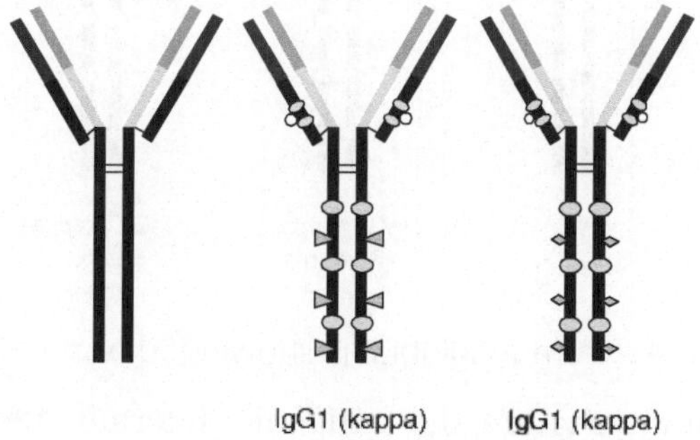

Figure 4.13 Immunoglobulins showing the allotype differences

The human Ig allotypes are named on the basis of the heavy or light chain on which they are located. Thus an allotype on a gamma heavy chain is given the name Gm system, on an alpha heavy chain is Am system and on kappa light chain is called Km system.

The following are the importance of allotypes

1. **Monitoring bone marrow grafts** Bone marrow grafts that produce a different allotype from the recipient can be used to monitor the graft.
2. **Forensic medicine** Km and Gm allotypes are detectable in blood stains and semen. Hence they are useful in forensic medicine.
3. **Paternity testing** The immunoglobulin allotypes are used in legal cases involving paternity.

Idiotypes

The idiotypes are localized on the Fab fragment of the Ig molecules and it is characteristic of an individual antibody molecule or of molecules of identical specificity (Figure 4.14).

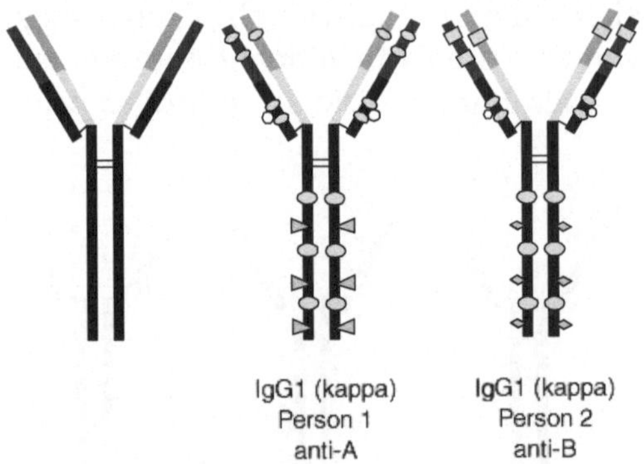

Figure 4.14 Immunoglobulins showing idiotype differences

Thus idiotypes are the antigenic determinants created by the hypervariable regions of an antibody and the antibodies that are directed against it are called as anti-idiotype antibodies. The idiotypes are useful markers for a particular variable region.

In some cases, anti-idiotypic antibodies stimulate B cells to produce antibody which may mimic the antigen. In this case the anti-idiotypic antibodies can be used as a vaccine.

IMMUNOGLOBULIN DEFICIENCY

All antibodies are made by B lymphocytes. Any disease that affects the development or function of B cells will cause a decrease in the amount of antibodies produced. Since antibodies are essential for fighting infectious diseases, people with immunoglobulin deficiency become ill more often. However, the cellular immune system will still be functional, so these patients are more prone to infection caused by organisms usually controlled by antibodies. Most of these invading microbes form capsules, a mechanism used to confuse the immune system. In a healthy body, antibodies can bind to the capsule and overcome the bacteria's defences. Bacteria that make capsules include the streptococci, meningococci and *Haemophilus influenzae*. These organisms cause such diseases as otitis, **sinusitis**, **pneumonia**, **meningitis**, **osteomyelitis**, septic arthritis and sepsis. Patients with immunoglobulin deficiencies are also prone to some viral infections, including echovirus, enterovirus and hepatitis B. They may also produce some reaction to the attenuated version of the **polio** virus vaccine.

There are two types of immunodeficiency diseases. They are primary and secondary. Secondary disorders occur in normally healthy people who suffer from an underlying disease. Once the disease is treated, the immunodeficiency is reversed. Immunoglobulin deficiency syndromes are called primary immunodeficiency diseases, if they occur due to defective B cells or antibodies. They account for 50% of all primary immunodeficiencies, and they are, therefore, the most prevalent type of immunodeficiency disorders.

X-linked Agammaglobulinaemia

It is an inherited disease. The **defect is on the X chromosome** and consequently, this disease is seen **more frequently in**

males than females (due to the lack of another X chromosome in male even a recessive mutation in any X-linked gene will be expressed). The defect results in a failure of B cells to mature. Mature B cells are capable of making antibodies and developing "memory", a feature in which the B cell will rapidly recognize and respond to an infectious agent the next time it is encountered. All classes of antibodies are decreased in agammaglobulinaemia.

Selective IgA Deficiency

It is also an inherited disease, resulting from failure of B cells to switch from making IgM, the early antibody, to IgA. Although the B cell numbers are normal (they can still make all other classes of antibodies), the amount of IgA produced is limited. This results in more infections of mucosal surfaces, such as the nose, throat, lungs and intestines.

Transient Hypogammaglobulinaemia of Infancy

It is a temporary disease of unknown cause. It is believed to be caused by a defect in the development of T helper cells (cells that recognize foreign antigens and activate T and B cells in an immune response). As the child ages, the number and condition of T helper cells improves and this situation corrects itself. Hypogammaglobulinaemia is characterized by low levels of gammaglobulin (antibodies) in the blood. During the disease period, patients have decreased levels of IgG and IgA antibodies. In laboratory tests, the antibodies that are present do not react well with infectious bacteria.

Common Variable Immunodeficiency

It is a defect in both B cells and T lymphocytes. It results in a near complete lack of antibodies in the blood.

Ig Heavy Chain Deletions

It is a genetic disease in which a part of the antibody molecule is not produced. It results in the loss of several antibody classes and subclasses, including most IgG antibodies and all IgA and IgE antibodies. The disease occurs because part of the gene for the heavy chain will be lost due to mutation.

Selective IgG Subclass Deficiencies

It is a group of genetic diseases in which some of the subclasses of IgG are not made. There are four subclasses in the IgG class of antibodies. As the B cell matures, it can switch from one subclass to another. In these diseases there is a defect in the maturation of the B cells that results in lack of class switching.

IgG Deficiency with Hyper-IgM

It is a disease that results when the B cell fails to switch from making IgM to IgG. This leads to increase in the amount of IgM antibodies and a decrease in the amount of IgG antibodies. This disease is a result of genetic mutation.

Causes and Symptoms

Immunoglobulin deficiencies are the result of congenital defects affecting the development and function of B lymphocytes (B cells). There are two main points that should be understood. First, B cells can fail to develop into antibody-producing cells. X-linked agammaglobulinaemia is an example of this kind. Second, B cells can fail to make a particular type of antibody or fail to switch classes during maturation. Initially, when B cells start making antibodies for the first time, they make IgM. As they mature and develop memory, they switch to one of the other four classes of antibodies. Failures in switching or failure to make subclass of antibody leads to immunoglobulin deficiency diseases. Another mechanism that results in decreased antibody production is a defect in T helper cells. Generally, defects in T helper cells are included under severe combined immunodeficiencies.

Symptoms are persistent and include frequent infections, diarrhoea, failure to thrive, and malabsorption of nutrients.

Diagnosis

Immunodeficiency disease is suspected when children become ill frequently, especially from infection by the same organism. The profile of organisms that cause infection in patients with immunoglobulin deficiency syndrome is unique and is preliminary evidence for this disease. Laboratory tests are performed to verify the diagnosis. Antibodies are found in the blood. Blood is collected and analysed for the content and types of antibodies present. Depending on the type of immunoglobulin deficiency the laboratory tests will show a decrease or absence of antibodies or specific antibody subclasses.

Treatment

Immunodeficiency diseases cannot be cured. Patients are treated with antibiotics and immune serum. Immune serum is a source of antibodies. Antibiotics are useful for fighting bacterial infections. There are some drugs that are effective against fungi, but very few drugs are effective against viral diseases.

Bone marrow transplantation can, in most cases, completely correct the immunodeficiency.

Prognosis

Patients with immunoglobulin deficiency syndromes must practice impeccable health maintenance and care, paying particular attention to optimal dental care, in order to stay in good health.

POINTS TO REMEMBER

- Antibodies are gammaglobulins.
- They are made up of four polypeptide chains, two heavy chains and two light chains.
- There are 5 classes of immunoglobulin molecules: IgG, IgA, IgM, IgD and IgE based on the presence of heavy chain.

- IgA is the predominant antibody in secretions.
- IgM is a pentamer. However membrane-bound IgM is a monomer.
- Immunoglobulins have 3 distinct antigenicities. They are isotype, allotype and idiotype.

REVIEW QUESTIONS

1. Write short notes on:
   i. Secretory IgA
   ii. Allotype
   iii. Isotype
   iv. Idiotype
   v. Domains of immunoglobulin
   vi. Hypervariable region
   vii. Hinge region
2. Write in detail the structure, types and functions of immunoglobulin.
3. Write a detailed account of immunodeficiency diseases.

5

ANTIBODY DIVERSITY

INTRODUCTION

For many years it remained a mystery as to how a B cell produces billions of different specific antibodies. If the concept of "one gene one polypeptide" is taken, then there should exist many millions of different genes in the cell which is quite impossible.

In 1965, Dreyer and Bennet proposed that there must be at least two genes for one polypeptide chain in immunoglobulin formation. They proposed that there must be hundreds or even thousands of different genes for the V region but only one constant region gene for each isotype of immunoglobulin.

In 1976, Tonegawa and Hozumi were the first to provide an experimental proof of the immunoglobulin formation by analysing the light chain genes from mouse myeloma cell DNA and comparing it to the light chain gene organization present in mouse embryonic DNA. They found that the V and C gene segments are brought together to form a combined V-C segment. Thus they concluded that the Ig gene undergoes gene rearrangement during B-cell development.

Susumu Tonegawa is a Japanese scientist who won the Nobel Prize for Physiology or Medicine in 1987 for his discovery of the genetic principle for generation of antibody diversity. Although he won the Nobel Prize for his work in Immunology, Tonegawa is a molecular biologist by training. In his later years, he turned his attention to the molecular and cellular basis of memory formation.

To achieve the diversity of antibodies needed to protect against any type of antigen, the immune system would require millions of genes coding for different antibodies, if each antibody was encoded by one gene. Tonegawa showed in a landmark series of experiments beginning in 1976 that genetic material can rearrange itself to form a vast array of available antibodies. Comparing the DNA of B cells in embryonic and adult mice, he observed that genes in the B cells of the older mice move around, recombine and are deleted to form the diversity of the variable region of antibodies.

Now it is well understood that the Ig gene rearrangement is responsible for the generation of Ig diversity in humans and other mammals. During the development of B cells, the Ig gene arrangement takes place and the surface immunoglobulin with a particular antigenic specificity is expressed on the surface of the B cell. Thus there are billions of B cells having different antigen-specific surface immunoglobulin. This surface immunoglobulin acts as the antigen receptor of the B cell and on binding with its specific antigen, B cells will get converted to plasma cells producing the specific antibodies.

IMMUNOGLOBULIN GENES

There are three gene groups that encode for immunoglobulin molecules. One for κ chain, one for λ chain (both are light chains) and one for the heavy chain, each on different chromosomes.

In humans, the heavy chain locus is found on chromosome 14, kappa chain locus on chromosome 2 and lambda chain locus on chromosome 22.

Heavy Chain Genes

The heavy chain gene locus shows basically two segments. They are variable and constant segments. The variable segment is again made up of three loci (Figure 5.1).

They are:

1. V locus (variable locus)
2. D locus (diversity locus)
3. J segment (joining segment)

The number of V segments in V locus is known to be 80 to 250 (differs from individual to individual). Each segment is accompanied by a leader sequence. The number of D segments in D locus is approximately 30 and J segments are 6 in number. The constant region contains various isotype genes that can determine the class of Ig.

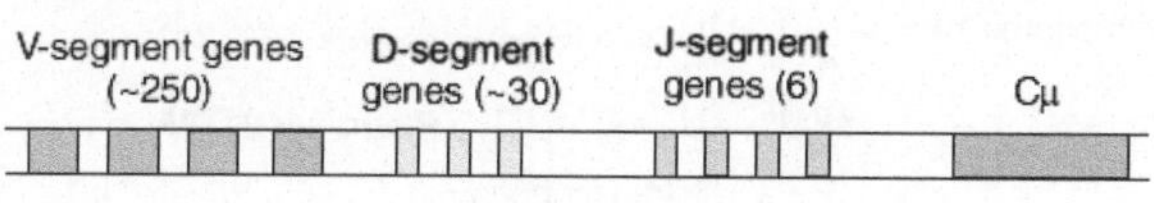

Figure 5.1 Heavy chain genes

Light Chain Genes

Like the heavy chain locus, light chain locus also contains both the constant and variable region.

The variable region of light chain consists of two loci. They are:

1. V locus
2. J locus

It does not possess a D locus like the heavy chain genes. It is believed that the V locus of kappa chain consists of 500 individual segments and 4 J segments of J locus. The V locus of lambda light chain consists of 100 V segments and several J segments for J locus.

GENE REARRANGEMENT

Heavy Chain Gene Rearrangement

During the development of B cell, the B cells randomly pick up some segments, form V segments, D segments and J segments, which are then combined together to form a gene segment that codes polypeptide chain of the variable region of heavy chain of the immunoglobulin. First the D and J segments are picked up and combined to form DJ segment. Then some V segments are picked up and it joins to DJ segment to form VDJ segment. Thus a variety of different V, D and J segments can be picked up in different B cell clones. As it is a random event, a B cell can make any combination of VDJ segment and attributes to antibody diversity. This is termed as combinational diversity.

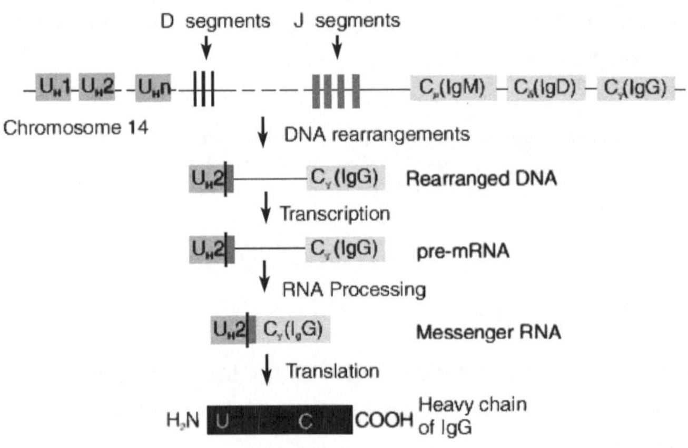

Figure 5.2 Heavy chain gene rearrangement and formation of heavy chain

After the formation of VDJ segment, some C segments from the C region will combine to form VDJC segment. The C segment will decide the isotype (the class of Ig) of the antibody.

The DNA thus formed will be the functional gene for the formation of the heavy chain of Ig (Figure 5.2).

> RSS (recombination signal sequences) flank all unarranged segments. They are the recognition sites for RECOMBINASE proteins (RAG-1/ RAG-2). The recombination signal sequences (RSS) is the essential element for the immunoglobulin gene rearrangement. The RSS consists of 3 elements: heptamer (7mer) 12 bp or 23 bp spacer and nanomer (9mer). The heptamer and nanomer are well conserved. The rearrangement occurs between an RSS with 12 bp spacer sequences and an RSS with 23 bp spacer sequences (the 12 23 rule). It ensures that V combines with D only. RSS are spliced out and coding sequences are joined.

Light Chain Gene Rearrangement

For the light chain, the V segments and J segments are picked up randomly and joined together to form the VJ segment. To this the C segment is joined to form VJC segment. Again this is a random event and the B cell can form any combination of VJ segment contributing to diversity of the antibody (Figure 5.3).

The DNA segment thus formed will be the functional gene for the light chain of Ig.

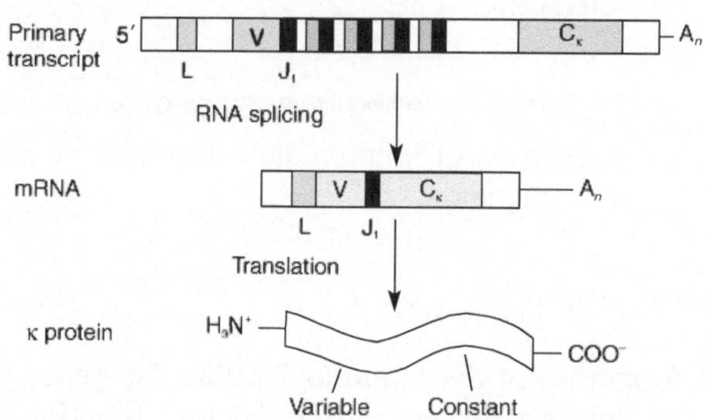

Figure 5.3 Light chain gene rearrangement

Figure 5.4 shows the formation of immunoglobulin molecule by Ig gene rearrangement.

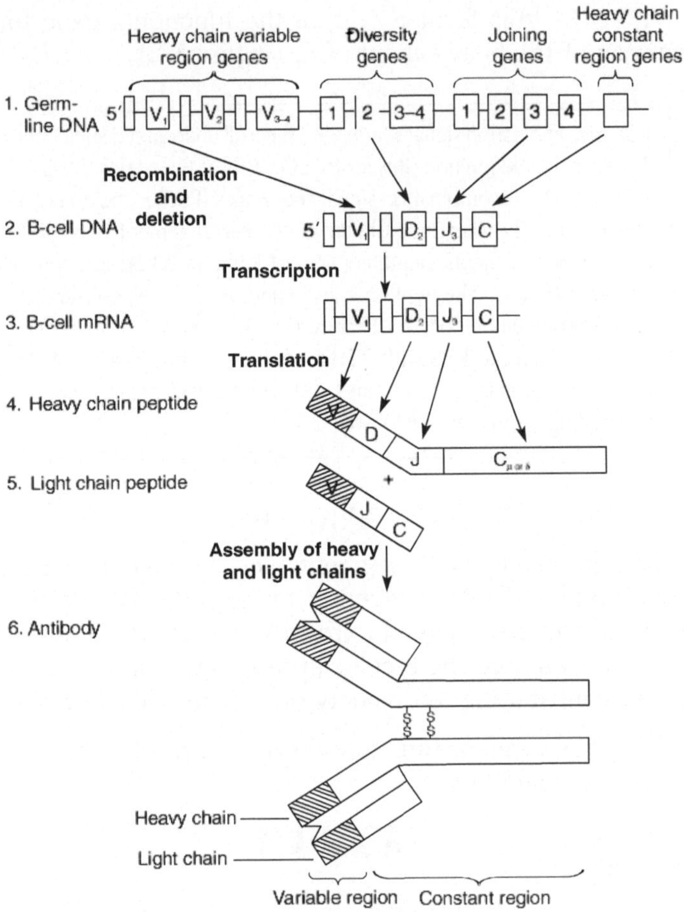

Figure 5.4 Formation of immunoglobulin molecule by Ig gene rearrangement

TRANSCRIPTION OF mRNA

For the formation of the immunoglobulin, the genes for heavy chain and light chain are transcribed into functional mRNA. During transcription of mRNA, only one segment of each V, D, J and C is transcribed for the heavy chain. For the light chain one segment of each V and J chain are transcribed to form mRNA.

The lymphocytes have two important genes called *RAG*-1 and *RAG*-2 (rearrangement activation gene). These genes produce two proteins that join together to form a dimer with enzymatic functions. These enzymes are responsible for the V region translocation. These enzymes break and rejoin the DNA during translocation and are thus important for the generation of diversity.

RAG mechanism: transesterification

1) Nick in one strand frees a hydroxyl
2) Hydroxyl attacks the opposite strand phosphate and leaves a hairpin (coding) and blunt end (RSS end)

ALLELIC EXCLUSION

As the successful rearrangement is completed for the V region of H chain and L chain, the rearrangement of the V region of the corresponding homologous chromosome is excluded. This is called as allelic exclusion. This may be probably due to the reason that *RAG*-1 and *RAG*-2 will be irreversibly inactivated after the successful translocation process.

But if there is any aberrant rearrangement in the first chromosome then the gene arrangement will be carried out by the corresponding chromosome.

As far as the light chain is concerned, first the kappa chain V region starts to rearrange. If both the homologous chromosomes fail to rearrange, the lambda chain will start to rearrange. Thus most of the immunoglobulins will have kappa chain as their light chain.

EXPRESSION OF SURFACE Ig

Once the translocation of gene is over, the antibodies are assembled in the endoplasmic reticulum. For the surface expression, H-chain gene thus transcribes mRNA that has coding sequence for its transmembrane domain. Once the Ig is formed, they get anchored in the plasma membrane by their C terminal region.

When the B cell is triggered by the antigen B-cell receptor binding, it will become an antibody-secreting cell called plasma cell. It is capable of secreting antibody of the same specificity of the membrane-bound Ig. The only difference is that the gene will produce the mRNA without the coding sequence of transmembrane domain. Thus the Ig produced will be secreted by the cell rather than getting anchored to the plasma membrane.

CREATION OF DIVERSITY

The following are the ways by which antibody diversity occurs.

1. Random rearrangement of V, D and J segments of heavy chain and the light chain.
2. Overlapping of the DNA segments during rearrangement.
3. Joining of segments sometimes creates gaps. The gaps may be filled with random nucleotides with the help of a special enzyme called terminal deoxyribonucleotidyl transferase (TdT).
4. Independent pairing of heavy and light immunoglobulin chain.

By the above process, the B cell clones express B-cell receptor (surface Ig) with different types of specificities. When an antigen attaches to any B-cell receptor due to its specificity they will get converted to plasma cell secreting the antibodies of the same specificity of the surface Ig. However, the isotype of the antibodies (classes and subclasses of Ig) is determined by the T_H cells and the cytokines that interact with the B cell.

Somatic Hypermutation (SHM) and Antibody Diversity

The diversifying mechanisms described above take place before the B cell encounters antigen. After a B cell encounters an antigen, it may begin mitosis, growing into a clone of cells synthesizing the same BCR (eventually secreting antibodies with the same binding site). Point mutations can occur while this goes on. Some of these may generate a binding site with increased affinity for its epitope. These are favourable mutations, and the "subclone" in which they occur tends to be favoured and may replace the ancestral clone. The result is affinity maturation—the production of antibodies with ever-increasing affinity for the antigen.

CLASS SWITCHING

During B-cell development, many B cells switch from making one class of antibody to making another—a process called class switching (Figure 5.5). All B cells begin to synthesize antibodies by making IgM molecules and inserting them into the plasma membrane as receptors for antigen. After the B cells leave the bone marrow and before interacting with antigen, they switch and make both IgM and IgD molecules as membrane-bound antigen receptors, both with the same antigen-binding sites. On stimulation by antigen and helper T cells, some of these cells are activated to secrete IgM antibodies, which dominate the primary antibody response. Later in the immune response, the combination of antigen and the cytokines that helper T-cells secrete induce many B cells to switch to making IgG, IgE or IgA antibodies. These cells generate both memory cells that express the corresponding classes of antibody molecules on their surface and effector cells that secrete the antibodies. The IgG, IgE and IgA molecules are collectively referred to as secondary classes of antibodies, both because they are produced only after antigen stimulation and because they dominate secondary antibody responses. As we saw earlier, each different class of antibody is specialized to attack microbes in different ways and in different sites.

The constant region of an antibody heavy chain determines the class of the antibody. Thus, the ability of B cells to switch the class of antibody they make without changing the antigen-binding site implies that the same assembled V_H region-coding sequence (which specifies the antigen-binding part of the heavy chain) can sequentially associate with different C_H-coding sequences (Figure 5.5). This has important functional implications. It means that, in an individual, a particular antigen-binding site that has been selected by environmental antigens can be distributed among the various classes of antibodies, thereby acquiring the different biological properties of each class.

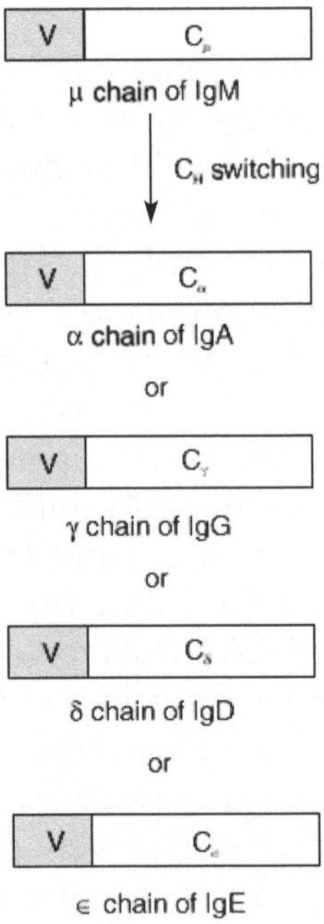

Figure 5.5 Class switching

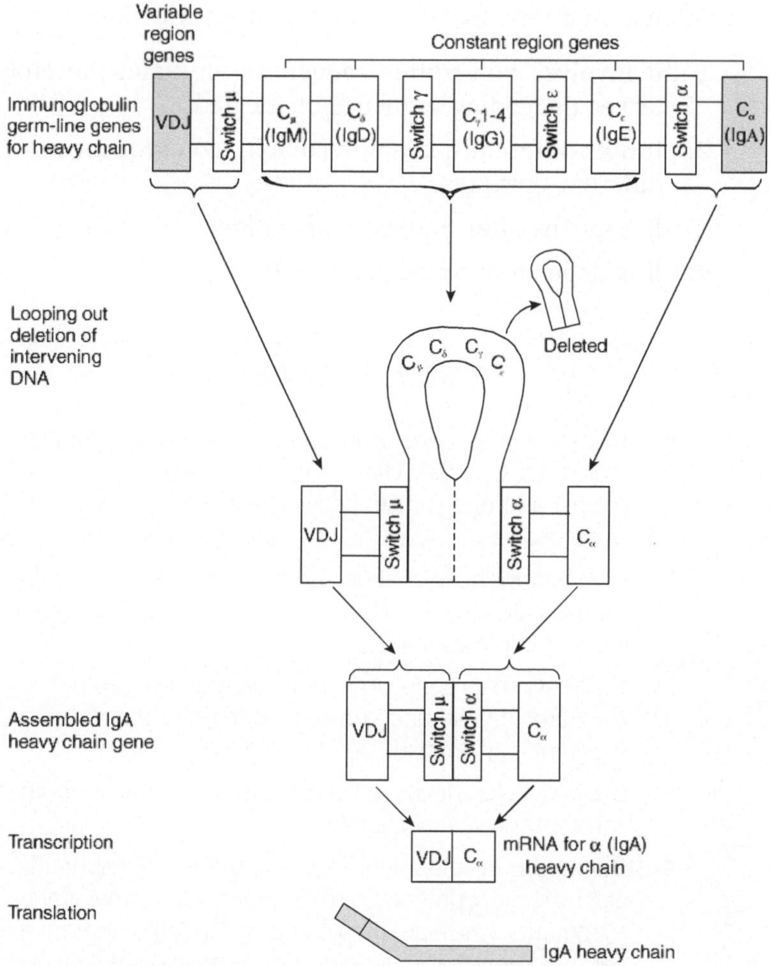

Figure 5.6 Mechanism of class switching

When a B cell switches from making IgM and IgD to one of the secondary classes of antibody, an irreversible change at the DNA level occurs—a process called class switch recombination (Figure 5.6). It entails deletion of all the C_H-coding sequences between the assembled VDJ-coding sequence and the particular

C_H-coding sequence that the cell is destined to express. Switch recombination differs from VDJ joining in several ways:

1. It involves non-coding sequences only and therefore leaves the coding sequence unaffected.
2. It uses different flanking recombination sequences and different enzymes.
3. It happens after antigen stimulation.
4. It is dependent on helper T cells.

POINTS TO REMEMBER

- There are three gene groups encoding immunoglobulin molecules. One for λ chain, one for κ chain (both are light chains) and one for the heavy chain, each on different chromosomes.
- In human, the heavy chain locus is found on chromosome 14, kappa chain locus on chromosome 2 and lambda chain locus on chromosome 22.
- The heavy chain gene locus shows basically two segments. They are variable and constant segments. The variable segment is again made up of three loci.
- Like heavy chain locus, light chain locus also contains both constant and variable region.
- During the development of B cell, the B cells randomly pick up some segments, form V segments, D segments and J segments, which are then combined together to form a gene segment that codes polypeptide chain of the variable region of heavy chain of the immunoglobulin.
- For the light chain, the V segments and J segments are picked up randomly and joined together to form the VJ segment.

REVIEW QUESTIONS

1. Write short notes on:
   i. Heavy chain genes
   ii. Light chain genes
   iii. Class switch
   iv. Allelic exclusion
   v. Somatic hypermutation
2. Write a detailed note on antibody diversity.

6

MAJOR HISTOCOMPATIBILITY COMPLEX

INTRODUCTION

When the tissues and organs are transplanted among the same species (allograft) they are recognized as foreign and are rejected. In the early 1930s, the availability of inbred mice marked the beginning of the experiments on transplantation.

Gorer identified two types of antigen system in mice. The one that was commonly found among species was named as H1 and the other antigen, restricted to some strains and found to be responsible for allograft rejection was called as H2 where H denotes histocompatibility. These antigens are expressed on the cells or tissues and induce an immune response for the graft rejection. These antigens are found to be produced by a group of multiallelic clusters of genes. This cluster of genes is called as major histocompatibility complex (MHC). In mice H2 is the MHC.

After the discovery of MHC in mice, various researches were carried out to identify the MHC of humans. The identification of MHC in humans in early stages was done with the help of sera of multiparous women (women who had delivered many children). When the woman becomes pregnant, the antigen of the foetus that is inherited from the father stimulates her immune system and produces antibodies against it. These antisera of the women were very helpful to identify the MHC in human.

George Davis Snell

Geneticist George Davis Snell (December 19, 1903 to June 6, 1996) is well known principally for his role in the discovery of H2, the major histocompatibility complex (MHC) of the mouse and the first known MHC. For this he was awarded Nobel Prize in Physiology or Medicine in the year 1980.

Gorer worked in parallel with Irwin and Coles, who had coined the term "immunogenetics" to describe their work on antigens of avian red cells. Gorer raised rabbit antisera to red blood cells of mice, which upon absorption distinguished two antigens present in different strains. He then joined George in demonstrating that his antigen II segregated in F2 mice together with the gene fused (Fu), which George had found to be linked to transplant rejection. Based on this the gene encoding the antigen was named H2 (H for histocompatibility and 2 for antigen II) and represents the first sighting of what later came to be called the major histocompatibility complex. It is worth noting that their 1947 paper wisely refrains from claiming that the antigen expressed on red blood cell antigen caused the rejection or that antibodies of the isoagglutinin type were responsible.

MAJOR HISTOCOMPATIBILITY IN HUMANS

It is found that the MHC complex antigens are present on the leucocytes. Thus these antigens are called as human leucocyte antigens (HLA) and MHC in human is called as HLA complex (Figure 6.1).

The HLA complex is found on the short arm of chromosome 6. It contains seven genetic loci organized into two classes namely class I and class II. The class I consists of three gene loci. They are HLA-A, HLA-B and HLA-C.

The class II consists of four gene loci, viz. HLA-D, HLA-DR, HLA-DQ and HLA-DP.

The class I gene locus produces molecules which are called as class I molecules or class I antigens and products of class II gene locus are called as class II molecules or class II antigens.

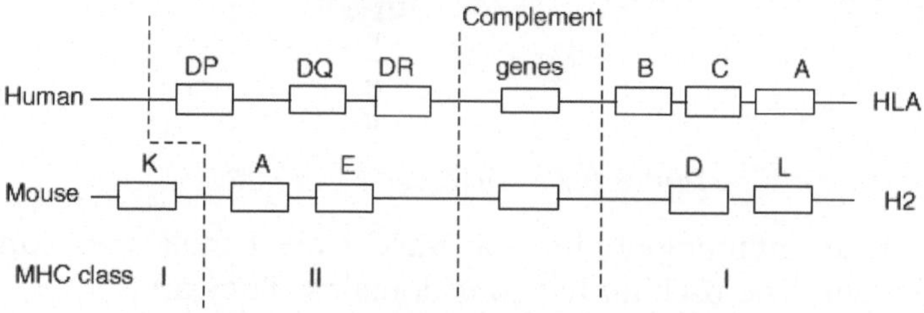

Figure 6.1 MHC in human and mouse

Nomenclature of the HLA System

Each gene locus of the HLA system is multiallelic (one of the several alternative forms of a gene locus). Thus an internationally accepted system of naming HLA system has been evolved. The antigens are named with their gene locus and a number (e.g. HLA-A3, HLA-B6, HLA-DR2, etc.). Those antigens which are identified and not confirmed are named "w" (w = workshop) along with their gene locus. The "w" will be deleted once the antigen is internationally accepted and recognized.

STRUCTURE OF MHC CLASS I MOLECULE

The class I molecules are made up of two polypeptide chains. The larger polypeptide chain is called as α-chain which has a molecular weight of 44 kDa and is encoded by an MHC class I gene. The smaller chain is called as β-2 microglobulin with a molecular weight of 12 kDa. The smaller chain is encoded by a gene that is present outside the MHC complex (Chromosome 15). The β-2 microglobulin is vital for the formation of MHC class I molecule (Figure 6.2).

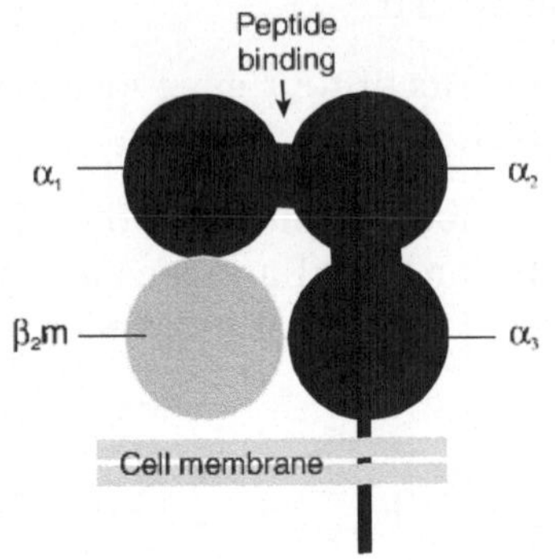

Figure 6.2 MHC class I molecule

Like immunoglobulin, the MHC class I molecules contain domains. The α-chain has four domains. They are:

1. Peptide-binding domain
2. Immunoglobulin-like domain
3. Transmembrane domain
4. Cytoplasmic domain

The peptide-binding domain is the only region that shows the allelic differences in the amino acid sequences whereas the other regions are common among all allelic antigens. As the name indicates, this domain is the site where the antigenic peptides are attached. The immunoglobulin-like domain resembles the domain of an immunoglobulin C-region. It contains the site for the binding of CD8 molecule of T lymphocyte. The transmembrane and the cytoplasmic domains are helpful for the attachment of MHC class I molecule to the plasma membrane and for the proper expression of the cell. The function of β-2 microglobulin is to stabilize the MHC class I molecule.

The important function of class I molecule is to bind to the processed endogenous antigen and to present it to the immunocompetent cells.

MHC class I expression is widespread on virtually every cell of the body. This is consistent with the protective function of cytotoxic T lymphocytes which continuously survey cell surfaces and kill cells harbouring metabolically active microorganisms.

MHC class I molecules bind peptide fragments derived from proteolytically degraded proteins endogenously synthesized by a cell. Small peptides are transported into the endoplasmic reticulum where they associate with nascent MHC class I molecules before being routed through the Golgi apparatus and displayed on the surface for recognition by cytotoxic T lymphocytes.

STRUCTURE OF MHC CLASS II MOLECULE

The MHC class II molecule contains two polypeptide chains. One chain is called as α-chain and another is β-chain. The molecular weight of α-chain is 32 to 34 kDa and that of β-chain is 29 to 32 kDa. Both are coded by different genes present in class II region (Figure 6.3).

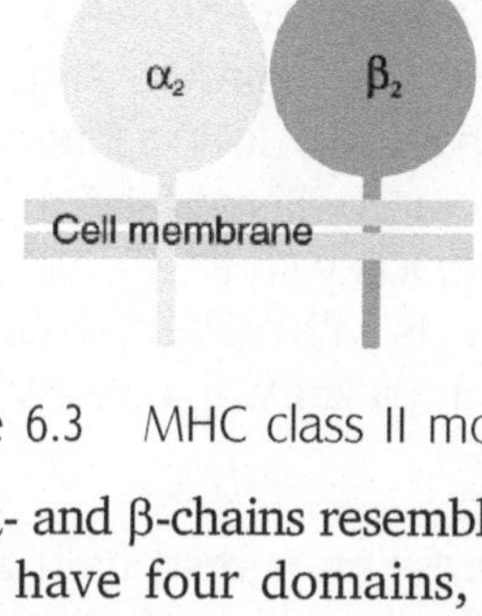

Figure 6.3 MHC class II molecule

The structure of α- and β-chains resemble the β-chain of class I molecule. They also have four domains, viz. peptide-binding, immunoglobulin-like, transmembrane and cytoplasmic domains.

The peptide-binding groove of the class II molecule is contributed by both the chains unlike class I molecule. The peptide binding site is deeper in class II molecule and can bind large peptide molecules. The immunoglobulin-like domain of the class II molecule binds to the CD4 molecule of T lymphocyte.

The functional significance of the MHC class II molecule is to present the exogenous antigen to the immunocompetent cells.

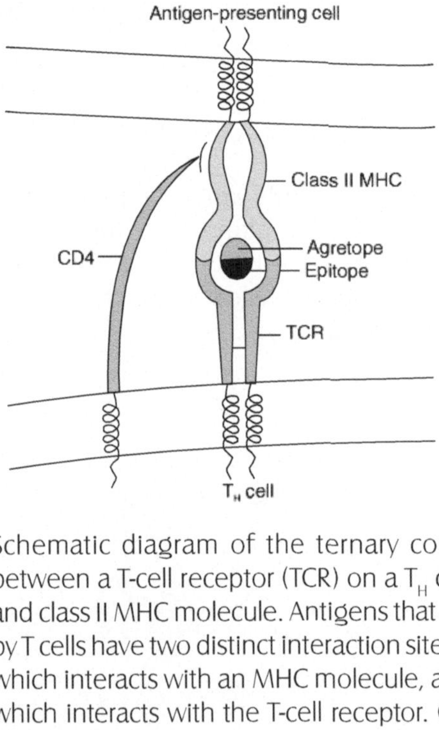

Figure 6.4 Schematic diagram of the ternary complex formed between a T-cell receptor (TCR) on a T_H cell, an antigen and class II MHC molecule. Antigens that are recognized by T cells have two distinct interaction sites: an agretope, which interacts with an MHC molecule, and an epitope, which interacts with the T-cell receptor. CD4 on T_H cells also interacts with MHC molecules. T_C cells form similar ternary complexes with class I MHC molecules on target cells.

MHC class II expression is restricted to "antigen-presenting cells". This is consistent with the functions of helper T_H

lymphocytes which are locally activated wherever these cells encounter macrophages, dendritic cells or B cells that have internalized and processed antigens produced by pathogenic organisms.

MHC class II molecules bind peptide fragments derived from proteolytically degraded proteins exogenously internalized by "antigen-presenting cells", including macrophages, dendritic cells and B cells. The resulting peptide fragments are compartmentalized in the endosome where they associate with MHC class II molecules before being routed to the cell surface for recognition by the helper T_H lymphocytes (Figures 6.4 and 6.5).

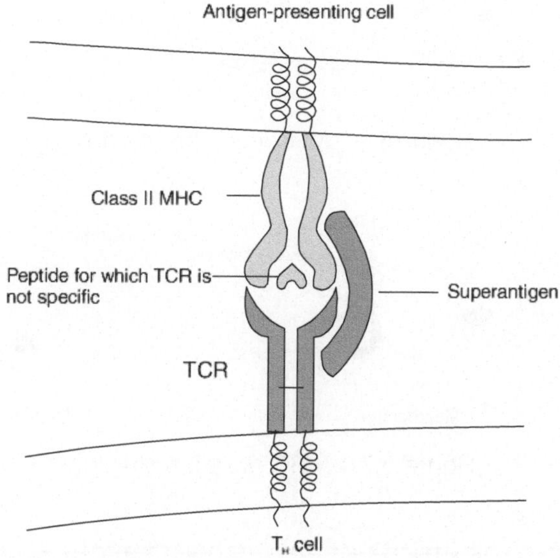

Figure 6.5 Schematic diagram of the ternary complex formed between a T-cell receptor (TCR), superantigen and class II MHC molecule. Superantigens bind to common sequences in class II MHC molecules and T-cell receptors that lie outside the normal antigen-binding sites. T-cell activation by superantigens is not limited by the antigenic specificity of the T cell.

MHC RESTRICTION

MHC restriction means that different T cells are restricted to either class I or class II MHC antigens. Cytotoxic T cells are restricted to class I antigens present on nucleated body cells and thus play a role in protecting against virus-infected cells or cancerous cells. Helper CD4 cells are restricted to class II MHC antigens present on immune cells, thus they play a role in increasing the humoral immune response (Figures 6.6 and 6.7).

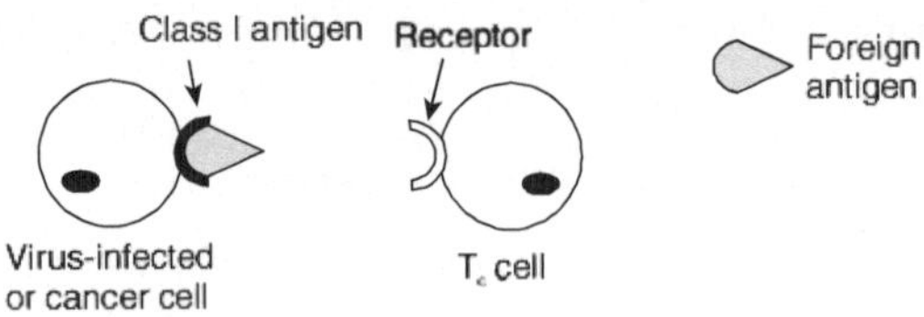

Figure 6.6 MHC class I restriction

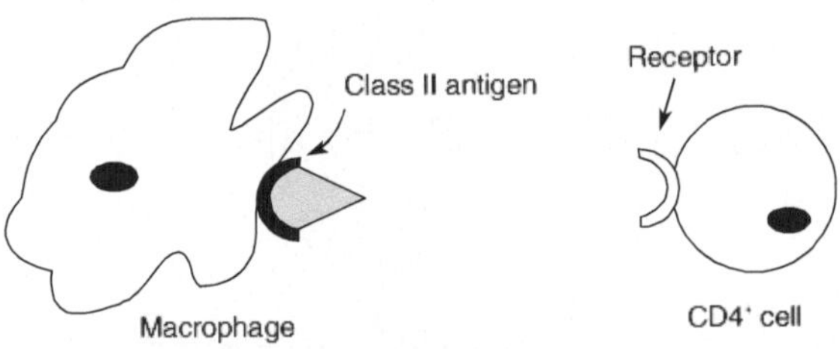

Figure 6.7 MHC class II restriction

Class III MHC genes encode several components of the complement system, a collection of soluble proteins found in the blood that targets foreign cells and break open their membranes. Adjacent to the class III region is a group of genes that control inflammation. Further genes with various immune and non-immune functions are distributed throughout the complex.

POINTS TO REMEMBER

- It is found that the MHC complex antigens are present on the leucocytes.
- The antigens are called as human leucocyte antigens (HLA) and MHC in human is called as HLA complex.
- The HLA complex is found on the short arm of chromosome 6.
- The MHC class I consists of three gene loci. They are HLA-A, HLA-B and HLA-C.
- The MHC class II consists of four gene loci, viz. HLA-D, HLA-DR, HLA-DQ and HLA-DP.
- The class I molecules are made up of two polypeptide chains. The larger polypeptide chain is called as α-chain. The smaller chain is called as β-2 microglobulin.
- The MHC class II molecule contains two polypeptide chains. One chain is called as α-chain and another is β-chain.

REVIEW QUESTIONS

1. Write short notes on:
   i. H2
   ii. HLA
   iii. MHC class I genes
   iv. MHC class II genes
   v. MHC class I molecules
   vi. MHC class II molecules
   vii. MHC restriction
   viii. Agretope
2. Explain in detail the structure and function of MHC class I and MHC class II molecules.
3. Write in detail on the genetics of MHC.

7

ANTIGENS

DEFINITIONS

Immunogen It is a substance that induces a specific immune response.

Antigen (Ag) They are the substances that react with the products (antibodies and immunocompetent cells) of a specific immune response.

Hapten It is a substance that is non-immunogenic but which can react with the products of a specific immune response (Figure 7.1). Haptens are small molecules which do not induce an immune response when administered alone but can do so when coupled to a carrier molecule. Free haptens, however, can react with products of the immune response after such products have been elicited. Haptens have the property of antigenicity but not immunogenicity.

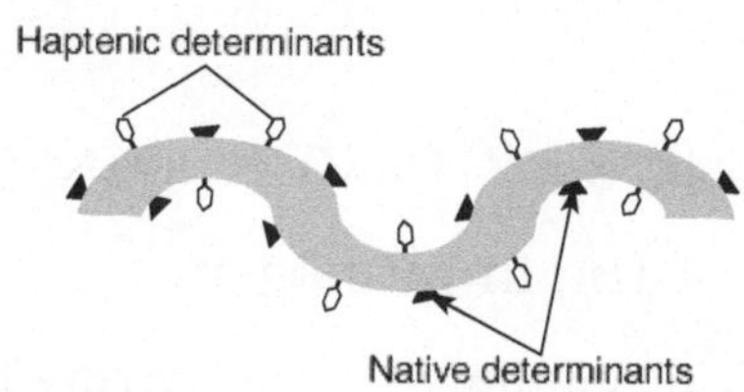

Figure 7.1 An antigen showing the haptenic determinant

Epitope It is that portion of an antigen that combines with the products of a specific immune response.

Antigenic determinant site It is the region of the antigen that can produce an antibody to which it can specifically bind. The simple type of antigen will have the same type of antigenic determinant site whereas the complex antigen will have different types of antigenic determinant sites. The antigenic determinant site for a protein antigen will be normally a pentapeptide or hexapeptide and for polysaccharide antigen it will be hexasaccharide.

Antibody (Ab) It is a specific protein which is produced in response to an immunogen and which reacts with an antigen.

PROPERTIES OF ANTIGENS

Immunogenicity It is the property that allows a substance to induce a detectable immune response (humoral or cellular) when introduced into an animal. Such substances are termed immunogens.

Antigenicity It is the property that allows a substance to combine specifically with antibodies or TCR (T-cell receptor), whether or not they are immunogenic. Therefore, all immunogens are antigens but all antigens are not immunogens.

Allerogenicity It is the property that allows a substance to induce an allergic response. Such substances are termed allergens.

Tolerogenicity It is the property that allows a substance to induce specific immunological non-responsiveness in either the humoral or cell-mediated branch. Such substances are termed tolerogens.

FACTORS INFLUENCING IMMUNOGENICITY

Contribution of the Immunogen

Foreignness The immune system normally discriminates between self and non-self such that only foreign molecules are immunogenic (Human albumin is not immunogenic in another human but would be immunogenic in a rabbit).

Molecular size Certain minimum size is required for immunogenicity. The most potent immunogens are macromolecular proteins with molecular weights greater than 100,000 daltons. Substances <1000 are not usually immunogenic.

Chemical complexity Homopolymers consisting of repeating units of a single amino acid are poor immunogens regardless of their size. Co-polymers of 2 or 3 amino acids may be good immunogens. A co-polymer of glutamic acid and lysine must be 30–40,000 daltons to be immunogenic. In general, immunogenicity increases with structural complexity. Aromatic amino acids contribute more to immunogenicity than non-aromatic (e.g. tyrosine or phenylalanine). This also explains why most of the polysaccharide antigens are poorly immunogenic. The polysaccharides are mostly homopolymers.

Degradability Macromolecules that cannot be degraded and processed by antigen-presenting cells are poor immunogens. For example, polymers of D-amino acids are not immunogenic. Proteolytic enzymes can only degrade proteins containing L-amino acids.

Contribution of the Biological System

Genetic factors Some substances are immunogenic in one species but not in another. Similarly, some substances are immunogenic in one individual but not in others. The species or individuals may lack or have altered genes that code for the receptors for antigen on B cells and T cells or they may not have the appropriate genes needed for the antigen-presenting cell to present antigen to the helper T cells.

Age Age can also influence immunogenicity. Usually the very young and the very old have a diminished ability to mount an immune response in response to an immunogen.

Method of Administration

Dose The dose of an immunogen can influence its immunogenicity. There is a dose of antigen above or below which the immune response will not be optimal.

Route Generally the subcutaneous route is better than intravenous or intragastric routes. The route of antigen administration can also alter the nature of the response.

Adjuvants Substances that can enhance the immune response to an immunogen are called adjuvants. The use of adjuvants, however, is often hampered by undesirable side effects such as fever and inflammation.

Chemical Nature of Immunogen

Proteins The vast majority of immunogens are proteins. These may be pure proteins or they may be glycoproteins or lipoproteins. In general, proteins are usually very good immunogens.

ANTIGEN vs. IMMUNOGEN

An immunogen is a substance capable of initiating an immune response. Immunogenicity is the ability to induce a humoral and/or a cell-mediated response. An antigen is any substance that is specific to an antibody or a T-cell receptor, and is often used as a synonym for immunogen, even though they are not the same by definition.

Antigenicity is the ability to combine specifically with antibodies or T-cell receptors. Though all immunogens are antigens, not all antigens are immunogens. Some small molecules called haptens, are antigenic but incapable, by themselves of inducing a specific immune response. In other words, they lack immunogenicity. However, if a hapten molecule binds to another antigen that also may lack immunogenicity, the resulting molecule may in fact initiate an immune response.

Polysaccharides Pure polysaccharides and lipopolysaccharides are good immunogens.

Nucleic acids Nucleic acids are usually poor immunogens. However, they may become immunogenic when they are single-stranded or when complexed with proteins.

Lipids In general lipids are non-immunogenic, although they may be haptens.

MITOGENS

Mitogens are agents that are able to induce cell division (mitosis) in a high percentage of T or B cells. This proliferation is described as polyclonal activation (that can act on all lymphocytes). There are T-cell and B-cell mitogens.

Most of the common mitogens are lectins. Lectins are proteins which bind to specific carbohydrate groups (moieties). They bind to glycoproteins on the surface of the lymphocytes and cause activation. Importantly they do not act via conventional TCR-epitope or Ig-epitope interactions.

Some examples of lectins are

Con A	T-cell mitogen
PHA	T-cell mitogen
PWM	T- and B-cell mitogen

LPS (lipopolysaccharide) is not a lectin but it does function as a mitogen for B cells.

TUMOUR ANTIGENS

Tumour antigens are those antigens that are presented by the MHC I molecules on the surface of tumour cells. These antigens can sometimes be presented only by tumour cells and never by the normal ones. In this case, they are called tumour-specific antigens and typically result from a tumour-specific mutation.

More common are antigens that are presented by tumour cells and normal cells, and they are called tumour-associated antigens. Cytotoxic T lymphocytes that recognize these antigens may be able to destroy the tumour cells before they proliferate or metastasize.

Tumour antigens can also be on the surface of the tumour in the form of, for example, a mutated receptor, in which case they will be recognized by B cells.

HETEROPHIL ANTIGENS

There are many examples of totally unrelated antigens that have small parts or epitopes (or antigenic determinants) in common. Thus an immune response to one such antigen can cause the production of antibodies that react with the second antigen. Such antigens can be called as heterophil antigens. Likewise, the antibodies produced against such antigens are often called heterophil antibodies. Some examples are as follows:

- Human blood group A antigen and pneumococcal capsule polysaccharide
- Human blood group B antigen and *E. coli* polysaccharide antigens
- *Streptococcus pyogenes* M-protein and human heart muscle protein
- Horse red blood cell antigens and Epstein–Barr virus

ISOANTIGENS

There are certain antigens that are same among specific groups in the same species. Based on these antigens, many groups can be identified in the species. These antigens can be called as **isoantigens**. An example of the isoantigen is the blood group antigen. The blood group antigens are called isoantigens or sometimes isoagglutinogens (as it can agglutinate the RBCs, these antibodies are called as isoagglutinins). Another good example for isoantigen is the HLA antigen.

EXOGENOUS AND ENDOGENOUS ANTIGENS

Exogenous Antigens

Exogenous antigens are extracellular antigens that have entered the body from outside, for example, by inhalation, ingestion or injection. By endocytosis or phagocytosis, these antigens are processed into fragments by antigen-presenting cells (APCs). APCs then present the fragments to T helper cells ($CD4^+$) by the use of MHC class II molecules on their surface. Some T cells are specific for the peptide–MHC complex. They become activated and start to secrete cytokines. Cytokines are substances that can activate cytotoxic T lymphocytes (CTL), antibody-secreting B cells, macrophages and other cells.

Endogenous Antigens

Endogenous antigens are antigens that are generated within the cell, as a result of normal cell metabolism, or because of viral or intracellular bacterial infection. The fragments are then presented on the cell surface in the complex with class I histocompatibility molecules. If activated cytotoxic $CD8^+$ T cells recognize them, the T cells begin to secrete different toxins that cause the lysis or apoptosis of the infected cell. In order to keep the cytotoxic cells from killing cells just for presenting self-proteins, self-reactive T cells are deleted from the repertoire as a result of central tolerance (also known as negative selection which occurs in the thymus). Only those CTLs that do not react to self-peptides that are presented in the thymus in the context of MHC class I molecules are allowed to enter the bloodstream.

SUPERANTIGENS

When the immune system encounters a conventional T-cell-dependent antigen, only a small fraction (1 in 104–105) of the T cell population is able to recognize the antigen and become activated (monoclonal/oligoclonal response). However, there are some antigens which polyclonally activate a large fraction of the T cells (up to 25%). These antigens are called superantigens.

Examples of superantigens include: staphylococcal enterotoxins (food poisoning), staphylococcal toxic shock toxin (toxic shock syndrome), staphylococcal exfoliating toxins (scalded skin syndrome) and streptococcal pyrogenic exotoxins (shock). Although the bacterial superantigens are the best studied, there are superantigens associated with viruses and other microorganisms as well.

The diseases associated with exposure to superantigens are, in part, due to hyperactivation of the immune system and subsequent release of biologically active cytokines by activated T cells.

T-CELL-INDEPENDENT ANTIGENS

T-cell-independent antigens are antigens which can directly stimulate the B cells to produce antibody without the requirement of T cell help. In general, polysaccharides are T-independent antigens. The responses to these antigens differ from responses to other antigens.

Properties of T-independent Antigens

- **Polymeric structure** These antigens are characterized by the same antigenic determinant repeated many times.
- **Polyclonal activation of B cells** Many of these antigens can activate B cell clones specific for other antigens (polyclonal activation). Thus T-cell-independent antigens can be subdivided into type 1 and type 2 based on their ability to polyclonally activate B cells. Type 1 T-independent antigens are polyclonal activators while type 2 antigens are not.
- **Resistance to degradation** T-cell-independent antigens are generally more resistant to degradation and thus they persist for longer periods of time and continue to stimulate the immune system.

T-CELL-DEPENDENT ANTIGENS

T-cell-dependent antigens are those that do not directly stimulate the production of antibody without the help of T cells. Proteins are T-dependent antigens. Structurally these antigens are characterized by a few copies of many different antigenic determinants.

ADJUVANT

An adjuvant is any substance that **enhances the response of the immune system** to an antigen. An adjuvant can also be any substance that enhances the effect of a drug in the body.

Adjuvants exert their effect in several different ways:

1. Some adjuvants retain the antigen and thus present the antigen to the immune system over a prolonged period of time. The immune response does not occur all at once, but rather is continuous over a longer time.
2. An adjuvant itself can react with some of the cells of the immune system. This interaction may stimulate the immune cells to heightened activity.
3. An adjuvant can also enhance the recognition and ingestion of the antigen by the phagocytes. This enhanced phagocytosis presents more antigens to the other cells that form the antibody.

The adjuvant selected typically depends on the animal being used to generate the antibodies. Different adjuvants produce different responses in different animals. Some adjuvants are inappropriate for certain animals, due to inflammation, tissue damage, and pain caused to the animal. Other factors that influence the choice of an adjuvant include the injection site, the manner of antigen preparation and the amount of antigen injected.

The most commonly used adjuvants are Alum and Freund's adjuvant.

Alum (aluminium potassium sulphate) forms precipitate with the antigens and when injected into the animal leads to the slow and prolonged release of antigen from the injection site. Alum also increases the size of the antigen thereby enhancing phagocytosis.

Freund's adjuvants also prolong the release of the antigen. There are two types of Freund's adjuvants. Freund's incomplete adjuvant contains water, mineral oil and an emulsifier. This adjuvant disperses the oil into small oil droplets into which the antigens are buried. This leads to the slow release of antigens. Freund's complete adjuvant additionally contains heat-killed mycobacteria. The muramyl dipeptide present in the cell wall of the mycobacteria activates macrophages and thus this adjuvant is more potent than the incomplete adjuvant.

Alum and Freund's adjuvants can also stimulate the local inflammatory reaction leading to the formation of granuloma (a dense macrophage-rich mass of cells). This will again help in the increase of the immunogenicity of the antigen.

EPITOPE RECOGNITION BY T CELL AND B CELL

T cells and B cells exhibit fundamental differences in epitope recognition.

B-cell Epitopes

The binding of antigen to immunoglobulin involves weak non-covalent interactions. So there must be complementarity between the two.

Properties of B-cell epitopes

1. Epitopes may be associated with soluble immunogens or particulate immunogens.
2. Epitopes tend to be accessible on the exposed surface of the immunogen.
3. Epitopes generally contain hydrophilic amino acids.

4. Often the epitopes are found where the molecule bends or where there is a high degree of segmental mobility (atomic mobility).
5. The epitope may consist of either sequential or non-sequential amino acids. If non-sequential, the 3D conformation of the epitope is vital.
6. Complex proteins contain multiple overlapping epitopes.
7. The size of a B-cell epitope is determined by the size of the antigen-binding site on the antibody molecule. Conformationally determined epitopes tend to require a larger epitope. Smaller ligands (carbohydrates, peptides, haptens, etc.) tend to fit within a deep concave pocket or crevice. Larger globular antigens tend to interact with Ig across a large planar face. Protrusions on the antigen-binding site would be complementary with depressions on the epitope and vice versa. This type of interaction is obviously highly dependent upon the 3D conformation of the globular antigen.

T-cell Epitopes

T cells recognize processed peptides associated with MHC on the surface of APCs (class II MHC) or altered self cells (class I MHC). In other words, T cells exhibit MHC restricted antigen recognition.

$CD4^+$ T cells are restricted to class II MHC and $CD8^+$ T cells are restricted to class I MHC.

More recent evidence suggests that a small population of T cells may possess TCRs capable of recognizing lipids or glycolipids. Little is known about this type of recognition and we will restrict our discussion to peptide epitopes.

1. The binding of the TCR to MHC-peptide represents a tri-molecular complex.
2. Only oligomeric peptides serve as epitopes. Epitopes which bind to class I MHC have an optimal size of 9 amino acids with a range of 8–11. Epitopes which bind

to class II MHC have an optimal size range of 12–25 amino acids.

3. Epitopes are often internal and are only exposed by processing within APCs or target cells.
4. Antigen processing is required to generate the peptides that interact specifically with MHC molecules.
5. T-cell epitopes must have two binding regions. The region of the peptide which binds to MHC is termed the agretope while the region which binds to the TCR is termed as epitope.
6. Complex proteins may contain multiple, overlapping epitopes.
7. Immunodominant T-cell epitopes are determined by the set of MHC molecules which are expressed by an individual.

Experimentally it has been demonstrated that there is a correlation between the ability of a peptide to bind to a particular MHC molecule and the T-cell response to that peptide.

WHY PENICILLIN ALLERGY?

Penicillin and other drug allergies can be induced by small doses of the drug, but they are not consequences of the pharmacological or physiological effects of the drugs. Penicillin, like most drugs, has a low molecular weight and is not capable of inducing an immune response unless it binds to another molecule to form a larger molecule. Penicillin can react to form a penicilloyl-protein derivative. The penicilloyl-protein behaves as a hapten-carrier conjugate, with penicillin-derived penicilloyl group acting as a hapten epitope. The epitope is recognized by the immune system and antibodies are produced against it. One of the antibodies produced is immunoglobulin E (IgE). Once produced, the IgE antibodies bind to mast cells and basophils. It is the presence of mast cells and basophils bound with penicillin-specific IgE that creates the allergic reaction.

POINTS TO REMEMBER

- Antigens are the substances that react with the products (antibodies and immunocompetent cells) of a specific immune response.
- Haptens are substances that are non-immunogenic but can react with the products of a specific immune response.
- Epitope is the portion of an antigen that combines with the products of a specific immune response.
- Exogenous antigens are extracellular antigens that have entered the body from the outside, for example by inhalation, ingestion or injection.
- Endogenous antigens are antigens that have been generated within the cell, as a result of normal cell metabolism, or because of viral or intracellular bacterial infection.
- T-cell-independent antigens are antigens which can directly stimulate the B cells to produce antibody without the requirement for T cell help.
- T-cell-dependent antigens are those that do not directly stimulate the production of antibody without the help of T cells. Proteins are T-dependent antigens.
- An adjuvant is any substance that enhances the response of the immune system to the antigen.

REVIEW QUESTIONS

1. Write short notes on:
   i. Antigens
   ii. Immunogen
   iii. Isotope
   iv. Hapten
   v. Allergen

vi. Adjuvant

vii. Antigenic determinant site

viii. Isoantigen

ix. Heterophil antigen

x. T-cell-independent antigen

xi. T-cell-dependent antigen

2. Write a detailed note on the factors that influence the antigenicity and immunogenicity.

8

T-CELL AND B-CELL MATURATION

INTRODUCTION

When the lymphocytes originate from the pluripotent stem cells they do not have the identity of lymphocytes. These stem cells upon entering into the primary lymphoid organs develop into lymphocytes. After encountering the antigen, they develop further and become complete immunocompetent cells.

B CELL

Generation of B Cells

B-cell development begins in the foetal liver and continues in the bone marrow throughout our lives. Once the cell expresses both the heavy and light chains of IgM on its membrane, it becomes a B cell. However, it is still immature and can be easily killed by contact with self-antigen unless it also expresses membrane IgD. The mature B cell that moves into the periphery can be activated by antigen and become an antibody-secreting plasma cell or a memory B cell, which will respond more quickly to a second exposure to antigen. B cells which fail to successfully complete B-cell development undergo apoptosis (programmed cell death).

B cells are produced through several stages, each stage representing a change in the genome content, and a variety of antibodies are produced (Table 8.1). The human antibody is composed of two light and two heavy chains, and there are genes specifying them, which are known as the H-chain loci and the L-chain loci. In the H-chain loci there are three regions: V, D and J, and combinations of these are drawn, in terms of rearrangement which results in deletions of bases between the two selected points, and results in the formation of a unique combination. In the L-chain loci there are only two regions, namely V and J, which undergo the same process.

The B-cell developmental stages are given below.

- *Progenitor B cells*—contain germ-line H genes, germ-line L genes.
- *Early pro-B cells*—undergo DJ rearrangement on the H chains.
- *Late pro-B cells*—undergo VDJ rearrangement on the H chains.
- *Large pre-B cells*—the H chain is VDJ rearranged, germ-line L genes.
- *Small pre-B cells*—undergo VJ rearrangement on the L chains.
- *Immature B cells*—VJ rearranged on L chains, VDJ rearranged on H chains. There is start of expression of IgM receptors.
- *Mature B cells*—there is start of expression of IgD.

When the B cell fails in any step of the maturation process, it will undergo apoptosis, and if it recognizes self-antigen during the maturation process, it will become suppressed (known as anergy) or undergo apoptosis.

B cells are continuously produced in the bone marrow, but only a small portion of newly made B cells survive in the long-lived peripheral B-cell pool.

Table 8.1 Stages in B-cell development

	Stem cell	Early pro-B cell	Late pro-B cell	Large pre-B cell	Small pre-B cell	Immature B cell	Mature B cell
H-chain genes	Germ line	DJ joining	VDJ joining	VDJ rearranged	VDJ rearranged	VDJ rearranged	VDJ rearranged
L-chain genes	Germ line	Germ line	Germ line	Germ line	VJ joining	VJ rearranged	VJ rearranged
Surface Ig	None	None	None	μ chain in pre-B receptor	μ chain in cytoplasm and on surface	Membrane IgM	Membrane IgM and IgD
RAG, TdT expression	No	Yes	Yes	No	Yes	Yes	No
Surrogate L chain expression	No	Yes	Yes	Yes	No	No	No
Ig-αβ expression	No	Yes	Yes	Yes	Yes	Yes	Yes
Membrane markers	CD34	CD34 CD45 (B220) Class II	CD45R Class II CD19 CD40	CD45R Class II pre-B-R CD19 CD40	CD45R Class II pre-B-R CD19 CD40	CD45R Class II IgM CD19 CD40	CD45R Class II IgM IgD CD19 CD21 CD40

Lymphoid progenitor cells receive signals from bone marrow stromal cells to begin B-cell development. Cytokines induce TdT (terminal deoxynucleotidyl transferase) and recombinase (RAG-1 and RAG-2) synthesis in $CD34^+$ lymphoid progenitors. The cells undergo DJ joining on the H chain chromosome to become early pro-B cells and also begin to express CD45 and class II MHC. Joining of a V segment to the DJ_H completes the late pro-B-cell stage.

Pro-B cells become pre-B cells when they express membrane μ chains with surrogate light chains in the pre-B receptor. Surrogate L chains resemble actual L chains but are the same on every pre-B cell. Signal transduction molecules Ig-alpha/Ig-beta are also part of the pre-B receptor complex. The cytoplasmic tails of Ig heavy chains are too short to enter the cytoplasm and transmit an antigen-binding signal. Ig-α/Ig-β signal transduction molecules have ITAMs (immunoreceptor tyrosine activation motifs) which become phosphorylated in response to antigen-BCR (B-cell receptor) binding. The phosphorylation initiates a cytoplasmic signalling cascade. The cell halts recombination of H chain and proliferates into a clone of B cells, all producing the same μ chain. Since dividing cells are larger than resting cells, this stage is called the large pre-B cell.

Following proliferation, small pre-B cells (no longer dividing) undergo VJ joining on one L chain chromosome. Once L chain has been successfully synthesized, it is expressed with μ chain on the cell membrane and the cell is called an immature B cell. Immature B cells are very sensitive to antigen-binding, so if they bind self-antigen in the bone marrow they die. B cells that do not bind self-antigen express δ chain and membrane IgD with their IgM about the time they leave the marrow and become mature naive (resting) B cells.

In the absence of specific antigen, mature B cells survive in the peripheral circulation for only a few days. Cells which do not encounter antigen within this period of time undergo apoptosis (programmed cell death). This is necessary in order to maintain an optimal number of B lymphocytes in the peripheral circulation.

Most B-cell antigens are T-dependent. In other words, the B cell requires direct contact with T-helper (T_H) lymphocytes as well as exposure to T_H lymphocyte cytokines in order to be fully activated.

There are a few T-independent antigens. One of the best known examples of a T-independent antigen is LPS (lipopolysaccharide). At low concentrations, LPS stimulates the production of specific antibodies (LPS-specific) but at high concentrations it can cause the polyclonal activation of B cells. The polyclonal activation of B cells leads to the proliferation and differentiation of large numbers of B cells, regardless of their antigen specificity.

Two-signal theory of B-cell activation

Early theories of B-cell activation proposed that two signals were required and that when one is lacking, cells would become "paralysed." Self-reactive cells would be deleted because of the low probability of two such faulty signals being generated simultaneously. It is now known that mature B cells that have received their first signal rapidly die unless they receive their second signal. The strength of the signal also affects time of death, with strong signals inducing rapid apoptosis in B cells.

Bacterial cell wall polysaccharides and bacterial flagellin can also serve as T-independent antigens. The cell wall polysaccharides are characterized as possessing repetitive monosaccharide subunits while the bacterial flagellin is a repetitive polymeric protein. It is believed that these repetitive antigens stimulate B cells by extensively cross-linking membrane-bound Ig. The antibody response to these antigens is specific. This process does not require direct T-cell contact but does require the presence of certain T-cell cytokines.

The antibody response to T-independent antigens is typically weaker, there are no memory B cells produced, and only IgM is produced (due to the lack of heavy chain class switching).

Regulation of B-cell Development

Progenitor cells receive signals from bone marrow stromal cells via cell–cell contacts and secreted signals. This bone marrow microenvironment is responsible for B-cell development. One set of CAMs (cell adhesion molecules) involved in both B- and T-cell development is SCF (stem cell factor) on the stromal cell membrane and CD117 on the lymphocyte membrane. A secreted cytokine important for both B- and T-cell development is IL-7, secreted by the stromal cell and which binds to IL-7R (IL-7 receptor) on the developing lymphocyte. Signals from these binding events initiate cytoplasmic cascades resulting in altered expression of proteins required for development. As the B cells develop in the marrow, they migrate from the outer part of the marrow towards the core.

Positive and Negative Selection of B Cells

Both B and T cells undergo positive and negative selection in the primary lymphoid organs. Positive selection requires signalling through the antigen receptor for the cell to survive. Developing B cells are positively selected when the pre-B receptor binds its ligand (developing T cells are positively selected based on their ability to bind MHC as well as peptide). Negative selection means that binding to the receptor results in cell death. Both immature B and T cells are negatively selected if they bind self-antigen.

Signalling for B-cell survival and movement through the appropriate stages of gene expression occurs through membrane pre-B receptor (a complete μ heavy chain associated with the surrogate light chain and the Ig-alpha/Ig-beta heterodimer) and membrane IgM expression. Once the B cell leaves the marrow, its survival appears to depend on further signals thought to be delivered in the lymphoid follicles of secondary lymphoid tissue. Competition between newly created B cells and older B cells for these signals probably maintains B cell homeostasis.

B cells which express only IgM are killed or inactivated (negatively selected) when they bind multivalent ligands, unlike mature B cells which are activated by cross-linking of their BCR. Binding to multivalent (cell-associated) self-ligands in the marrow leads to B-cell apoptosis and clonal deletion. Binding to soluble self-antigen does not kill the B cell. The cell can move to the periphery and express IgD but little IgM. These cells are anergic and cannot respond to antigen. They have a short life span. Cells which do not bind, self-express normal levels of IgM and IgD. If they successfully enter the lymphoid follicles, they can survive for a few weeks until they either encounter their specific antigen or die.

B-cell Heterogeneity

During foetal life, bone marrow stem cells give rise to B cells with different properties. They are B-1 and B-2 cells. B-1 cells have membrane CD5. They are self-renewing, meaning they can produce more mature naive cells themselves by division in the peripheral lymphoid tissues. Conventional B-2 cells can only divide in response to antigen and give rise to memory or plasma cells in the periphery. B-1 BCR is much less diverse than that of B-2 cells.

B-1 BCR is produced preferentially from only some Ig gene segments, does not have additional N nucleotides at the joints between segments, and is specific for mainly common bacterial carbohydrate antigens. B-1 cells secrete predominantly IgM and undergo very little somatic hypermutation. Since they respond to antigens found on multiple pathogens and bind many antigens with low affinity, B-1 cells and their secreted antibodies are called polyreactive. B-1 cells produced after birth have more diverse Ig than those produced during foetal life, but not as diverse as that on B-2 cells. Eventually, bone marrow stem cells stop producing B-1 cells. An analogous T-cell type produced early in development is the gamma/delta T cell.

Location of B Cell

B cells change their location with their stages of maturation, each location providing the microenvironment suitable for the B cell at that life stage. Stem cells produce lymphoid progenitors and pro-B cells in the bone marrow just under the bone. Developing B cells move towards the centre of the marrow as they mature. Mature naive B cells leave the marrow and use selectins to bind addressins on blood vessel endothelium to enter peripheral lymphoid tissues, T-cell areas and B-cell areas (follicles). Peyer's patches, tonsils and appendix are predominantly composed of large follicles. The microenvironment in the MALT follicles signals the B cells to produce IgA, while that in the lymph nodes and spleen signals the B cells to make IgG.

B cells encountering antigen and receiving appropriate T cell help in the T-cell areas form germinal centres in the follicles, where they divide rapidly, and undergo somatic hypermutation. Selection for B cells with higher affinity receptors also occur in these germinal centres. Antibody-secreting plasma cells, short-living ones that do not pass through the follicles and long living ones that undergo hypermutation and class switching in the follicles, are found primarily in the medullary cords of the lymph nodes, the red pulp of the spleen, and in the bone marrow (primarily IgG-secreting plasma cells) or mucosal lamina propria (IgA-secreting plasma cells). Memory B cells are found predominantly in the marginal zone of the spleen, the sub-capsular sinus of the lymph nodes, under the intestinal epithelium in the Peyer's patches, crypt epithelium of the tonsils and a few are also found in the blood.

Burkitt's Lymphoma

B-cell tumours arise from different maturation stages of normal B cells, and these naturally-occurring tumours have helped immunologists understand B-cell development. Each tumour type has its characteristic Ig gene recombination state and homing properties. In nearly every case these tumours are monoclonal, arising from a single B cell which become a cancer cell. Monoclonality allows physicians to identify the tumour cells and track their responses to treatment.

DNA translocations resulting in activation of oncogenes are found in some B-cell tumours. Translocation is the movement of a chromosome segment to another chromosome. Oncogenes are genes usually associated with regulated cell division. When their function is disrupted by translocation, unregulated growth can result. Epstein–Barr Virus (EBV), which usually causes a mild childhood disease or a more debilitating infectious mononucleosis in young adults in the United States, is associated with a B-cell cancer called Burkitt's lymphoma in Africa. In Burkitt's lymphoma cells, the oncogene *myc* is translocated under the control of an H or L chain promoter. Because these promoters are active in B cells, unregulated growth can occur in a B cell with this translocation along with other mutations. There seems to be a link between the occurrence of Burkitt's lymphoma and malaria. Another gene activated by translocation to Ig loci is *bcl-2*. Bcl-2 protein protects B-lineage cells from programmed cell death, so cells carrying translocated *bcl-2* survive beyond their normal lifespan and may become cancerous.

B-CELL RECEPTOR COMPLEX

All isotypes of membrane immunoglobulin (mIg) have very short cytoplasmic tails. Both mIgM and mIgD have a cytoplasmic domain which is only 3 amino acids in length. The cytoplasmic tails of mIg are too short to be able to associate with intracellular signalling molecules. Thus it raises a question as to how this can evoke activating signals after binding to the antigen.

The mIg is always associated with the Ig-alpha/Ig-beta heterodimer collectively termed as the B-cell receptor complex (BCR complex).

Two molecules of this heterodimer associate with one mIg to form a single BCR. The Ig-alpha/Ig-beta heterodimer carries out the signal transducing function of the complex. The Ig-alpha chain has a long cytoplasmic domain containing 61 amino acids while the Ig-beta chain has a long cytoplasmic domain containing 48 amino acids (Figure 8.1).

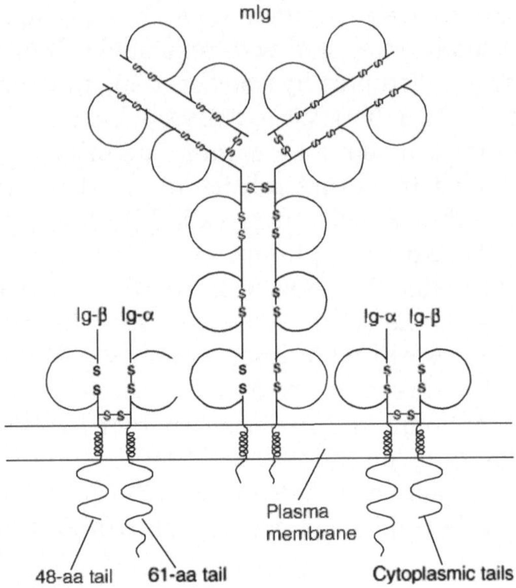

Figure 8.1 B-cell receptor

Significant progress has been made towards delineation of the intrinsic molecular processes that regulate B-lymphocyte immune function. Recent observations have provided a clear picture of the interactive signalling pathways that occur in the mature B-cell antigen receptor (BCR) complex and the different precursor complexes that are expressed during development. Studies have also revealed that the net functional response to a given antigenic challenge is affected by the combined action of BCR-dependent signalling pathways, as well as those originating from various co-receptors expressed by B cells (e.g. CD19, CD22, FcgRIIb and PIR-B). It is now well established that reversible tyrosine phosphorylation plays an important role in regulating B-cell biology. In particular, binding of antigen to the BCR promotes the activation of several protein tyrosine kinases (PTK), which in conjunction with protein tyrosine phosphatases (PTP), alter the homeostasis of reversible tyrosine phosphorylation in the resting B cell. The net effect is a transient increase in protein tyrosine phosphorylation that facilitates the phosphotyrosine-dependent formation of effector protein complexes, promotes targeting of effector proteins to specific microenvironments within the B cell and initiates the catalytic activation of downstream effector proteins.

Studies have demonstrated that Src (***src***, a gene which encodes a **tyrosine kinase**, an enzyme that attaches phosphate groups to tyrosine residues on a variety of host cell proteins) family PTKs are activated initially and serve to phosphorylate CD79a and CD79b thereby creating phosphotyrosine motifs that recruit downstream signalling proteins. In particular, phosphorylation of the BCR complex leads to the recruitment and activation of the **PTK Syk** (Syk-Spleen tyrosine kinase), which in turn promotes phosphorylation of **PLC**γ (Phospholipase C gamma, an enzyme), **Shc** (Shc are the adapter proteins that link tyrosine kinase receptors to Ras and make tyrosine kinase functionally associated with receptors) and **Vav** (Vav is a guanine nucleotide exchange factor implicated in cell proliferation and cytoskeletal organization). Additionally, the **Tec** family (non-receptor protein tyrosine kinases) member **Btk** (Bruton's tyrosine kinase) is recruited to the plasma membrane where it is involved in activation of PLCγ. Initiation of B-lymphocyte activation is dependent on the tyrosine phosphorylation-dependent formation of multimolecular effector protein complexes that activate downstream signalling pathways. The formation of such complexes was initially hypothesized to occur primarily via effector protein binding to the BCR complex itself. However, recent studies have demonstrated that productive signalling via the BCR is in fact dependent on tyrosine phosphorylation of one or more adapter proteins that play a crucial role in recruitment and organization of effector proteins at the plasma membrane. The **SLP-65/BLNK** adapter protein has recently been shown to play a crucial role in recruitment and activation of key signal transducing effector proteins in the B cell. After the BCR has been engaged by antigen and the activation response has been initiated, numerous second messengers and intermediate signal transducing proteins are activated. These include the production of lipid second messengers by phosphatidylinositol 3-kinase, and the **PLC** (phospholipase C)-dependent hydrolysis of phosphatidylinositol 4,5-bisphosphate to yield diacylglycerol and 1,4,5-inositol trisphosphate (IP3). DAG is important for activation of PKC whereas IP3 promotes release of calcium from the endoplasmic reticulum and the subsequent influx of Ca^{2+} from the extracellular space. Numerous intermediate signalling proteins are also activated including the **Ras** and **Rap1**, which are small molecular weight GTPases and these ultimately lead to the activation of **MAP kinases** including **Erk**, **JNK** and **p38**. The net effect of second messenger production and activation of intermediate signalling proteins is the concerted regulation of several transcription factors that mediate gene transcription in the B cell.

T CELL

Generation of T Cell

Immature thymocytes begin life as emigrants from the bone marrow. When they arrive in the thymus, these precursors do not express CD4, CD8 or T-cell receptor (TCR). CD4 and CD8 are very useful markers in thymocyte development as the early thymocytes never express both and are said to be "double-negative" cells.

T-cell development has many parallels with B-cell development but there is one very important difference which is that the T-cell repertoire is profoundly influenced by the MHC allotype of an individual through selection. Developing T cells (thymocytes) in fact undergo three different types of selection.

The first critical stage in T-cell development is the rearrangement of the T-cell receptor beta chain. This stage is analogous to the rearrangement of the immunoglobulin heavy chain locus. As with the Ig heavy chain locus, the cell needs to make an in-frame functional rearranged TCR beta chain to avoid death by apoptosis and to be given the signal to continue its development. Again, as with B cells, the cell senses when a functional TCR beta chain is made by expressing surrogate partners which can pair with the rearranged chain to form a **pre-T-cell receptor complex**. When a functional TCR beta chain is made, these signals switch off beta chain rearrangement (thus providing allelic exclusion at the TCR beta locus) and initiate a programme of differentiation which includes arrangement of the TCR alpha locus and expression of both the CD4 and CD8 molecules. Thus the first selective step in T-cell development selects for a functional TCR beta rearrangement and cells which successfully pass this scrutiny become $CD4^+CD8^+$ "double-positive" cells.

T-cell development is greatest during foetal development and before puberty. After puberty the thymus shrinks and T-cell production declines. In adult humans, removal of the thymus impairs T-cell function. Children born without a thymus because of an inability to form a proper third pharyngeal pouch during embryogenesis (DiGeorge Syndrome) were found to be deficient in T cells. Of several different T-cell deficiencies that have been identified in mice, two complementary defects are found in SCID and nude mice. Nude mice, also called athymic, have a defective thymic epithelium and lack T cells. They are called nude because the defect also affects skin epithelium and results in lack of body hair as well as T cells. **SCID mice** (severe combined immune deficiency) have a thymus but cannot produce lymphocytes because of defects in enzymes (such as RAG and TdT) required for somatic recombination. Lymphoid progenitors from nude mice can develop normally when transferred into SCID mice with a normal thymus microenvironment.

Positive Selection of T Cells

Double positive αβ TCR cells must successfully undergo positive and negative selection before they can leave the thymus. Cells which have successfully rearranged αβ TCR will die in the thymus cortex if they do not bind self MHC within 3–4 days. Positive selection occurs when double positive T cells bind cortical epithelial cells expressing class I or class II MHC plus self-peptides with a high affinity to get the survival signal. Negative selection occurs when double positive T cells bind to bone marrow-derived APC (macrophages and dendritic cells) expressing class I or class II MHC plus self-peptides with a high affinity to receive an apoptosis signal. Note that selection occurs on self-peptides in the thymus and MHC presents self-peptides in the absence of pathogen.

Positive selection also determines whether the T cell will become a helper or a cytotoxic T cell. Positive selection on class I MHC will produce a CD8 T_C cell, while positive selection on class II MHC will yield a CD4 T_H cell.

Negative Selection of T Cells

T cells that survive positive selection migrate further into the cortico-medullary junction of the thymus where they encounter macrophages and dendritic cells, bone marrow-derived APC with high expression of MHC–self-peptide complexes. T cells which bind self-peptide–MHC with high affinity at this stage undergo negative selection and die by apoptosis.

The signals received during positive and negative selection must differ; otherwise all developing T cells would die before they leave the thymus.

The differential avidity hypothesis proposes that the same peptide–MHC complex delivers both signals, but the avidity of positive selection is lower (less signal is required to save the cells from death), while the avidity of the negative selection signal is higher (more signal is required to kill them).

The differential signalling hypothesis proposes that qualitatively different signals are delivered during positive and negative selection. Experiments to study this hypothesis use agonist peptides which stimulate T cells and slightly different antagonist peptides which deliver partial signals that interfere with T-cell activation by agonist peptides. In this model, antagonist peptides could deliver signals leading to positive selection, but only agonist peptides could deliver strong enough signals for negative selection. This result was obtained with CD8 cells in thymus organ culture, but with CD4 cells antagonist peptides could not positively select. Differences were also observed in positive selection of CD4 and CD8 cells by directly cross-linking the TCR and co-receptors. Cross-linking TCR and CD8 (in the absence of peptide on MHC) results in the production of a partial signal and positive selection of CD8 T cells. CD4 T cells could be positively selected by cross-linking TCR with either CD4 or CD8, which produce signals similar to normal activating signals.

Structure of T-cell Receptor

The structure of the T-cell receptor was not elucidated until the 1980s. It was much more difficult to isolate the TCR than Ig because the T cell does not secrete its receptor and the receptor is specific for both antigen and MHC. Monoclonal antibodies and nucleic acid probes were both vital to the purification and isolation of the TCR.

The T-cell receptor is a heterodimer composed either of alpha and beta or gamma and delta polypeptide chains. Amino acid sequencing analysis shows a surprising similarity to the domain structure of the Igs. Each chain has a variable region domain and a constant region domain (designated V_α and C_α, V_β and C_β) (Figure 8.2).

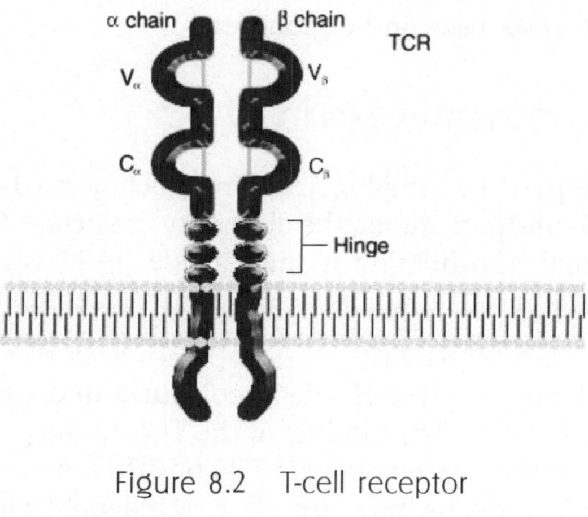

Figure 8.2 T-cell receptor

Three complementarity-determining regions which appear to be equivalent to the CDRs in Ig heavy and light chains have been identified in both the alpha-beta and gamma-delta chains. The variable region domains of alpha and beta or gamma and delta come together to form the antigen-binding cleft.

The TCR has been visualized by X-ray crystallography and is quite similar in the way in which the antigen-binding site is formed by the CDRs of Ig variable region domains.

The TCR heterodimer has a MW of approximately 85,000–90,000 daltons. The two polypeptide subunits are approximately 40,000 daltons each, have different *pI* points and can be separated by isoelectric focusing.

The alpha/beta TCR is present on more than 95% of peripheral T cells and the vast majority of TCR+ thymocytes. The gamma/delta TCR is abundant in various epithelia (intestinal epithelium and epithelium of uterus and tongue).

Current evidence indicates that the gamma/delta TCR may recognize bacterial peptides presented perhaps on non-classical MHC proteins or may recognize HSPs (heat shock proteins) produced in areas of infection.

The gamma/delta T cells may be our first line of defence, limiting the extent of infection until an MHC-restricted alpha/beta T-cell response can be mounted.

T-CELL RECEPTOR COMPLEX

The TCR is part of a complex signalling machinery which includes the TCR alpha/beta dimer, the accessory molecules CD4 or CD8 and a signal transduction module made up of various chains (CD3). CD4 and CD8 are associated in the membrane with the TCR. Both CD4 and CD8 function as adhesion molecules.

CD4 binds to class II MHC molecules and CD8 binds to class I MHC molecules. Binding of the TCR to the peptide/MHC complex is greatly augmented if CD4 or CD8 are assisting. Their cytoplasmic domains may also allow for signal transduction to occur. The signal transduction property of CD4 and CD8 is mediated through their cytoplasmic domains. Both CD4 and CD8 are noncovalently associated with the protein kinase lck (Figure 8.3).

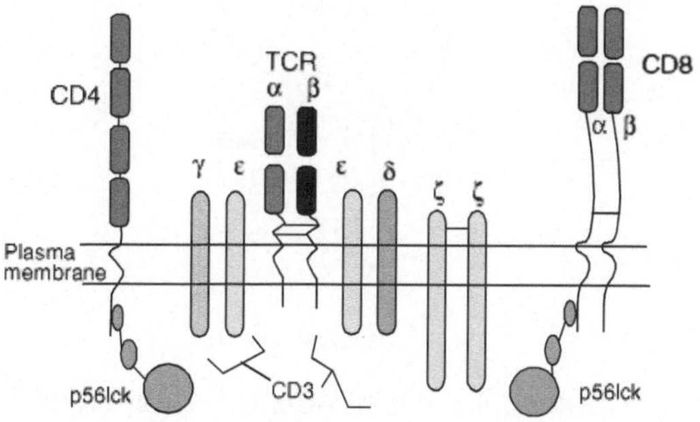

Figure 8.3 T-cell receptor complex

Engagement of the TCR by peptide antigen, in association with MHC gene products, leads to a series of intracellular biochemical events culminating in the transcription of new genes and cellular activation. The earliest identifiable intracellular change documented at present is the activation of one or more tyrosine kinases that first phosphorylate the CD3 chains and subsequently other substrates. Subsequent to tyrosine kinase activation a series of secondary events have been observed to follow TCR engagement, including activation of serine/threonine kinases, activation of the GTP-binding protein p21Ras and activation of transcription factors for receptors and growth factors such as the major T-cell growth factor interleukin-2 (IL-2). The CD4 and CD8 co-receptors bind a tyrosine kinase (p56lck) via their intracytoplasmic tail which plays a critical role in T-cell signalling.

Co-stimulation

However TCR receptor binding is not sufficient to activate T cells. The two-signal hypothesis state that a second signal is required in addition to that through the antigen receptor. This model may even be too simple. There is evidence of a critical role for two distinct co-stimulator molecules on T cells which interact with specific ligands on the surface of antigen-presenting cells (APC).

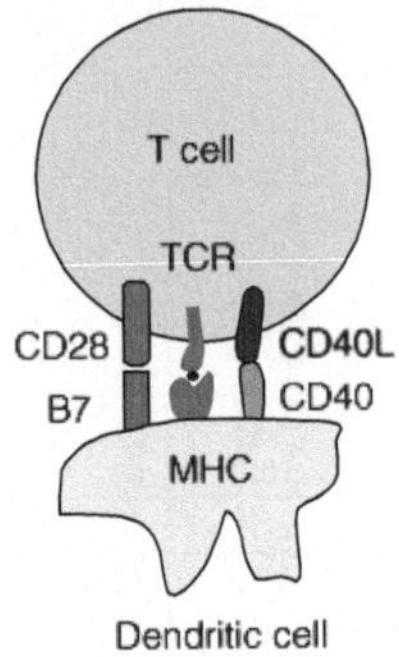

Figure 8.4 Co-stimulatory molecules

The CD28 molecule on T cells delivers a co-stimulatory signal upon engaging either of its ligands, B7.1 (CD80) or B7.2 (CD86). A distinct signal is transduced by the CD40L (for ligand) molecule on the T cell when it is ligated to CD40 (Figure 8.4). The expression of B7.1/B7.2 and CD40 are among the properties which distinguish specialized antigen-presenting cells (termed professional APC) from other MHC-positive cells. A number of other molecules on the surface of APC may serve some role in co-stimulation, although their full role or mechanism of action is not clear.

Mechanisms For Generating TCR Diversity

Recombinatorial diversity This is generated by multiple V gene segments and J gene segments for the alpha and gamma chain and by multiple V, D and J gene segments for the beta and delta chain.

Diversity in D region usage Due to the arrangement of the heptamer 12 nanomer and heptamer 23 nanomer sequences in the beta and delta genes, differential D region usage has been observed. In beta chain gene rearrangement, both direct VJ rearrangement and VDJ rearrangement occur. In delta chain gene rearrangement, VJ, VDJ and VDDJ rearrangements are observed.

Combinatorial diversity This is due to the random combination of alpha and beta chains or gamma and delta chains.

Junctional diversity N-nucleotide addition occurs at the VJ, DJ, and VDJ junctions due to the action of terminal deoxynucleotidyl transferase (TdT). This mechanism occurs with all four of the TCR gene families. P-nucleotide addition also occurs following rearrangement with all four of the TCR gene families. Little to no somatic mutation occurs within V region gene segments, so TCRs do not show affinity maturation.

POINTS TO REMEMBER

- B-cell development begins in the foetal liver and continues in the bone marrow throughout our lives.
- When the B cell fails in any step of the maturation process, it will undergo apoptosis, and if it recognizes self-antigen during the maturation process, it will become suppressed (known as anergy) or undergo apoptosis.
- Developing B cells are positively selected when the pre-B receptor binds its ligand.
- Positive selection occurs when double positive T cells bind cortical epithelial cells expressing class I or class II MHC plus self-peptides with a high enough affinity to get the survival signal.
- Both immature B and T cells are negatively selected if they bind self antigen.
- The mIg is always associated with the Ig-alpha/Ig-beta heterodimer collectively termed as the B-cell receptor complex (BCR complex).
- The T-cell receptor is a heterodimer composed either of alpha and beta or gamma and delta polypeptide chains.

REVIEW QUESTIONS

1. Write short notes on:
    i. Positive and negative selection of B cell
    ii. Positive and negative selection of T cell
    iii. Location of B cell
    iv. Burkitt's lymphoma
    v. B-cell receptor
    vi. T-cell receptor
    vii. T-cell receptor complex
    viii. TCR diversity
2. Write a detailed account on B-cell maturation.
3. Explain in detail, T-cell development and maturation.

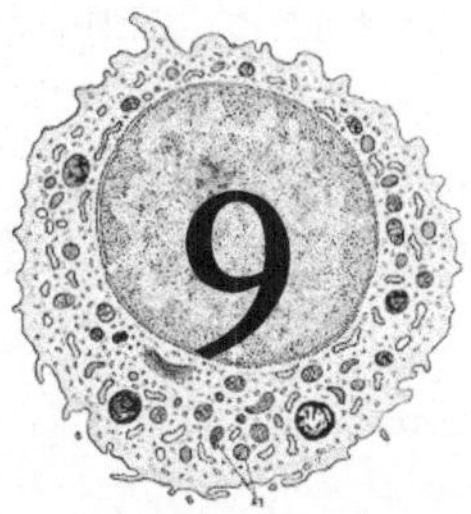

ANTIGEN PROCESSING AND PRESENTATION

INTRODUCTION

Both $CD4^+$ and $CD8^+$ T cells can recognize antigen only when it has been processed and presented on the cell membrane of an antigen-presenting cell (APC) or target cell respectively in association with self-MHC. This phenomenon is termed as **restriction to self-MHC**.

In this terminology, $CD4^+$ T cells are class II MHC-restricted while $CD8^+$ T cells are class I MHC-restricted. Table 9.1 shows the different types and properties of antigen-presenting cells.

By convention, the cells which present antigen to CD4 T_H cells are termed antigen-presenting cells while the cells which present antigen to CD8 T_C cells are termed target cells.

Processing of antigen involves degradation of protein antigen into peptides and presentation is the association of peptide with MHC and transportation of MHC/peptide complex to the cell membrane. Figure 9.1 gives a brief outline of presentation of exogenous and endogenous antigens.

Table 9.1 Different types of antigen-presenting cells and their properties

Properties	Macrophage	Dendritic cell	B cell
MHC-II expression	Low levels induced by bacteria and/or cytokines	Always expressed	Always expressed inducible upon activation
Antigen type and presentation by MHC	Extracellular antigens: presentation via MHC II	Intracellular and extracellular antigens: presentation via MHC I and II	Extracellular antigen binds to specific Ig receptors: presentation via MHC II
Co-stimulation (B7 expression)	Low levels induced by bacteria and/or cytokines	Always expressed in high levels	Low levels inducible upon activation
Location	Lymphoid tissue, connective tissue body cavities	Lymphoid tissue, connective tissue, epithelium	Lymphoid tissues

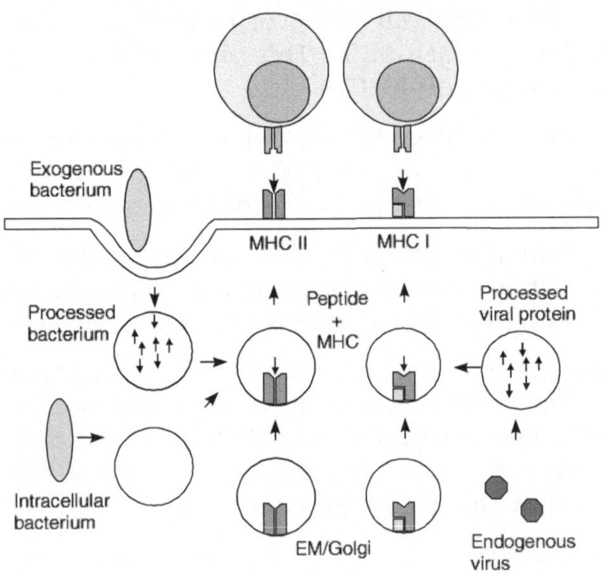

Figure 9.1 Presentation of endogenous and exogenous antigen

ENDOGENOUS ANTIGENS

The endogenous antigens are the peptides that are derived from proteins that are synthesized within the cytoplasm of the cell. So this pathway is also termed the cytosolic pathway.

Some examples of endogenous antigens are viral proteins, tumour-specific antigens and proteins synthesized by intracellular protozoa or bacteria.

Protein levels are carefully regulated in the cell. Misfolded, aged or damaged proteins are naturally degraded within the cytoplasm of all nucleated cells. Proteins targeted for proteolysis often have a small protein known as **ubiquitin** associated with them. Such conjugates are known to be degraded within structures known as **proteasomes** (basically a cylinder composed of hydrolytic enzymes which functions as a large protease complex). Degradation is thought to occur within the hollow centre of the cylinder.

Two alternative subunits of the proteasome (termed **LMP2** and **LMP7**) are encoded within the MHC complex and their synthesis is triggered by the T_H cell cytokine, interferon-gamma. The proteolytic activities of these two subunits apparently generate peptides which preferentially bind to class I MHC proteins.

Peptides generated in the cytoplasm are transported across the membrane of the rough endoplasmic reticulum by a transporter protein known as the **TAP protein** (transporter associated with antigen processing). The TAP protein is a heterodimer composed of two subunits known as **TAP 1** and **TAP 2**. Interestingly, the genes for TAP 1 and TAP 2 map to the class II region of the MHC complex. TAP 1 and 2 are classified as **ABC proteins** (ATP-binding cassette proteins) because such proteins are involved in the ATP-dependent transport of ions, sugars, amino acids, peptides, etc. across membranes.

Meanwhile, the class I MHC alpha chains are being synthesized by the cell at ribosomes associated with the rough endoplasmic reticulum. An alpha chain must associate with a β-2 microglobulin molecule and a processed peptide before it is transported to the Golgi complex and then to the plasma membrane. A molecular chaperone known as **calnexin** binds to the alpha chain and retains it within the rough endoplasmic reticulum until β-2 microglobulin can associate. At this point, calnexin is released and calreticulin binds. Then, tapasin acts like a bridge between the TAP protein and the class I MHC and allows the peptide binding cleft to bind to peptide (Figure 9.2).

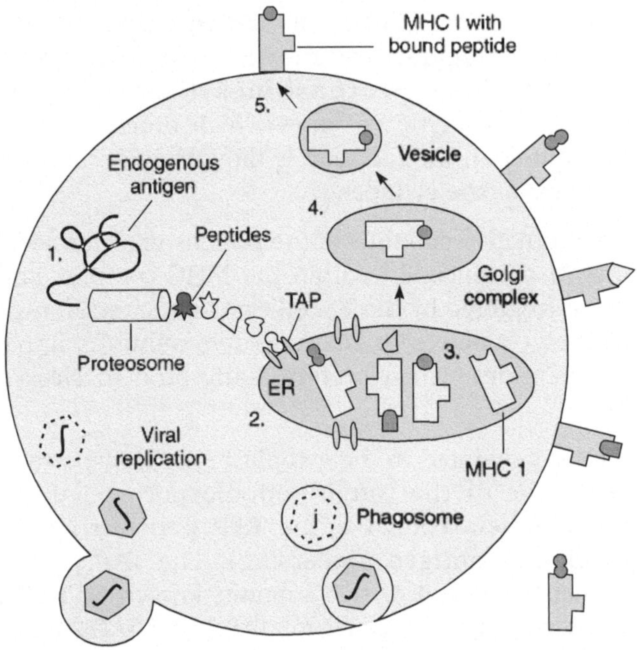

Figure 9.2 Processing of endogenous antigen

The "completed" class I MHC protein/peptide complex can now be transported to the Golgi in membrane-bound vesicles for further modification, packaging and sorting to the plasma membrane.

It recently became clear that peptides in the endoplasmic reticulum (ER) might require further processing before loading onto MHC class I molecules, and an ER-resident amino peptidase ERAAP (ER aminopeptidase associated with antigen processing, known as ERAP1 in humans), was recently identified in mice and implicated in this process. Now, two studies show that ERAAP is required for the generation of the normal repertoire of MHC class I peptides, and that ERAP1 trims precursor peptides by a "molecular ruler" mechanism.

The presentation of class I MHC/ peptide by a target cell to a CD8$^+$ T_C cell results in the proliferation and subsequent differentiation of a T_C into a CTL (killer/effector cell). The CTL can then participate in target cell killing. Target cell killing occurs due to induction of apoptosis (Figure 9.3).

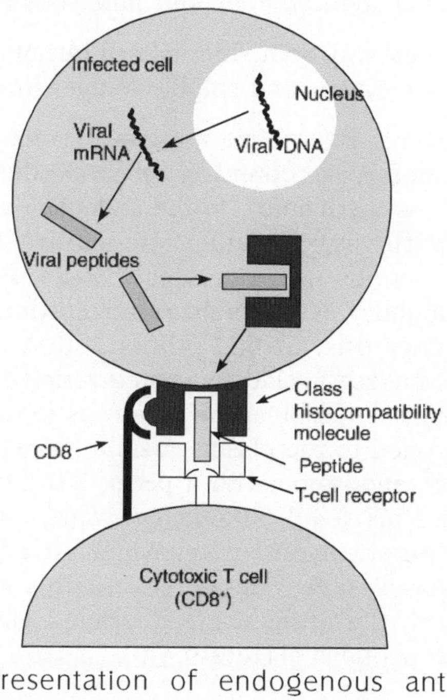

Figure 9.3 Presentation of endogenous antigen by the class I molecule to cytotoxic T cell

EXOGENOUS ANTIGENS

The exogenous antigens are processed by specialized antigen-presenting cells. The following are the three important antigen-presenting cells—macrophages, B lymphocytes and dendritic cells.

Mature dendritic cells are the most effective antigen-presenting cells in that they constitutively produce a high level of class II MHC protein and the co-stimulatory protein, B7. B cells constitutively express the class II MHC protein but must be activated to produce B7. Macrophages must be induced (activated) by the process of phagocytosis before expressing class II MHC or B7.

In addition, there is a category of cells known as non-professional APCs that can act as APCs for short periods of time, particularly during periods of sustained inflammatory activity. Such cells must be induced to express both class II MHC and B7.

Macrophages and dendritic cells internalize particulate antigen by phagocytosis and soluble antigens by endocytosis.

B lymphocytes internalize antigen by receptor-mediated endocytosis (membrane-bound Ig serves as the receptor and antigen serves as the ligand). In the endocytic pathway or the phagocytic pathway, protein antigens are processed by proteolytic enzymes into peptides (13–24 amino acids in length). B lymphocytes internalize antigen through receptor-mediated endocytosis. In this process, surface Ig bound to antigen is internalized in clathrin-coated vesicles. As this vesicle becomes an early endosome, the pH drops to 6.0–6.5 and the clathrin is recycled to the plasma membrane. The vesicle then becomes a late endosome with a pH of 5.0–6.0 and finally a lysosome with a pH of 4.5–5.0. The lysosome contains a battery of over 50 different hydrolytic enzymes including proteases, nucleases, lipases, glycosidases, etc. At this stage, proteins associated with the antigen are degraded (hydrolysed) by proteases into peptides (Figure 9.4).

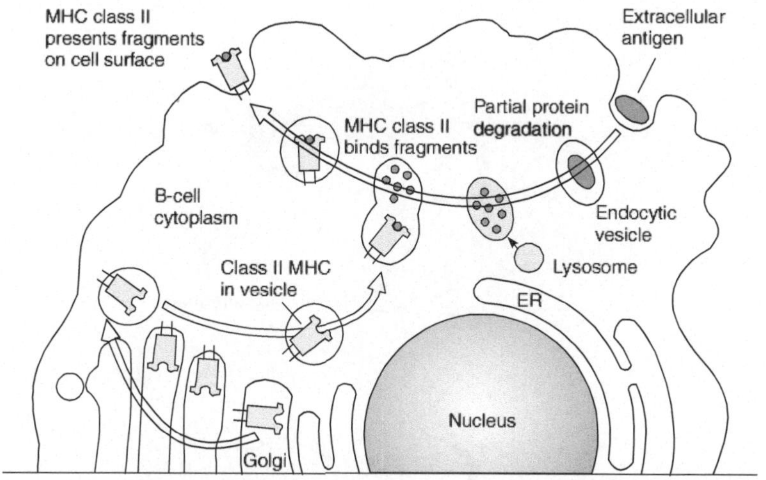

Figure 9.4 Processing of exogenous antigen

Meanwhile, the class II MHC molecules are translated in the ribosomes associated with rough endoplasmic reticulum. The two polypeptide chains (alpha and beta) enter the rough endoplasmic reticulum where they attach to each other and bind to a third protein known as Ii—the invariant chain. The presence of the invariant chain prevents the binding of endogenously synthesized peptides from binding to the class II MHC molecule. Calnexin participates in this process as a molecular chaperone. Membrane-bound vesicles traffic the MHC class II (Ii serves as a targeting protein in this process) to the endosomal pathway. Under acidic conditions within this compartment, Ii is first cleaved leaving a fragment known as CLIP (class-II-associated invariant chain peptide).

The CLIP must either dissociate or be displaced to allow peptides to bind to the peptide-binding cleft. HLA-DM (a non-classical class II-like protein) catalyses the release of CLIP and the binding of other peptides to the class II MHC. HLA-DM also catalyses the release of unstably bound peptides in a process known as "peptide editing". The peptide–MHC complex must be stable on the cell surface. It may take a long time for an appropriate T_H lymphocyte to encounter the APC.

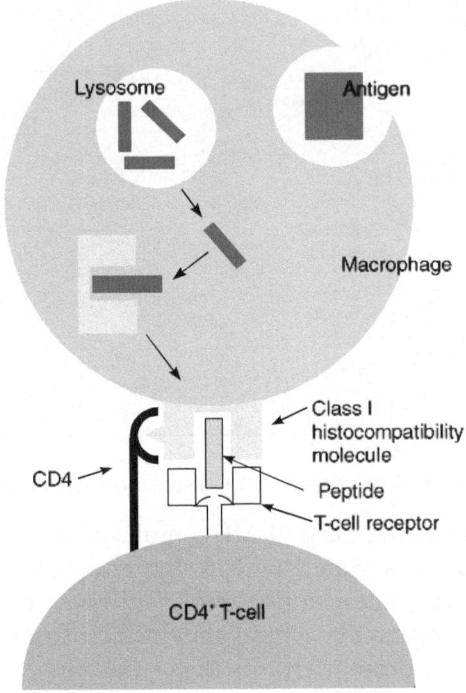

Figure 9.5 Presentation of exogenous antigen by the class II molecule to T helper cell

Once a stable peptide–MHC interaction occurs, the complex is released and the class II MHC/peptide complex is then transported to the Golgi and ultimately to the plasma membrane where it is inserted.

The class II MHC/peptide complex is now presented to a $CD4^+$ T_H cell (Figure 9.5). The result of this interaction is the clonal expansion (proliferation) and differentiation of the T_H cell to effector or memory T_H cell.

As with class I, class II proteins can also bind and present self-peptides (many derived from MHC proteins themselves). These presented "self-peptides" would be ignored since self-reactive T_H cells are deleted during thymic education.

Peptide-binding Site

In both class I and class II MHC, the greatest polymorphic variability in the amino acids is in those facing the binding groove. Thus, the polymorphism among MHC gene products creates variation in the chemical surface of the peptide-binding groove.

For any given MHC molecule, binding of a peptide usually requires the peptide to have one or more specific amino acids at a fixed position, frequently the terminal or penultimate amino acid of the peptide. Binding of the specific amino acid in the groove of the MHC molecule occurs in what is termed the anchor site. The other amino acids can be variable so that each MHC molecule can bind many different peptides. Other polymorphic residues of the MHC molecule are those in contact with the TCR, which interact with both peptide and the MHC molecule itself. Fragments of self, as well as non-self, proteins associate with MHC molecules of both classes and are expressed at the cell surface. Which protein fragments bind is a function of the chemical nature of the groove for that specific MHC molecule.

Interestingly, there are some pathogens which actually reside within the endosomal or phagosomal pathway of antigen-presenting cells. Some examples include *Mycobacterium tuberculosis* and *Mycobacterium leprae* (the bacteria which cause tuberculosis and leprosy respectively), and the protistan parasite, *Leishmania* sp. (the causative agent of Leishmaniasis).

Peptides derived from these pathogens can also be processed and presented in association with class II MHC proteins on the surface of the infected APC. In this way, pathogen-specific T_H cells can be activated. Cytokines (especially interferon gamma) produced by these activated T_H cells can in turn activate the APC to destroy the internalized pathogen through the production of the highly toxic nitrogen intermediate, nitric oxide.

Figures 9.4 and 9.5 show the steps involved in the processing and presentation of exogenous antigens.

POINTS TO REMEMBER

- Both $CD4^+$ and $CD8^+$ T cells can recognize antigen only when it is processed and presented on the cell membrane of an antigen-presenting cell (APC) or target cell respectively in association with self-MHC.
- The cells which present antigen to $CD4^+$ T_H cells are termed antigen-presenting cells while the cells which present antigen to $CD8^+$ T_C cells are termed target cells.
- The endogenous antigens are peptides derived from proteins that are synthesized within the cytoplasm of the cell. They are presented by MHC class I molecule to cytotoxic T cell.
- The exogenous antigens are processed by specialized antigen-presenting cells. They are presented by MHC class II molecule to T helper cell.

REVIEW QUESTIONS

1. Write a short note on antigen-presenting cell.
2. Explain endogenous antigen presentation.
3. Explain exogenous antigen presentation.

10

CYTOKINES

INTRODUCTION

"Cytokine" is a word that comes from *cyto,* a combining form meaning "cell" and *kinin,* a combining form used in naming hormones, especially peptide hormones (e.g. bradykinin).

Cytokines are low molecular weight, soluble proteins that are produced in response to an antigen and function as chemical messengers for regulating the innate and adaptive immune systems. They are produced by virtually all cells involved in innate and adaptive immunity, but especially by T-helper (T_H) lymphocytes. The activation of cytokine-producing cells triggers them to synthesize and secrete their cytokines. The cytokines, in turn, are then able to bind to specific cytokine receptors on other cells of the immune system and influence their activity in some manner. They act by binding to specific membrane receptors, which then signal the cell via **second messengers**, often tyrosine kinases, to alter its behaviour (gene expression). Responses to cytokines include increasing or decreasing expression of membrane proteins (including cytokine receptors), proliferation and secretion of effector molecules.

For many immunologists, migration inhibition factor (MIF) was the first of what came to be known as lymphokines. This was an activity in supernatants from antigen-activated lymphocytes that inhibited the movement of macrophages in *in vitro* assays. It was identified simultaneously in 1966 by John David and Barry Bloom, working in two separate laboratories. The next of the lymphocyte-derived factors to be described was lymphotoxin (LT) (Nancy Ruddle and Byron Waksman). Others followed and in 1969 Dudley Dumonde proposed the term "lymphokine" to describe these factors. Subsequently, factors derived from macrophages and monocytes in culture were naturally called "monokines". These lymphokines and monokines were first described in antigen- or mitogen-activated cell cultures. Following the discovery of a lymphokine activity in virus-infected kidney cell cultures, it was suggested that the various soluble substances represented a broad class of mediators of host defence secreted by cells and should more properly be called "cytokines".

TYPES OF CYTOKINES

The **different types of cytokines** are:

- **lymphokines**—cytokines made by lymphocytes
- **monokines**—cytokines made by monocytes
- **chemokines**—cytokines with chemotactic activities
- **interleukins**—cytokines made by one leucocyte that act on other leucocytes

Cytokines may act on the cells that secrete them (autocrine action), or on nearby cells (paracrine action), or in some instances on distant cells (endocrine action).

Cytokines are pleiotropic, redundant and multifunctional.

- **Pleiotropic** means a particular cytokine can act on a number of different types of cells rather than a single cell type.

- **Redundant** refers to the ability of a number of different cytokines to carry out the same function.
- **Multifunctional** means the same cytokine is able to regulate a number of different functions.

There are three functional categories of cytokines:

1. Cytokines that regulate innate immune responses
2. Cytokines that regulate adaptive immune responses
3. Cytokines that stimulate haematopoiesis

Cytokines that Regulate Innate Immune Response

Cytokines that regulate innate immunity are produced primarily by mononuclear phagocytes such as macrophages and dendritic cells. They can also be produced by T lymphocytes, natural killer cells (NK cells) and other cells. They are produced primarily in response to pathogen-associated molecules such as lipopolysaccharide, peptidoglycan monomers, teichoic acids and double-stranded DNA. They mostly act on leucocytes and the endothelial cells of blood vessels to promote and control early inflammatory responses.

Examples Tumour necrosis factor, interleukin-1(IL-1), chemokines, interleukin-12 (IL-12), type I interferons, etc.

Cytokines that Regulate Adaptive Immune Responses

Cytokines that regulate adaptive immunity (Humoral immunity and cell-mediated immunity) are produced primarily by T lymphocytes that have recognized an antigen specific for that cell. These cytokines function in the proliferation and differentiation of B lymphocytes and T lymphocytes after antigen recognition and in the activation of effector cells.

Examples Interleukin-2 (IL-2), interleukin-4 (IL-4), interleukin-6 (IL-6), interferon-gamma (IFN-γ), etc.

CYTOKINES THAT STIMULATE HAEMATOPOIESIS

These cytokines are produced by bone marrow stromal cells. They stimulate the growth and differentiation of immature leucocytes.

Examples Colony stimulating factor (CSF), interleukin-3 (IL-3), interleukin-4 (IL-4), etc.

INTERLEUKINS

They have a variety of functions, but most are involved in directing other immune cells to divide and differentiate. Each IL acts on a specific, limited group of cells that express the correct receptor for that interleukin.

- **Interleukin-1 (IL-1)** It is produced by activated macrophages, endothelial cells, B cells and fibroblast cells. IL-1 induces inflammatory responses, oedema, promotes the production of prostaglandins, IL-2, the growth of leucocytes, and induces acute phase reaction. It stimulates the synthesis of collagen and collagenase for scar tissue formation; stimulates the synthesis of adhesion factors on endothelial cells and leucocytes for diapedesis and activates macrophages. IL-1 also augments corticosteroid release, induces fever and shivering which are useful responses because elevated body temperature reduces bacterial growth.
- **Interleukin-2 (IL-2)** It is also known as T-cell growth factor (TGF). It is secreted by stimulated helper T cells ($CD4^+$), cytotoxic T cells ($CD8^+$) and large granular lymphocytes (LGL). IL-2 promotes proliferation (clonal expansion) and differentiation of additional $CD4^+$ cells, B cells, and activates macrophages and oligodendrocytes.
- **Interleukin-3 (IL-3)** It is produced by activated T cells. It stimulates the proliferation of precursors in all haematopoietic cells (red cells, granulocytes, macrophages and lymphocytes).

- **Interleukin-4 (IL-4)** It is produced by T_H2 cells and mast cells. It stimulates the production of antibody-producing B cells, leading to the production of IgG and IgE. IL-4 also promotes $CD8^+$ cell growth and promotes T_H2 cell differentiation.

  IL-4 induces MHC class II expression, on macrophages, but inhibits production of the proinflammatory cytokines (IL-1) and tumour necrosis factor alpha (TNF-α). It is important in the defence against helminthes and arthropods.

- **Interleukin-5 (IL-5)** It is produced mainly by T_H2 cells. It is chiefly a growth and activation factor for eosinophils. It contributes to defence against helminthes and arthropods. It also stimulates the proliferation and differentiation of antigen activated B cells and the production of IgA.

- **Interleukin-6 (IL-6)** It is produced by many cell types, including T cells, macrophages, B cells, fibroblasts and endothelial cells. IL-6 stimulates several types of leukocytes, and the production of acute phase proteins in the liver. IL-6 is particularly important in inducing B cells to differentiate into antibody forming cells (plasma cells).

- **Interleukin-7 (IL-7)** It is made by bone marrow stromal cells and acts on thymocytes. IL-7 is a T-cell growth and activation factor, and a macrophage activation factor.

- **Interleukin-8 (IL-8)** It is produced by many cells of the body, especially by macrophages and endothelial cells. IL-8 enhances inflammation, by enabling immune cells to migrate into tissue and is a powerful inducer of chemotaxis for neutrophils.

- **Interleukin-9 (IL-9)** It up-regulates T_H1 responses (enhancing inflammation) by inhibiting T-cell apoptosis.

- **Interleukin-10 (IL-10)** It down-regulates anti-viral responses by inhibiting the production of interferon gamma (IFN-γ). It also inhibits antigen presentation, and macrophage production of IL-1, IL-6, and TNF-α. IL-10 is also very important in B-cell activation. IL-10 inhibits the production of IL-12, a co-stimulator molecule.

- **Interleukin-12 (IL-12)** It acts in a contrasting manner to IL-10. It promotes T_H1 type response in macrophages, NK cells and induces IFN-γ production.
- **Interleukin-13 (IL-13)** It has structural and functional similarities to IL-4 and promotes B-cell differentiation, inhibits T_H1 cells and the production of macrophage inflammatory cytokines.
- **Interleukin-15 (IL-15)** It shares several biological activities with IL-2 and is produced by both epithelial cells and monocytes. IL-15 also induces T-cell proliferation, enhances NK-cell cytotoxicity and stimulates B cells to proliferate and secrete immunoglobulins.
- **Interleukin-17 (IL-17)** It induces production of inflammatory cytokines.
- **Interleukin-18 (IL-18)** It induces interferon-gamma (IFN-γ) production.

TUMOUR NECROSIS FACTOR (TNF)

TNF is the principal cytokine that mediates acute inflammation. When present in excessive amounts it becomes the principal cause of systemic complications such as the shock cascade. Functions include acting on endothelial cells to stimulate inflammation and the coagulation pathway; stimulating endothelial cells to produce selectins and ligands for leucocyte integrins during diapedesis; stimulating endothelial cells and macrophages to produce chemokines that contribute to diapedesis, chemotaxis and the recruitment of leucocytes; stimulating macrophages to secrete interleukin-1 (IL-1) for redundancy; activating neutrophils and promoting extracellular killing by neutrophils; stimulating the liver to produce acute phase proteins, and acting on muscles and fat to stimulate catabolism for energy conversion.

In addition, TNF is cytotoxic for some tumour cells; interacts with the hypothalamus to induce fever and sleep; stimulates the synthesis of collagen and collagenase for scar tissue formation and activates macrophages. TNF is produced by monocytes, macrophages, dendritic cells, T_H1 cells, and other cells.

This is one of the first cytokines that appear during an inflammatory response. There are two types of TNF. They are TNF-α and TNF-β. TNF-α is sometimes referred to as cachectin. TNF-β is sometimes referred to as lymphotoxin (LT).

COLONY STIMULATING FACTORS (CSFs)

CSFs are cytokines that stimulate the proliferation of specific pluripotent stem cells of the bone marrow in adults. Granulocyte-CSF (G-CSF) is specific for proliferative effects on cells of the granulocyte lineage. Macrophage-CSF (M-CSF) is specific for cells of the macrophage lineage.

Granulocyte-macrophage-CSF (GM-CSF) has proliferative effects on both classes of lymphoid cells. GM-CSF binds to receptors on neutrophils, eosinophils and monocytes, it activates these cells and inhibits their apoptosis. GM-CSF increases adhesion of these cells to capillary walls during diapedesis, enhances their phagocytosis and extracellular killing and increases both superoxide anion generation and antibody-dependent cytotoxicity.

IL-3 (secreted primarily from T cells) is also known as multi-CSF, since it stimulates stem cells to produce all forms of haematopoietic cells.

CHEMOKINES

Chemokines are a group of cytokines that enable the migration of leucocytes from the blood to the tissues at the site of inflammation. Chemokines are a family of structurally related glycoproteins with potent leucocyte activation and/or chemotactic activity. They are 70 to 90 amino acids in length and approximately 8 to 10 kDa in molecular weight. Most of them fit into two subfamilies with four cysteine residues. These subfamilies are based on whether the two amino terminal cysteine residues are immediately adjacent or separated by one amino acid.

They increase the affinity of integrins on leucocytes for ligands on the vascular wall during diapedesis, regulate the polymerization and depolymerization of actin in leucocytes for movement and migration and function as chemoattractants for leucocytes. In addition, they trigger some WBCs to release their killing agents for extracellular killing and induce some WBCs to ingest the remains of damaged tissue. Chemokines also regulate the movement of B lymphocytes, T lymphocytes and dendritic cells through the lymph nodes and the spleen. Certain chemokines have also been shown to suppress HIV, probably by binding to the chemokine receptors serving as the second binding factor for HIV on $CD4^+$ cells. When produced in excess amounts, chemokines can lead to damage of healthy tissue as seen in such disorders as rheumatoid arthritis, pneumonia, asthma, adult respiratory distress syndrome (ARDS) and septic shock.

Examples of chemokines include IL-8, MIP-1a, MIP-1b, MCP-1, MCP-2, MCP-3, GRO-a, GRO-b, GRO-g, RANTES, and eotaxin. Chemokines are produced by many cells including leucocytes, endothelial cells, epithelial cells, and fibroblasts.

INTERFERONS (IFN)

Type I Interferon

There are two important type I interferons. They are IFN-α and IFN-β. Type I IFN has several sources and effects. The Type I IFN inhibits viral replication in virus-infected cells. IFN-α has been used to treat HIV, some forms of cancer and multiple sclerosis. IFN-α is made predominantly by neutrophils, and IFN-β is made predominantly by fibroblasts.

Type II Interferon

There is one important type II interferon. It is called as interferon gamma (IFN-γ). It is produced by T helper cells. It is an important cytokine for activating macrophages. An activated macrophage is more phagocytic. It processes and presents antigen more efficiently, produces more cytokines and becomes more bactericidal than resting macrophages. Activated macrophages are an important mediator of cellular immunity (which is distinguished from antibody-mediated humoral immunity). IFN-γ acts antagonistically against other cytokines such as IL-4.

TRANSFORMING GROWTH FACTOR (TGB-β)

TGF-β is produced by a variety of cells. It is also produced by platelets. TGF-β tends to inhibit the proliferation of lymphocytes in experimental systems. It also tends to promote cellular differentiation, IgA and IgG2b production. TGF-β functions to inhibit the macrophage function. It also promotes tissue repair.

CYTOKINE RECEPTORS

Cytokines act on their target cells by binding specific membrane receptors. The receptors and their corresponding cytokines have been divided into several families based on their structure and activities (Figure 10.1).

- Haematopoietin family receptors are dimers or trimers with conserved cysteines in their extracellular domains and a conserved Trp-Ser-X-Trp-Ser sequence. Examples are receptors for IL-2 to IL-7 and GM-CSF.
- Interferon family receptors have the conserved cysteine residues but not the Trp-Ser-X-Trp-Ser sequence, and include the receptors for IFN-α, IFN-β, and IFN-γ.

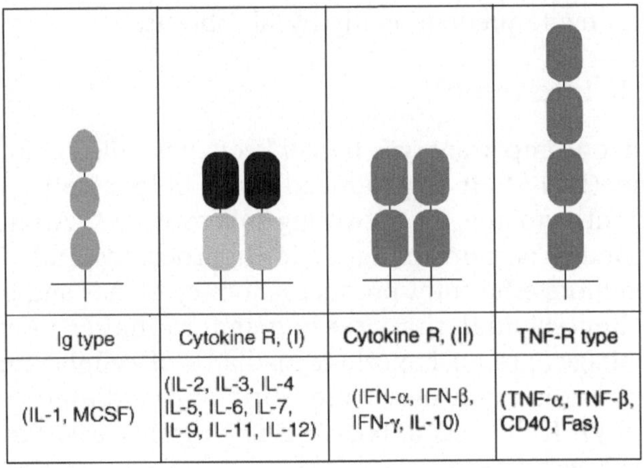

Figure 10.1 Different types of cytokine receptors

- Tumour necrosis factor family receptors have four extracellular domains; they include receptors for soluble TNF-α and TNF-β as well as membrane-bound CD40 (important for B cell and macrophage activation) and Fas (which signals the cell to undergo apoptosis).
- Chemokine family receptors have seven transmembrane helices and interact with G protein. This family includes receptors for IL-8, MIP-1 and RANTES. Chemokine receptors CCR5 and CXCR4 are used by HIV to preferentially enter either macrophages or T cells.

Cytokine activity can be blocked by antagonists, molecules which bind cytokines or their receptors. IL-1 has a specific antagonist that blocks binding of IL-1α and IL-1β to their receptor. During immune responses, fragments of membrane receptors may be shed and then compete for cytokine binding. Microbes also influence cytokine activities. For example, vaccinia virus (smallpox and cowpox) encodes soluble molecules which bind IFN-γ, while Epstein–Barr virus (infectious mononucleosis) encodes a molecule homologous to IL-10 that suppresses immune function in the host. The different cytokines and their functions are presented in Table 10.1.

Table 10.1 Cytokines and their functions

Cytokine	Producing cell	Target cell	Function
GM-CSF	T_H cells	Progenitor cells	Growth and differentiation of monocytes and DC
IL-1α IL-1β	Monocytes Macrophages, B cells, DC	T_H cells	Co-stimulation
		B cells	Maturation and proliferation
		NK cells	Activation
		Various cells	Inflammation, acute phase response, fever
IL-2	T_H1 cells	Activated T and B cells, NK cells	Growth, proliferation, activation
IL-3	T_H cells NK cells	Stem cells	Growth and differentiation
		Mast cells	Growth and histamine release
IL-4	T_H2 cells	Activated B cells	Proliferation and differentiation IgG_1 and IgE synthesis
		Macrophages	MHC class II
		T cells	Proliferation
IL-5	T_H2 cells	Activated B cells	Proliferation and differentiation IgA synthesis
IL-6	Monocytes, macrophages, T_H2 cells stromal cells	Activated B cells	Differentiation into plasma cells
		Plasma cells	Antibody secretion
		Stem cells	Differentiation
		Various cells	Acute phase response
IL-7	Marrow stroma thymus stroma	Stem cells	Differentiation into progenitor B and T cells
IL-8	Macrophages, endothelial cells	Neutrophils	Chemotaxis
IL-10	T_H2 cells	Macrophages	Cytokine production*
		B cells	Activation

(Contd.)

Table 10.1 (Continued)

Cytokine	Producing cell	Target cell	Function
IL-12	Macrophages, B cells	Activated Tc cells	Differentiation into CTL (with IL-2)
		NK cells	Activation
IFN-α	Leucocytes	Various cells	Viral replication* MHC I expression
IFN-β	Fibroblasts	Various cells	Viral replication* MHC I expression
IFN- γ	T_H1 cells, T_C cells, NK cells	Various cells	Viral replication
		Macrophages	MHC expression
		Activated B cells	Ig class switch to IgG_{2a}
		T_H2 cells	Proliferation*
		Macrophages	Pathogen elimination
		Tumour cells	Cell death
MIP-1α	Macrophages	Monocytes, T cells	Chemotaxis
MIP-1β	Lymphocytes	Monocytes, T cells	Chemotaxis
TGF-β	T cells, monocytes	Monocytes, macrophages	Chemotaxis
		Activated macrophages	IL-1 synthesis
		Activated B cells	IgA synthesis
		Various cells	Proliferation*
TNF-α	Macrophages, mast cells, NK cells	Macrophages	CAM and cytokine expression
		Tumour cells	Cell death
TNF-β	T_H1 and T_C cells	Phagocytes	Phagocytosis, no production
		Tumour cells	Cell death

* Inhibited activities

CTL = cytotoxic T lymphocytes; DC = dendritic cells; GM-CSF= granulocyte-macrophage colony-stimulating factor; IL = interleukin; IFN = interferon; TGF = tumour growth factor; TNF = tumour necrosis factor.

POINTS TO REMEMBER

- Cytokines are low molecular weight, soluble proteins that are produced in response to an antigen and function as chemical messengers for regulating the innate and adaptive immune systems.
- Cytokines may act on the cells that secrete them (autocrine action), or on the nearby cells (paracrine action), or in some instances on distant cells (endocrine action).
- Cytokines are pleiotropic, redundant and multifunctional.
- Cytokines that regulate innate immunity are produced primarily by mononuclear phagocytes such as macrophages and dendritic cells.
- Cytokines that regulate adaptive immunity are produced primarily by T lymphocytes.
- Interleukin acts on a specific, limited group of cells that express the correct receptor for that interleukin.
- TNF is the principal cytokine that mediates acute inflammation.
- Colony-stimulating factors are cytokines that stimulate the proliferation of specific pluripotent stem cells of the bone marrow in adults.
- Chemokines are a group of cytokines that enable the migration of leucocytes from the blood to the tissues at the site of inflammation.
- TGF-β is produced by a variety of cells. TGF-β functions to inhibit the macrophage function. It also promotes tissue repair.

REVIEW QUESTIONS

1. Write short notes on:
    i. Tumour necrosis factor
    ii. Types of cytokines

    iii. Colony-stimulating factor
    iv. TGF-β
    v. Cytokine receptors
2. Explain the role of cytokines in immunity.
3. Write a detailed account on interleukins.
4. Write an essay on interferons.

11

IMMUNOLOGICAL SYNAPSE

INTRODUCTION

The helper T cells must first become activated before they can help other immune cells to respond to a foreign protein or pathogenic organism. This process occurs when an antigen-presenting cell submits a fragment of a foreign protein bound to a class II MHC molecule (endogenous antigenic fragments are bound to class I MHC molecules) to the helper T cell. Antigen-presenting cells are derived from bone marrow, and include both dendritic cells and Langerhans' cells, as well as other specialized cells like macrophages. Because T cell responses depend upon direct contact with their target cells, their antigen receptors (unlike antibodies made by B cells), are bound to the membrane only. In the intercellular gap between the T cell and the antigen-presenting cell, a special pattern of various receptors and complementary ligands forms, which is several microns in size. This patterned collection of receptors is called the immune synapse.

STRUCTURE OF IMMUNE SYNAPSE

The immunological synapse is characterized by the ordered organization of proteins at the interface. In the T cell, the TCR is clustered in the centre, along with the co-stimulatory receptor,

CD28, in a structure known as the central supramolecular activation cluster (cSMAC). This in turn is surrounded by a ring of the β2 integrin lymphocyte function-associated antigen-1(LFA-1), in an area known as peripheral supramolecular activation cluster (pSMAC). Integrins are a family of cell-surface proteins that are involved in binding to extracellular matrix components.

This arrangement of protein is in turn mirrored in the antigen-presenting cell. It has a central cluster of MHC-peptide and CD80 (ligand for CD28), surrounded by a ring of intracellular adhesion molecules-1 (ICAM-1). The ICAM-1 acts as ligand for the LFA-1 (Figure 11.1).

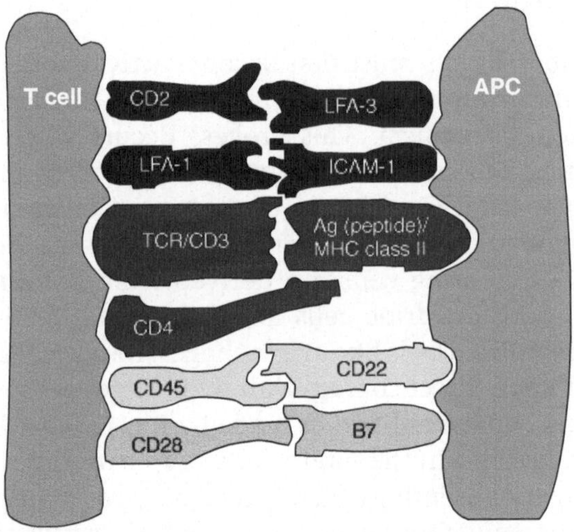

Figure 11.1 The immune synapse showing various receptors and adhesion molecules present on the antigen-presenting cells and the T cells

This specialized cell–cell junction was named as the immunological synapse because it is thought to be involved in the transfer of information across the T-cell–APC junction.

Specifically, the immune synapse appears to play an essential role in organizing the immune response, the level of control and the nature of that response. The formation of the synapse requires several minutes and it appears to be stable for several hours. The structural protein actin seems to have an important role in that stability as T-cell activation is blocked by disruption of actin filaments. There also appears to be a temporal spatial component in that signals that modulate T-cell maturity and functions are received in a serial manner as well as simultaneously (Figure 11.2).

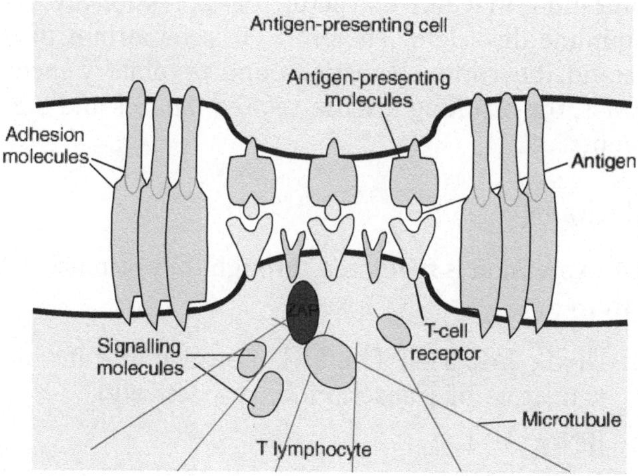

Figure 11.2 The immune synapse showing the actin filaments and signalling molecules

ICAM-1

Intercellular adhesion molecule-1 (ICAM-1, also CD54) is a transmembrane glycoprotein molecule of the immunoglobulin superfamily. Each molecule is characterized by five distinct immunoglobulin-like domains, a transmembrane domain and a cytoplasmic tail. The entire protein is coded by seven exons and six introns on chromosome 19. Each immunoglobulin domain is coded by a different exon. While the final protein is only 505 amino acids long, the molecule weighs between 80 and 114 kDa depending on the level of glycosylation, which varies among cell types and environments.

ICAM-1 is a fundamental component in many immune-related processes. ICAM-1 associates with receptors of the adhesion, thereby mediating cell–cell interactions and allowing for signal transduction. ICAM-1 interacts specifically with its receptors to induce a reversible adhesion interaction. Since normal immune function relies on ICAM-1 for processes like T-cell activation and leucocyte recruitment, it is understandable that alterations in ICAM-1 structure or expression are associated with immune disorders. Therefore, it is important to properly understand the various functions and regulatory mechanisms of ICAM-1, the resulting disease-related failures and the various treatments.

Regulation

ICAM-1 expression is regulated through four primary pathways. They are:

1. NFκB, JAK/STAT (Janus kinase/signal transducers and activators of transcription, respectively)
2. IFN-γ, AP-1
3. MAP kinase
4. PKC (protein kinase C)

It is understood that ICAM-1 is regulated at the level of transcription by one of these signalling cascades.

Function

The primary receptors for ICAM-1 are integrins which mediate cell–cell interactions and allow for signal transduction. Integrins are characterized by two subunits, α and β, of which there are several α and β families which can combine in different ways to make a wide array of integrins, each with a specific function.

Specifically, ICAM-1 is targeted to two integrins of the β_2 subunit family: LFA-1 (also $\alpha_L\beta_2$ or CD11α/CD18) and Mac-1 (also $\alpha_M\beta_2$ or CD11β/CD18). The interaction with these two molecules gives ICAM-1 a role in its two most important immune-related functions:

1. T-cell function and activation and
2. Leucocyte–endothelial cell interaction.

CD8 T cells utilize non-specific interactions between ICAM-1 and LFA-1 as the primary step in antigen recognition. The momentary adhesion interaction induced by the ICAM-1 gives the T cell time to align the T-cell receptor with the MHC class I–peptide complex. A successful TCR interaction will increase the adhesive force and commence the effector function. Conversely, an unsuccessful TCR interaction will not provide adequate adhesive forces and the T cell will leave the cell and repeat the process elsewhere. Antigen-presenting cells (APC) also express ICAM-1 alongside MHC class II. By a mechanism analogous to the CD8 target cell ICAM-1 interaction, APCs also use ICAM-1 to pause CD4 T cells allowing time to prompt activation via interaction of the TCR and the MHC II–peptide complex.

The first stage of leucocyte–endothelial cell interaction is mediated by the weaker non-specific selectin molecules, including P, E and L selectin. Inflammatory responses will up-regulate the expression of ICAM-1 thereby increasing the adhesive nature of leucocytes and endothelial cells. While the selectins instigate a rolling behaviour over the endothelial layer, the ICAM-1 interaction with leucocyte LFA-1 or Mac-1 actually stabilizes the leucocyte for extravasation. The arrested leucocytes then begin diapedesis, the process of crossing the endothelial layer, which is mediated by CD31, a protein expressed both on leucocytes and the intercellular junctions of endothelial cells. Though the ICAM-1–integrin interaction is not specific in the same way as TCR is specific for a certain MHC-protein, ICAM-1 is specifically regulated by cytokines and other factors as described above thus controlling the nature of the inflammatory response.

CONCLUSION

T-cell activation requires interaction of T-cell antigen receptors with proteins of the major histocompatibility complex (antigen). This interaction takes place in a specialized cell–cell junction referred to as an immunological synapse. The immunological synapse contains at least two functional domains: a central cluster of engaged antigen receptors and a surrounding ring of adhesion molecules. The segregation of the T-cell antigen receptor (TCR) and adhesion molecules is based on size, with the TCR interaction spanning 15 nm and the lymphocyte-function-associated antigen-1 (LFA-1) interaction spanning 30–40 nm between the two cells. Therefore, the synapse is not an empty gap, but a space populated by both adhesion and signalling molecules.

POINTS TO REMEMBER

- When a T cell meets an antigen-presenting cell (APC), a specialized structure known as an immunological synapse—or a supramolecular activation cluster (SMAC)—forms at the point of cell–cell contact.
- This consists of a central cluster of T-cell receptors (TCRs), forming the central SMAC (cSMAC).
- They are surrounded by a peripheral ring of adhesion molecules (the pSMAC). It was thought that the cSMAC increases the concentration of TCRs and therefore the probability of activation.
- The cSMAC increases both TCR activation and TCR degradation. This could be a mechanism to ensure that the immune system can respond appropriately to antigenic stimulation over a wide range of magnitudes.
- At low levels of stimulation, TCR clustering in the cSMAC enhances T-cell activation, whereas at high levels of stimulation, TCR degradation in the cSMAC protects against T-cell death due to over-stimulation.

REVIEW QUESTIONS

1. Write a short note on ICAMs.
2. Write a detailed account of immunological synapse.

12

COMPLEMENT

INTRODUCTION

The complement system is a complex system of serum proteins which interact in a cascade and activate each other sequentially. The complement system plays an essential role in host defence against infectious agents and also in the inflammatory process. The role of complement in host defence has been established through genetic deficiencies of certain complement components, which may result in life-threatening recurrent bacterial infections or immune complex diseases. The role of complement in inflammation and tissue injury has become apparent through clinical investigations and discoveries that the pathogenesis of certain experimental inflammatory diseases is complement-dependent.

The complement consists of about twenty plasma proteins that function either as enzymes or as binding proteins. In addition to these plasma proteins, the complement system includes multiple distinct cell-surface receptors that exhibit specificity for the physiological fragments of complement proteins and those that occur on inflammatory cells and cells of the immune system. There are also several regulatory membrane proteins that function to prevent autologous complement activation and protect host cells from accidental complement attack.

There are two pathways by which complement activation is initiated:

1. The classical pathway
2. The alternative pathway

The classical pathway is activated by antibody–antigen complexes. The classical pathway is activated by the binding of antibody molecules (specifically IgM and IgG1, IgG2 and IgG3) to a foreign particle. This pathway is antibody-dependent.

The alternative pathway is initiated when a previously activated complement component binds to the surface of a pathogen, where it is protected. Activation of complement has a number of important biological effects. The alternative pathway seems to be of major importance in host defence against bacterial infections because, unlike the classical pathway, it is activated by invading microorganisms and does not require antibody. This pathway is antibody-independent.

Jules Bordet is the discoverer of complement. He was born in Soignies, Belgium in 1870. In 1895, Bordet proved that two elements have to be present in the serum in order to destroy the bacterial wall.

1. One of these elements is an antibody that can be found only in animals that are already immunized against the bacteria.
2. The other element called alexine or complement can be found in any animal.

At that time, Jules Bordet set the basis of serology or the study of humoral immunity because it is contained in the "humors" (body fluids) as opposed to cellular immunity.

The alternative pathway constitutes the humoral component of natural defence against infections, which can operate without antibodies. The six proteins C3, B, D, H, I and P together perform the functions of initiation, recognition and activation of this pathway which results in the formation of activator-bound C3/C5 convertase.

C3—THE KEY COMPONENT OF COMPLEMENT

The central component of the complement system is C3. C3 is an abundant serum protein (1.2 mg/ml) which contains an unusual internal thiol–ester bond (Figure 12.1). In native C3 this bond is stable but this thiol–ester bond can become highly reactive as a result of conformational changes in the C3 protein structure. Generally the activation of C3 occurs as a result of proteolytic cleavage of the C3 molecule into 2 biologically active fragments. The activation of C3 is achieved either through the classical or alternative pathway. The activation process leads to the membrane attack pathway for the eventual cell lysis.

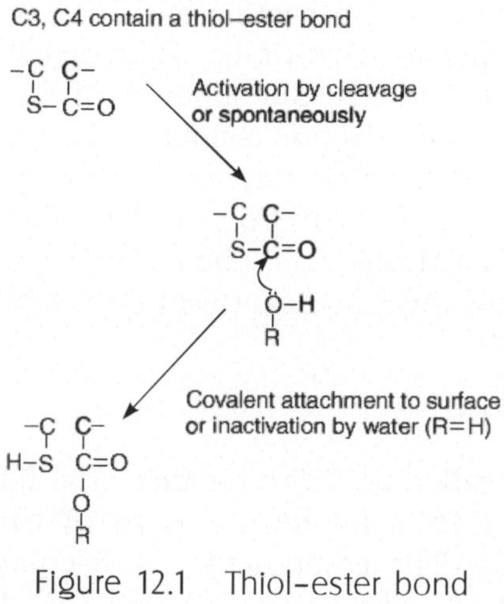

Figure 12.1 Thiol–ester bond

THE CLASSICAL COMPLEMENT PATHWAY

The classical pathway of complement activation is mediated by the specific antibody response. It is triggered by antigen-bound antibody molecules. It is the binding of a specific part of the antibody molecule to the C1 component of the complement that initiates this pathway.

C1 is a complex formed through a calcium-dependent association. The complex consists of three subunits namely C1q, C1r and C1s. C1 occurs in serum as a proenzyme which tends to undergo autoactivation but which is strictly controlled by C1 inhibitor (C1-In). A C1q molecule resembles a bunch of tulips and has an affinity for the complement-binding domain ($C_\gamma 2$ of IgG and $C\mu 3$ of IgM) of immunoglobulin. Upon the binding of C1 to immune complexes by virtue of the affinity of C1q for immunoglobulins (specifically IgM and IgG and the C1q molecule should interact with the two adjacent domains), the controlling action of C1-In is overcome and C1q effects the further activation (Figure 12.2). C1q possesses no intrinsic catalytic activity, but when any of the several activators binds to the C1q subcomponent of C1, the homologous C1r and C1s subcomponents are converted into catalytically active species, namely C1r* and C1s*, triggering the first step of the classical pathway of complement activation. Thus, on binding to immune complexes through C1q, the subunits of C1 become firmly associated and autoactivation commences even in the presence of the Cl-In. Initially, a conformational change in C1r occurs, followed by proteolytic activation which results in the activation of C1s. The two activated C1s subunits are then able to catalyse the assembly of the C3 convertase, C4b2a, which is formed from C2 and C4.

C3 Convertase

The C4b2a is called as C3 convertase. The initial step in the assembly of C4b2a is the cleavage of the complement component C4. The component C4 is composed of three polypeptide chains, alpha (93 kDa), beta (75 kDa) and gamma (33 kDa). C4 is a very sensitive substrate of C1s*. It is composed of two isotypes, C4A and C4B. Both these isotypes show clear differences in function, chemical reactivity and antigenicity despite the high sequence identity between their respective genes (about 99%).

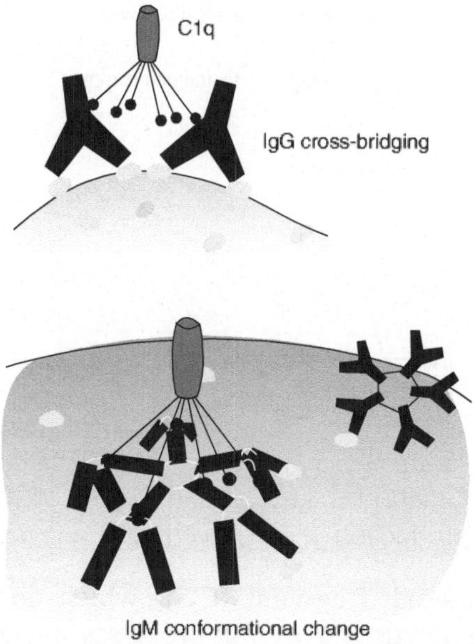

Figure 12.2 The binding of C1q molecule to the two adjacent antibody domains

It seems that gene duplication may have conferred on C4 the ability to react with a wide range of different substances. C4 undergoes cleavage, with the loss of a small fragment, C4a. The larger fragment formed by the cleavage reaction, C4b, develops a labile binding site allowing it to attach to antigens nearby. The participation of C4 in the complement cascade takes place only when the cleavage of C4 occurs in the presence of acceptor sites for C4b. In the absence of such acceptor sites for C4b, the labile binding site on C4b is no longer available, and C4 becomes C4bi incapable of further proceeding with the next step in the complement cascade.

The formation of a haemolytically active C4b site represents the binding site for C2, the second natural substrate of C1s*. C2, like C4, is encoded for within the major histocompatibility complex (MHC). Cleavage of C2 results in the formation of two

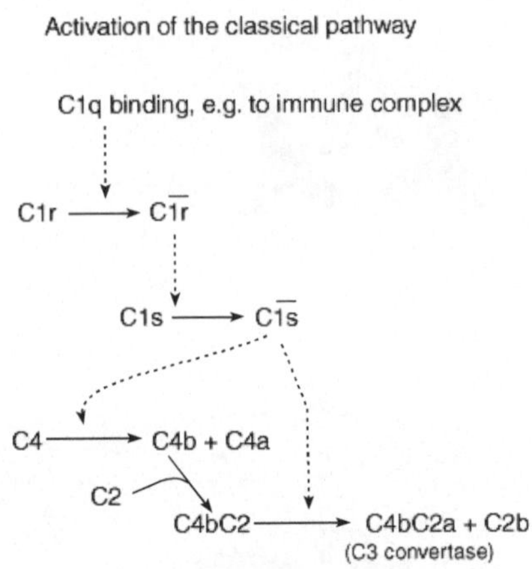

Figure 12.3 Outline of classical pathway of complement for the production of C3 convertase

fragments; a small fragment (C2b) which does not proceed further in haemolytic and opsonic processes, and a large fragment (C2a) which goes on to become part of the C3 convertase enzyme. Upon cleavage by C1s*, the C2a fragment becomes firmly associated with C4b, a reaction dependent on the presence of magnesium ions, and the C3 convertase, C4b2a, is generated (Figure 12.3). The formed C4b2a complex is now able to cleave the next component of the cascade, C3. In contrast to the C4bC2 complex, the newly formed C4b2a complex is no longer dependent on magnesium ions. The enzymatic site of the C3 convertase is located in the C2a molecule and has substrate specificity for C3. This unstable enzyme undergoes a time- and temperature-dependent decay, lasting only few minutes, unless there is a sufficient quantity of C3 in the vicinity of the cell-bound complex to mediate the next site in the sequence. Decay is associated with the release of the C2a fragment, in a functionally inactive form, into the fluid phase. The remaining C4b site is now able to take up new naive C2 and a new C4b2a enzyme is formed. Upon cleavage of the C3 complement component, two fragments are produced; a small fragment C3a which is released and appears to be important in many inflammatory responses, as increased serum levels of C3a are found as a sign of complement activation in various inflammatory skin diseases particularly in psoriasis and a larger fragment, C3b, which becomes covalently bound to the cell or bacterial surface and appears to be of great importance in the process of opsonization. After the binding of the C3b component to the C4b component, the C3 convertase, C4b2a, becomes the C5 convertase, C4b2a3b (Figure 12.4).

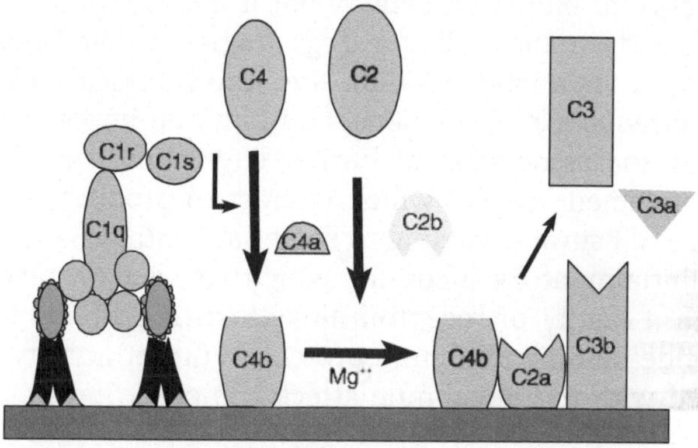

Figure 12.4 Classical complement pathway showing the cross-linking of C1q to domains of two adjacent IgG molecules followed by formation of the C5 convertase

> CR3 and CR4 are related receptors which bind iC3b and are found on monocytes/macrophages and neutrophils. They trigger phagocytosis of opsonized particles, either in concert with Fc receptors or independently. Phagocytosis of microorganisms via CR3 and CR1+FcR is the major defence mechanism against bacterial and fungal infections.

C5 Convertase

The C5 convertase is capable of cleaving the next component of the cascade, C5. The change in substrate specificity (from C3 to C5) of C4b2a is accomplished by C3b and not by modulation of the enzyme. High-affinity C5 binding sites have been demonstrated on C3b4b dimers in which C3b is linked to the chain of C4b through an ester bond. It is possible that this is the prevalent mode of C3b association with C4b2a within the C5 convertase structure on target cells. By implication, C5 may bind to both C3b and C4b before it is cleaved by the enzyme. A single cleavage in the C5 molecule leads to the formation of two fragments namely C5a and C5b. C5a is a small fragment that has important biological activity but it does not associate with the cell surface and C5b is a large fragment that binds to the cell surface via a labile binding site. C5b is critical in initiating the lytic sequence of reactions, and plays an important role in directing the association of further late-acting components of the complement system which interact to produce a lesion in the bacterial surface leading to bacterial death. C5a is found to be an inflammatory mediator which can act on target cells through a family of receptors linked to Gi proteins (guanine nucleotide-binding inhibitory protein). Further activity of C5 is involved with the membrane attack complex.

THE ALTERNATIVE COMPLEMENT PATHWAY

The alternative complement pathway constitutes the humoral component of natural defence against infections which can operate without antibody participation. Six proteins, C3, B, D, H, I, and P, by themselves perform the functions of initiation, recognition and amplification of the pathway, which results in the formation of the activator-bound C3/C5 convertase. A variety of activators of this pathway have been described. They are: particulate polysaccharides, for example, bacterial (LPS), yeast (zymosan) or plant (inulin) polysaccharides, fungi, bacteria, viruses, certain mammalian cells and aggregates of immunoglobulins, for example, the Fab portions of IgA or IgE. It is not yet known what is common in the structures of these activators as far as recognition by the pathway is concerned. Likewise, the mechanisms by which these heterogeneous groups of substances can initiate the activation of this pathway are not fully understood. It is clear, however, that recognition involves C3b. The microenvironment of the particle-bound C3b determines whether C3b prefers the binding of B, which causes activation of the pathway, or of H, which cancels the progression of the reaction. Since C3b becomes covalently bound to receptive surfaces, interaction between the putative recognition site in C3b and the recognized structure in the immediate environment may be quite weak. This strategy would allow a wide spectrum of different but related substances to be recognized by C3b. The degree of specificity of the pathway is low, but by no means non-specific (Figure 12.5).

C3 has been shown to contain a thiol–ester bond which reacts with almost anything that exposes –OH or $-NH_2$ groups, by spontaneously "ticking over". The spontaneous slow hydrolysis of the thiol–ester converts inactive native C3 to a functionally active C3b-like molecule. This form of C3, referred to as $C3(H_2O)$, constitutes a subunit of the initial C3 convertase, and its continuous production is the chemical basis of what is called the "tick-over" phenomenon.

The studies of mixtures of the six isolated proteins have shown that these six proteins are, indeed, sufficient to generate all known biological activities ascribed to the pathway. One such mixture behaved qualitatively and quantitatively like the alternative pathway in whole human serum. In addition to these six proteins, the five precursor proteins of the membrane attack complex (MAC) reconstituted the cytolytic and bactericidal alternative pathway.

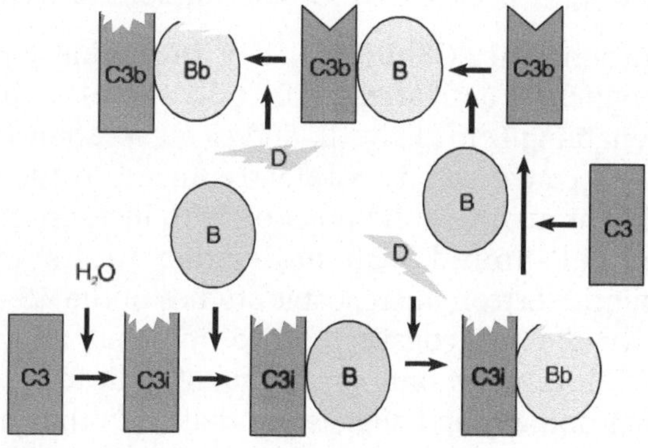

Figure 12.5 Alternative pathway of complement activation

Initial C3 Convertase

Without the participation of enzymes, initiation of the alternative pathway is safeguarded by spontaneous low-rate hydrolysis of the thiol–ester in C3 and the resultant continuous supply of C3(H_2O). When the Mg^{2+}-dependent binding of factor B to C3(H_2O) occurs, the activation of the proenzyme complex by factor D is triggered. The resultant cleavage of B causes the release of Ba, a 30 kDa fragment and the formation of C3(H_2O)Bb(Mg), the initial C3 convertase, which is confined to the fluid phase. Bb, a 60 kDa fragment, is formed second as a result of cleavage of B. The initial C3 convertase is under positive regulation by properdin (P), whose function in the alternative pathway is to bind to cell-bound C3b and to stabilize the C3/C5 convertase and is under negative regulation by factors H and I. It has been shown that for modified C3, the acquisition of H-binding capacity is slower than that of B-binding sites. Thus, modified C3 has temporarily a greater chance to form the fluid-phase C3 convertase than to become enzymatically degraded by H and I.

Target Cell-bound C3/C5 Convertase

The function of the initial C3 convertase is to produce metastable C3b and to deposit C3b on the surface of surrounding particles. The thioester in metastable C3b is more reactive than that in native C3. Since C3b deposition is catalysed by a fluid-phase enzyme, it is expected to be a random process.

The proenzyme, C3bB(Mg), is a reversible trimolecular complex which is activated by D. C3b serves as a substrate modifier enabling D to cleave B, and as an acceptor for Bb. The activated C3 convertase is covalently linked to the surface of target cells through C3b. Its function is to increase the number of target cell-bound C3b molecules in its immediate environment. Electron microscopic studies of the C3 convertase have shown that Bb consists of two domains, and that Bb is linked to C3b through only one domain, indicating that this is the binding domain and suggesting that the other may be the

C-terminal catalytic domain. In the formation of the enzyme, Ni^{2+} can replace Mg^{2+}, yielding a seven-fold more stable complex—C3bBb(Ni). The metal is released on decay-dissociation of the enzyme.

The C3 convertase can function as a C5 convertase provided, an additional C3b molecule is available in close proximity. The role of this second C3b molecule is to bind C5 and to modify it for cleavage by Bb. At this stage, it should be noted that factor I cannot inactivate C3b in the bimolecular C3 convertase, C3bBb, although it is able to inactivate the additional C3b which converts the C3bBb complex to the C5 convertase, C3bBbC3b, also, that factor H augments the rate of inactivation of C3b by dissociating Bb from the complex C3bBb. The C3/C5 convertase is physically stabilized by the cyclic protein, properdin.

> The presence of bound C3d molecules enhances the response to antigen by about 20-fold for each molecule of C3d bound (at least up to 3). In this way the innate immune system guides the antibody response onto the most "dangerous" antigens and greatly lowers the threshold of antigen required to generate a response. This effect is mediated by CR2 cross-linking with surface IgM. CR2 recruits a signalling chain which both amplifies signals and triggers a distinct signalling pathway.

Feedback Control of Alternative Pathway

The C3b-dependent positive feedback is a unique feature of the alternative pathway. C3b, the product of the reaction catalysed by the C3 convertase, forms a subunit of the C3 convertase itself. Each newly produced C3b molecule has the potential to form the enzyme together with B, D and Mg^{2+} and thus to produce more C3b and more enzyme. This process occurs rapidly in solution or on the surface of cells. In the latter case, cells become covered with C3b. The progression of amplification is controlled by H and I, both in the fluid-phase and on non-activating particles. C3b, instead of binding B, binds H and is subsequently cleaved to C3bi (inactivated), therefore H restricts the formation of C3 convertase and accelerates decay-dissociation of C3bBb. The action of factor I requires that C3b is in complex with H so that the actual substrate for I is C3bH. Factors H and I together constitute a very efficient scavenger system for C3b and also for C3(H_2O). Table 12.1 gives a comparison of classical and alternative pathways.

Table 12.1 Comparison of classical and alternative pathway

	Classical pathway	**Alternative pathway**
Initiated by	Antibody bound to antigen which activates C1	Microbial surface molecules bind C3b
C3 convertase	C4b2a	C3bBb
C5 convertase	C4b2a3b	C3bBb3b

Role of Antibody

There are evidences to show that antibody, independent of its role in activating the classical complement pathway, is able to function in the alternative complement pathway also. Metastable C3b is capable of binding directly to immunoglobulin G (IgG). The role of immunoglobulin in alternative pathway function is important because C3b covalently bound to IgG displays relative resistance to inactivation by H and I when compared to free C3b. The resistance appears to be entirely due to the reduced affinity for H, and this confers on the complex an enhanced capacity to activate C3 in serum. This complex of C3b with bactericidal IgG was found to be much more effective than IgG alone in the killing of *Escherichia coli* by serum.

It seems that different classes of immunoglobulins are able to activate the alternative complement pathway and play an important role in host defence in the infective process. Aggregated IgG and aggregated IgM both activate the pathway, as well as some aggregated IgA myeloma proteins, and some

IgE myelomas can mediate the alternative pathway although the immunoglobulin concentration required for such activation is actually relatively high.

Regulatory Proteins

Both C3 and C5 convertase enzymes are regulated and rigidly controlled by fluid-phase and membrane-regulatory proteins. The following regulatory proteins function in both activation pathways with the exception of C4bp which is only involved in the classical pathway.

1. C4bp (C4 binding protein)
2. DAF (decay-accelerating factor)
3. CRI (C3b receptor) and
4. MCP (membrane cofactor protein)

C4bp and CR1 allow the degradation of C4b by factor I (a glycosylated, disulphide-linked, two-chain serine protease of high substrate specificity found in serum and plasma as an active enzyme.

The only activity that DAF expresses is directed towards C3b, Bb (alternative pathway components) and C4b2a. It prevents the assembly of the complexes and it disassembles the formed enzymes. DAF has no cofactor activity for I action on C3b or C4b and it does not function as a receptor, although it does possess a measurable affinity for C3b but not for C4b. DAF can be inhibited by C3b located on the same cell, where just a few thousand molecules per cell have a significant effect.

Solubilized MCP has potent I-cofactor activity for fluid-phase C3b or C3b bound to solubilized molecules, but very weak cofactor activity for cell- or particle-bound C3b.

The Membrane Attack Complex (MAC)

Both the classical and the alternative complement pathways eventuate in the cleavage and activation of the C5 complement component. This process leads to the assembly of the membrane attack complex.

Cell injury by complement occurs as a consequence of activation of either the classical or the alternative pathway on the surface of a cell. The "killer" molecule is the MAC. The MAC constitutes a supramolecular organization that is composed of approximately twenty protein molecules and has a molecular weight of approximately 1.7 million. The fully assembled MAC contains one molecule each of C5b, C6, C7, and C8 and one or more molecules of C9.

There are the five precursors present in the MAC, all of which are hydrophilic glycoproteins. When C5 is cleaved by C5 convertase, a nascent C5b (C5b*) is produced and this will be followed by self-assembly of the MAC. C5b* and C6 form a stable and soluble bimolecular complex which binds to C7. This induces it to express a metastable site through which the nascent trimolecular complex (C5b-7*) can insert itself into membranes, when it occurs on or in close proximity to a target lipid bilayer. Insertion is mediated by hydrophobic regions on the C5b-7 complex that appear following C7 binding to C5b-6. Membrane-bound C5b-7 commits MAC assembly and forms the receptor for C8. The binding of one C8 molecule to each C5b-7 complex gives rise to small transmembrane channels of less than 1 nm functional diameter that may disturb target bacterial and erythrocyte membranes.

Each membrane-bound C5b-8 complex acts as a receptor for multiple numbers of C9 molecules and appears to facilitate insertion of C9 into the hydrocarbon core of the cell membrane. The concentration of C9 in human serum is such that there are only about two molecules of C9 for every C8 molecule. Therefore C5b-9 complexes generated on target membranes display a degree of heterogeneity with regard to C9 content. Binding of one molecule of C9 initiates a process of C9 oligomerization at the membrane attack site and after at least 12 molecules are incorporated into the complex, a discrete channel structure is formed. Therefore the end product consists of the tetramolecular C5b-8 complex (approximately 550 kDa) and tubular poly-C9 (approximately 1100 kDa) (Figure 12.6). This form of the MAC, once inserted into the cell membranes, creates complete transmembrane channels leading to osmotic lysis of the cell (Figure 12.7 and 12.8). The transmembrane channels formed vary in size depending on the number of C9 molecules incorporated into the channel structure.

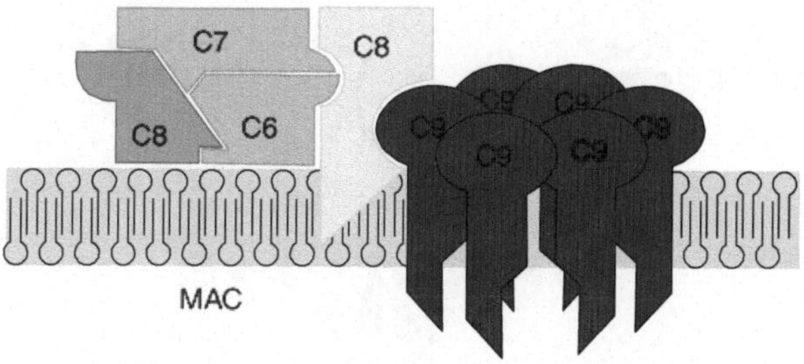

Figure 12.6 Membrane attack complex of complement

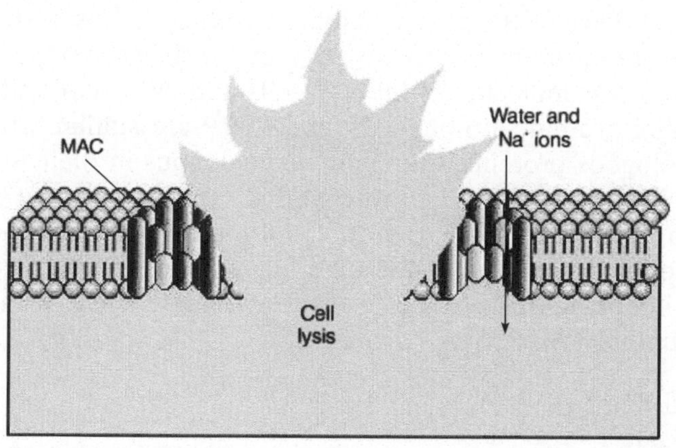

Figure 12.7 Cell lysis caused by membrane attack complex

The terminal components of the complement system, i.e., all five proteins, C6, C7, C8, C9, and C9RP are called as "cytolysin", "perforin", or "pore-forming protein". C9-related protein (C9RP) is the protein responsible for pore-forming

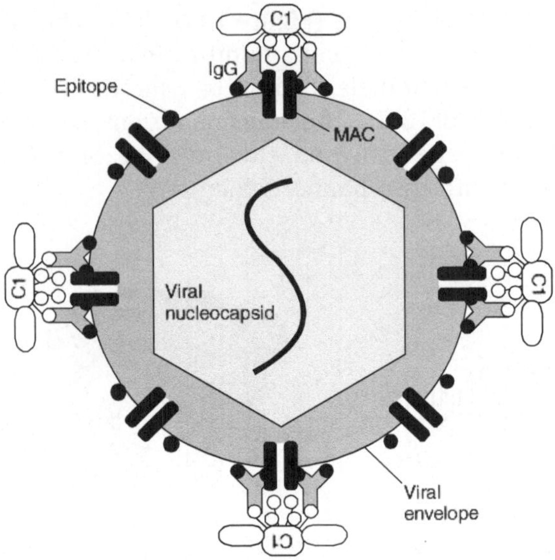

Figure 12.8 Virus attacked by MAC of complement

activity. Because the protein interacts with C9, it has been called C9-related protein, a term synonymous with cytolysin, perforin (an effector molecule of killer T cells and NK cells) and pore-forming protein. Although C9 and C9RP are similar, probably homologous proteins, they may be analogous in their function and differ in that, isolated C9RP is cytotoxic by itself, whereas isolated C9 is not. Under conditions promoting homopolymerization, C9RP kills cells without the participation of other proteins. For C9 to exert its cytotoxic effect, it requires cell-bound C5b-8.

Inhibitors of MAC

S protein is the primary inhibitor of MAC. It competes with membrane lipids for the metastable binding sites of C5b-7 and allows the binding of C8 and C9, but prevents C9 polymerization. Its function appears to be to protect the cells adjacent to sites of complement activation from accidental attack.

The cell-surface antigen, CD59, has been found to be an inhibitor of complement-mediated lysis. The function of CD59 was first suggested by the finding that the purified antigen inhibits complement-mediated lysis by binding in a glycosylation-dependent manner to C5b-8 and/or C5b-9, preventing the formation of MAC.

LECTIN PATHWAY

C4 activation can be achieved without antibody and C1 participation by the lectin pathway (Figure 12.9). This pathway is initiated by three proteins: a mannan-binding lectin (MBL), also known as mannan-binding protein (MBP) which interacts with two mannan-binding lectin-associated serine proteases (MASP and MADSP2), analogous to C1r and C1s. This interaction generates a complex analogous to C1qrs and leads

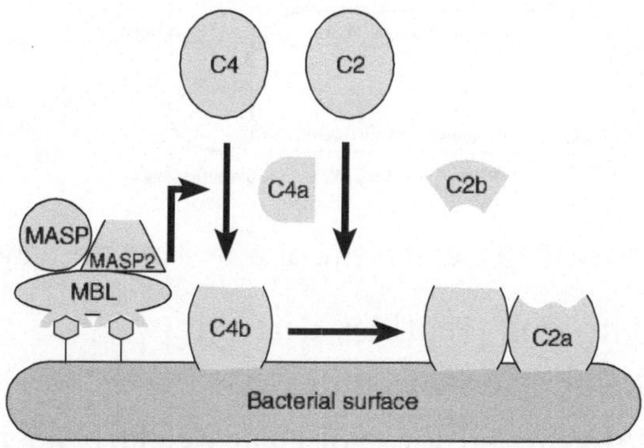

Figure 12.9 Lectin pathway of complement

to antibody-independent activation of the classical pathway. C1q can also bind to a number of agents including some retroviruses, mycoplasma, poly-inosinic acid and aggregated IgG, and initiate the classical pathway.

BIOLOGICAL EFFECTS OF COMPLEMENT

The complement system is one of the most important humoral systems. It mediates many activities that contribute to host defence (Figure 12.10). It initiates and amplifies inflammation, even before the immunity is produced where specific antibodies and lymphocytes are not available. Therefore it is not surprising that the complement cascades can be initiated by multiple ways in addition to antibody–antigen reactions. Activation of the complement cascade leads to the fragmentation of C3, C4 and C5 into low-molecular-weight hormone-like peptides, C3a, C4a and C5a (Table 12.2).

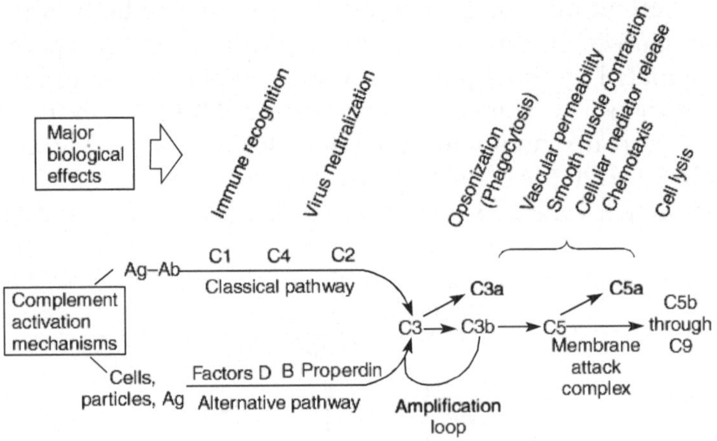

Figure 12.10 Complement activation mechanism

Anaphylatoxins (Peptides derived from C3, C4 and C5)

Anaphylatoxins are low-molecular-weight, biologically active peptides that are defined functionally by their action on small blood vessels, smooth muscle, mast cells and peripheral blood leucocytes.

C3a was the first anaphylatoxin to have its complete primary structure elucidated. C4a has a pentapeptide structure and contracts smooth muscle, although it is approximately 500-fold less active in this respect than the C3a pentapeptide.

C5a functions also as a chemotactic factor, inducing the migration of leucocytes into an area of complement activation. These molecules induce smooth muscle contraction and enhance vascular permeability. They bind to specific receptors and induce the release of vasoactive amines such as histamine from mast cells and basophils, and lysosomal enzyme release from granulocytes (particularly C3a and C5a).

Table 12.2 Biological properties of C activation products and their regulatory molecules

Component	Biological activity	Effect	Controls
C2b (Prokinin)	Accumulation of body fluid	Oedema	C1-INH
C3a (Anaphylatoxin)	Basophil and mast cell degranulation; enhanced vascular permeability; smooth muscle contraction	Anaphylaxis	Carboxy-peptidase-B (C3a-INA)
	Induction of suppressor T cells	Immuno-regulation	
C3b and its products	Opsonization; Phagocyte activation	Phagocytosis	Factors H and I
C4a (Anaphylatoxin)	Basophil and mast cell activation; smooth muscle contraction; enhanced vascular permeability	Anaphylaxis	C3a-INA
C4b	Opsonization	Phagocytosis	C4bp, Factor I
C5a (Anaphylatoxin; Chemotactic factor)	Basophil and mast cell activation; enhanced vascular permeability; smooth muscle contraction	Anaphylaxis	C3a INA
	Chemotaxis; neutrophil aggregation; oxidative metabolism stimulation	Inflammation	
	Stimulation of leukotriene release	Delayed anaphylaxis	
	Induction of helper T cells	Immuno-regulation	
C5b67	Chemotaxis; attachment to other cell membranes	Inflammation; lysis of bystander cells	Protein-S

The activities of the anaphylatoxins, C3a and C5a, are abolished by anaphylatoxin inactivator (AI), which removes the carboxy-terminal arginine from both molecules yielding C3a des Arg and C5a des Arg, respectively. The anaphylatoxin inactivator simply functions as a regulator of anaphylatoxin activity. C5a is also a potent activator of neutrophils, basophils and macrophages and causes induction of adhesion molecules on vascular endothelial cells.

Kinin Production

C2b generated during the classical pathway of C activation is a prokinin which becomes biologically active following enzymatic alteration by plasmin. Excess C2b production is prevented by limiting C2 activation by C1 inhibitor (C1-INH) also known as serpin which displaces C1rs from the C1qrs complex (Figure 12.3). A genetic deficiency of C1-INH results in an overproduction of C2b and is the cause of hereditary angioneurotic oedema. This condition can be treated with Danazol, which promotes C1-INH production or with ε-amino caproic acid, which decreases plasmin activity.

Opsonins

C3b and C4b on the surface of microorganisms attach to C-receptor (CR1) on phagocytic cells and promote phagocytosis.

OTHER BIOLOGICALLY ACTIVE PRODUCTS OF C ACTIVATION

Degradation products of C3 (iC3b, C3d and C3e) also bind to different cells via distinct receptors and modulate their functions.

COMPLEMENT RECEPTORS

CR1 binds C3b and C4b. CR1 is found on erythrocytes, where it plays a vital role in removing immune complexes from circulation. It is also found on macrophages and neutrophils and can trigger phagocytosis (only after activation of cells by other mediators including C5a). CR1 also has a regulatory role.

CR2 binds iC3b and C3d and is found on B cells where it plays an important role in activating class switching and memory formation. The different complement receptors and their functions are given in Table 12.3.

Table 12.3 Complement receptors and their functions

	Complement receptors	
Receptor	**Ligands**	**Functions**
CR1 (CD35)	C3b, C3bi, C4b, C4bi	Opsonization and antigen clearance (phagocytes) Antigen persistence (FDC) Immune complex clearance (RBC) Complement regulation
CR2 (CD21)	C3d, C3dg, C3bi, EBV	B-cell activation (B cells, FDC)
CR3 (MAC-1, CD11b/CD18)	C3bi	Adhesion, extravasation, phagocytosis (macrophages, neutrophils)
CR4 (CD11c/CD18)	C3bi	Adhesion, extravasation, phagocytosis (macrophages, neutrophils)
C1qR	C1q	Immune complex binding to phagocytes (macrophages, neutrophils)
C5aR	C5a	Adherence, phagocytosis, CR1 and CR3 expression (macrophages, neutrophils)
C3aR, C4aR, C5aR	C3a, C4a, C5a	Mast cell degranulation, smooth muscle contraction

POINTS TO REMEMBER

- The complement system is a complex system of serum proteins which interact in a cascade and activate each other sequentially.
- The complement consists of about twenty plasma proteins that function either as enzymes or as binding proteins.
- The classical pathway is activated by antibody–antigen complexes. The classical pathway is activated by the binding of antibody molecules (specifically IgM and IgG1, IgG2 and IgG3) to a foreign particle. This pathway is antibody-dependent.
- The alternative pathway is initiated when a previously activated complement component binds to the surface of a pathogen, where it is protected. This pathway is antibody-independent.
- The central component of the complement system is C3.
- The C4b2a is called as C3 convertase.
- The C5 convertase (C4b2a3b) is capable of cleaving the next component of the cascade, C5. Activity of C5 is involved with the membrane attack complex.
- Both the classical and the alternative complement pathways eventuate in the cleavage and activation of the C5 complement component. This process leads to the assembly of the membrane attack complex.
- The fully assembled MAC contains one molecule each of C5b, C6, C7, and C8 and one or more molecules of C9.

REVIEW QUESTIONS

1. Write short notes on:
    i. C1
    ii. C3 convertase
    iii. C5 convertase

   iv. Membrane attack complex
   v. Complement receptor
2. Write a detailed account of complement components.
3. Write an essay on alternative pathway of complement activity.
4. Explain the steps involved in classical pathway of complement activity.

13

INNATE IMMUNITY

INTRODUCTION

The mechanism of protection against various infectious diseases are diverse. There are two important mechanisms.

1. *Non-specific or innate immunity* This is of the pre-existing defence mechanism of an animal.
2. *Specific or adaptive immunity* This is the response to a specific immune stimulus (antigen) that involves cells of the immune system and most often leads to a state of immune memory.

Table 13.1 gives a comparison of innate and adaptive immunity.

The term innate immunity refers to the basic resistance to disease that a species possesses, i.e., the first line of defence against infection. It refers to antigen–non-specific defence mechanisms that a host uses immediately or within several hours after exposure to almost any antigen. This is the immunity one is born with and is the initial response of the body to eliminate microbes and prevent infection.

The following are the characteristics of innate immune mechanism:

- Broad spectrum of response

- No immunological memory or long-lasting protectiveimmunity
- Limited repertoire of recognition molecules
- Responses are phylogenetically ancient

Table 13.1 Comparison of innate immunity and adaptive immunity

Innate immunity	**Adaptive immunity**
Pathogen recognized by receptors encoded in the germ line	Pathogen recognized by receptors generated randomly
Receptors have broad specificity, i.e., recognize many related molecular structures called PAMPs (pathogen-associated molecular patterns)	Receptors have very narrow specificity, i.e., recognize a particular epitope
PAMPs are essential polysaccharides and polynucleotides that differ little from one pathogen to another but are not found in the host	Most epitopes are derived from polypeptides (proteins) and reflect the individuality of the pathogen
Receptors are PRRs (pattern recognition receptors)	In jawed vertebrates, the receptors are B-cell (BCR) and T-cell (TCR) receptors for antigen
Immediate response	Slow (3–5 days) response (because of the need for clones of responding cells to develop)

TYPES OF INNATE IMMUNITY

There are many elements of innate immunity (Figure 13.1) that can be broadly classified into two types.

Anatomical barriers They are

- Skin
- Mucous membrane
- Broncho-pulmonary cilia

Secretory molecules These include

- Transferrin and lactoferrin—deprives organisms of iron
- Interferon—inhibits viral replication and activates other cells to kill pathogens

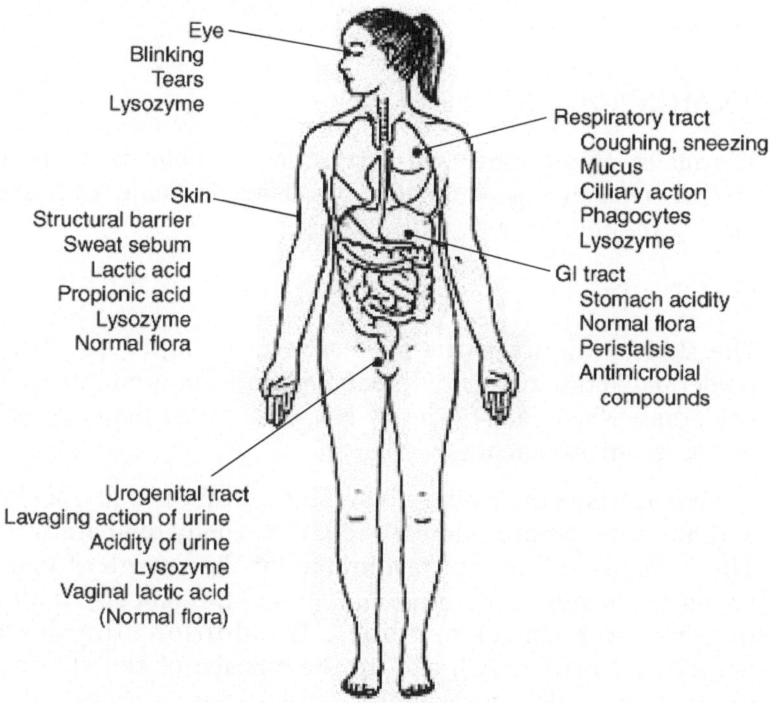

Figure 13.1 The various innate immunity mechanisms in human

- Lysozyme—present in serum and tears. It breaks down the bacterial cell wall (peptidoglycan)
- Fibronectin—coats (opsonizes) bacteria and promotes their rapid phagocytosis

- Complement—components and their products cause destruction of microorganisms directly or with the help of phagocytic cells. Acute-phase proteins (such as CRP) interact with the complement system proteins to combat infections
- TNF-alpha—suppresses viral replication and activates phagocytes

Cellular components They include various phagocytic cells. Neutrophils (PMN), macrophages and monocytes are the most important cellular components of the non-specific immune system.

ANATOMICAL BARRIER

The major physico-chemical barriers which help the body to protect itself against infections are listed in Table 13.2 and include the following.

Skin

The skin by nature is the protective covering of our body. Epidermal keratinocytes differentiate and form a multilayered epidermis, which is the primary barrier between the body and the outer environment.

It consists of the epidermis and the dermis. It is dry, acidic, and has a temperature lower than 37°C (body temperature). These conditions are not favourable for the growth of many bacteria. Further resident normal flora of the skin also inhibit potentially harmful microbes. In addition, the dead, keratinized cells that make up the surface of the skin are continuously being sloughed off so that microbes that colonize these cells are constantly being removed. Hair follicles and sweat glands produce lysozyme and toxic lipids that can kill bacteria. Finally, beneath the skin surface is skin-associated lymphoid tissue (SALT) that contains cells for killing microbes.

As the epidermis is constantly exposed to a variety of microbial pathogens, its function of resisting microbial pathogens is vital. This characteristic feature is formed during differentiation. Immunohistochemical analysis revealed that the upper epidermis of normal human skin expresses beta-defensins 1 to 3 and LL37.

The three human β-defensins, HBD-(1–3), are 33–47 residues long, cationic, antimicrobial proteins expressed by epithelial cells. All three proteins have broad-spectrum antimicrobial activity, with HBD-3 consistently being the most potent. Additionally, HBD-3 has significant bactericidal activity against gram-positive *Staphylococcus aureus* at physiological salt concentrations.

LL37 is an important peptide in the human body's first line of defence against infection. LL37 is the only human antimicrobial peptide in the cathelicidin family. It is widely expressed in both neutrophils and epithelial tissues. Infection and inflammation of epithelial tissues due to a variety of causes have been shown to induce a marked increase in its expression. Although LL37 is an amphipathic peptide, it is suggested that it is salt-resistant, with significant resistance to proteolytic degradation. The mast cell is one of the major effector cells in inflammatory reactions and can be found in most tissues throughout the body. During inflammation, an increase in the number of mast cells in that particular area occurs and such accumulation requires directed migration of this cell population. LL-37 stimulates the degranulation of mast cells and it has been hypothesized that LL-37 could be a mast cell chemotaxin (a chemical that promotes movement of a cell or microorganism in the process of chemotaxis).

Mucous Membranes

Mucous membranes line body cavities that open to the exterior, such as the respiratory tract, the gastrointestinal tract, and the genito-urinary tract. Mucous membranes are composed of epithelial cells which secrete mucus and can have cilia which act to mechanically remove microbes. In addition, these cells can secrete degradative enzymes, like lysozyme, which degrade bacterial peptidoglycan which makes up the bacterial cell walls. They can also secrete defensins which constitute a diverse family with a broad spectrum of antimicrobial activity including the ability to kill bacteria, fungi and even HIV. Mucous membranes can also have lactoferrin which binds iron to keep it from being used by harmful microbes as well as lactoperoxidase which generates toxic superoxide anion radicals which kill infectious microbes. Mucous membranes are not foolproof. Pathogens like the influenza virus have a surface molecule that enables it to attach to cells in the mucous membrane. *N. gonorrheoae* has surface projections that allow it to bind to mucous membrane epithelial cells in the urogenital tract.

Mucus also contains an antibody called secretory IgA that prevents microbes from attaching to mucosal cells and traps them in the mucus. Resident normal flora of the mucosa also inhibit potentially harmful microbes. Beneath the mucosal membrane is mucosa-associated lymphoid tissue (MALT) that contains cells for killing microbes and sampling antigens on the mucosa to start adaptive immune responses against them.

Mechanical Removal

It is the process of physically flushing microbes from the body. Methods include:

Cilia Cilia on the surface of the epithelial cells propel mucus and trapped microbes upwards towards the throat where it is swallowed. This is sometimes called the tracheal toilet. In lungs mucociliary clearance provides an efficient system to clear pathogenic particles and is assisted by aerodynamic filtering and airway reflexes, such as coughing and sneezing. Mucus secreted by mucous gland and goblet cells of the large airways entraps particles that are then propelled by movement of cilia.

Cough and sneeze reflex Coughing and sneezing removes mucus and trapped microbes.

Vomiting and diarrhoea These processes remove pathogens and toxins from the gastrointestinal tract.

Physical flushing action of body fluids The body fluids such as urine, tears, saliva, perspiration and blood from injured blood vessels also flush out microbes from the body.

Intraepithelial T lymphocytes and B1 cells

Intraepithelial T lymphocytes are found in the epidermis of the skin and the mucosal epithelia. These T lymphocytes are known as gamma delta T cells. They have a limited diversity of antigen receptors for microbes often encountered on the skin and mucous membranes. Thus a T cell of this type can interact with many diverse antigens. As such they function more as effector cells for innate immunity rather than adaptive immunity.

B1 cells are B lymphocytes with a limited diversity of antigen receptors that initially produce a class of antibodies called IgM against common polysaccharide and lipid antigens of microbes. As such they function more as effector cells for innate immunity rather than adaptive immunity. Antibodies produced by B1 cells are often called natural antibodies.

Table 13.2 Physico-chemical barriers to infections

System/Organ	Active component	Effector mechanism
Skin	Squamous cells, sweat	Desquamation, flushing, organic acids
Gastrointestinal tract	Columnar cells	Peristalsis, low pH, bile acid, flushing
Lung	Tracheal cilia	Mucociliary elevator, surfactant
Nasopharynx and eye	Mucus, saliva, tears	Flushing, lysozyme

(Contd.)

Table 13.2 (Continued)

System/Organ	Active component	Effector mechanism
Circulation and lymphoid organs	Phagocytic cells	Phagocytosis and intracellular killing
	NK cells and K cells	Direct and antibody-dependent cytolysis
	LAK	IL-2-activated cytolysis
Serum	Lactoferrin and transferrin	Iron binding
	Interferons	Antiviral proteins
	TNF-alpha	Antiviral, phagocyte activation
	Lysozyme	Peptidoglycan hydrolysis
	Fibronectin	Opsonization and phagocytosis
	Complement	Opsonization, enhanced phagocytosis, inflammation

Bacterial Antagonism by Normal Flora

About three million bacteria are present in the normal human body. These normal body flora keep potentially harmful opportunistic pathogens in check and also inhibit the colonization of pathogens. The mechanisms involved are

- producing metabolic products (fatty acids, bacteriocins, etc.) that inhibit the growth of many pathogens,
- adhering to target host cells thus covering them and preventing pathogens from colonizing,

- depleting nutrients essential for the growth of pathogens and
- non-specifically stimulating the immune system.

SECRETORY MOLECULES

The various secretory molecules involved in innate immunity have been depicted in Table 13.3.

Hydrochloric Acid

The hydrochloric acid and enzymes found in gastric secretions destroy microbes that are swallowed along with the food.

Lysozyme

It is found in tears, mucus, saliva, plasma, tissue fluid, etc. It breaks down peptidoglycan in bacteria causing osmotic lysis. Lysozyme is an enzyme (EC 3.2.1.17), commonly referred to as the "body's own antibiotic" since it kills bacteria. It is abundantly present in a number of secretions such as tears (except bovine tears). This protein is present in cytoplasmic granules of the polymorphonuclear neutrophils (PMN) (except for bovine neutrophils) and released through the mucosal secretions (such as tears and saliva). They can also be found in high concentration in egg white.

> Alexander Fleming (1881–1955), who discovered penicillin, described lysozyme in 1922. Its structure was described by David Chilton Phillips (1924–1999) in 1965 when he got the first two angstrom resolution image. This work led Phillips to provide an explanation of how enzymes speed up a chemical reaction in terms of its physical structures.

Lysozyme (also called muramidase) from hen's egg white is a polypeptide of 129 amino acid residues with a molecular weight of 14,400 dalton. It is a basic protein (positively charged) with isoelectric point of 10.7–11.0. In the hen's egg white, lysozyme accounts for 3.5% of the total egg white protein. Lysozyme is an enzyme which has the ability to lyse certain gram-positive bacteria by hydrolysing the beta-linkage between *N*-acetylmuramic acid (NAM) and *N*-acetylglucosamine (NAG) of the peptidoglycan layer in the bacterial cell wall. Lysozyme molecule is ovoid and consists of two domains or lobes linked by a long α-helix, between which lies the active site of the enzyme. The lower N-terminal lobe (residues 40–88) consists of some helices and is mostly antiparallel β-sheets. The second lobe is made up of residues 1–39 and 89–129 and its secondary structure is largely α-helical.

Most of the bacteria affected by lysozyme are not pathogenic. However, it could be considered that lysozyme is the primary reason for these organisms not developing into pathogenic strains.

Lysozyme serves as a non-specific innate opsonin by binding to the bacterial surface to reduce the negative charge and facilitate phagocytosis of this bacterium before opsonins from the acquired immune systems enter the scene.

Lysozyme has been used in pharmaceutical and food applications for many years, due to its lytic activity on the cell wall of gram-positive microorganisms. These organisms are responsible for many infections of the human body as well as the spoilage of various foods.

Lysozyme is used in certain drugs in order to increase the natural defence of the body against bacterial infections. The pharmaceutical use of lysozyme encompasses applications such as otorhinolaryngology (lozenges for the treatment of sore throats and of canker sores) and in ophthalmology (eye drops and solutions for the decontamination of contact lenses). Lysozyme is also added to infant formulae in order to make them more closely resemble human milk (cow's milk contains very low levels of a lysozyme).

Defensins

Defensins, are found in all animals and plants. They are generally short (12–50 amino acids long), positively charged and have hydrophobic or amphipathic domains in their folded structure. They constitute a diverse family with a broad spectrum of antimicrobial activity, including the ability to kill or inactivate gram-negative and gram-positive bacteria, fungi (including yeasts), parasites (including protozoa and nematodes) and even enveloped viruses like HIV. Defensins are also the most abundant protein type in neutrophils which use them to kill phagocytosed pathogens.

It is still uncertain how defensins kill pathogens. One possibility is that they use their hydrophobic or amphipathic domains to insert into the membrane of their victims, thereby disrupting membrane integrity. Some of their selectivity for pathogens over host cells may come from their preference for membranes that do not contain cholesterol. After disrupting the membrane of the pathogen, the positively charged peptides may also interact with various negatively charged targets within the microbe, including DNA. Because of the relatively non-specific nature of the interaction between defensins and the microbes they kill, it is difficult for the microbes to acquire resistance to the defensins. Thus, in principle, defensins might be useful therapeutic agents to combat infection, either alone or in combination with more traditional drugs.

Human beta-defensins are short peptides found in blood plasma and mucus. They forms pores in the cytoplasmic membrane of a variety of bacteria causing leakage of cellular needs. Certain defensins also block the fusion of viral envelopes with host cell membranes.

Lactic and Fatty Acids

It is found in perspiration and sebaceous secretions. They inhibit microbes on the skin.

The relative efficacy of lactic acid against gram-negative bacteria is not unexpected, considering that, as a small water-soluble molecule lactic acid gains access to the periplasm through the water-filled porin proteins of the outer membrane. The outer membrane functions as an efficient permeability barrier that is able to exclude macromolecules (such as bacteriocins or enzymes) and hydrophobic substances (i.e., hydrophobic antibiotics). The permeability barrier property of the outer membrane is largely due to the presence of a specific lipopolysaccharide (LPS) layer on the membrane surface. LPS molecules consist of a lipid part, termed lipid A, and a hydrophilic heteropolysaccharide chain protruding outward and providing the cell with a hydrophilic surface. Certain external agents that either release LPS and other components from the outer membrane or intercalate in the membrane can abolish the integrity of the outer membrane. Lactic acid is able to cause cell injury involved in the disruption of the LPS layer. This may enable other compounds like hydrophobic antibiotics, detergents, lysozyme or bacteriocins to penetrate.

Fatty acids are simple lipids made up of a hydrophilic carboxylate group attached to a long hydrocarbon chain. Short- and medium-chain fatty acids have antimicrobial properties and are thought to help the immune response and protect the digestive tract from harmful microorganisms.

Lactoferrin and Transferrin

It is found in body secretions, plasma, and tissue fluid. Lactoferrin is a glycoprotein that belongs to the iron transporter or transferrin family. It was originally isolated from bovine milk, where it is found as a minor protein component of whey proteins. Lactoferrin contains 703 amino acids and has a molecular weight of 80 kilodaltons. In addition to its presence in milk, it is also found in exocrine secretions of mammals and is released from neutrophil granules during inflammation.

The possible antibacterial activity of supplemental lactoferrin might be accounted for by its ability to strongly bind iron. Iron is essential to support the growth of pathogenic bacteria. Lactoferrin may also inhibit the attachment of bacteria to the intestinal wall. A breakdown product of lactoferrin is the peptide lactoferricin. Lactoferricin, classified as a bioactive peptide, also has antibacterial, as well as antiviral activity. The possible antiviral activity of supplemental lactoferrin may be due to its inhibition of virus–cell fusion and viral entry into cells.

A few mechanisms are proposed for lactoferrin's possible immunomodulatory activity. Lactoferrin may promote the growth and differentiation of T lymphocytes. Lactoferrin appears to bind uniquely in the region of major histocompatibility (MHC) proteins and the CD4 and CD8 determinants on T4 (helper) and T8 (suppressor) lymphocytes; it bears sequence homologies with the MHC class II determinant. Lactoferrin also appears to play a role in the regulation of cytokines and lymphokines, such as tumour necrosis factor-alpha (TNF-α) and IL-6. Lactoferrin's possible antioxidant activity may also contribute to its possible immunomodulatory activity.

Transferrins are found in the mucosa and bind iron, thus creating an environment low in free iron, where only few bacteria are able to survive.

Peroxidase Enzymes

It is found in saliva and milk as well as in tissue fluids and phagocytic granules. It breaks down hydrogen peroxide to produce potent oxidizing agents. Thus it is potent against catalase-negative organisms.

Fibronectin

Fibronectin is a large multidomain glycoprotein found in connective tissue, on cell surfaces and in plasma and other body fluids. It interacts with a variety of macromolecules including components of the cytoskeleton and the extracellular matrix, circulating components of the blood clotting, fibrinolytic, acute-phase and complement systems, and with cell surface receptors on a variety of cells including fibroblasts, neurons, phagocytes and bacteria. Thus it can bind to the bacteria and can act as opsonins for the engulfment of the bacteria by the phagocytes.

Table 13.3 Details of the secretory molecules involved in innate immunity

Substance	**Common sources**	**Chemical composition**	**Activity**
Lysozyme	Serum, saliva, sweat, tears	Protein	Bacterial cell lysis
Basic proteins and polypeptides (histones, lysins and other cationic proteins, tissue polypeptides)	Serum or organized tissues	Proteins or basic peptides	Disruption of bacterial plasma membrane
Lactoferrin and transferrin	Body secretions, serum, organized tissue spaces	Glycoprotein	Inhibit microbial growth by binding (withholding) iron
Peroxidase	Saliva, tissues, cells (neutrophils)	Protein	Acts with peroxide to cause lethal oxidations in cells
Fibronectin	Serum and mucosal surfaces	Glycoprotein	Binds to bacteria and assists in clearance (opsonization) by phagocytes

CELLULAR COMPONENTS IN INNATE IMMUNITY

Neutrophils

Neutrophils are the most abundant leucocytes and the second most important phagocytes in the blood. These are the first cells to enter a tissue during infection. Further, the neutrophils constitute the principal cells of acute inflammation. Their entrance is stimulated by various chemotactic factors secreted from injured cells, resident tissue macrophages and complement activation. Neutrophils can also be activated by Fc region of antibodies and by T-cell-derived cytokines. A very potent chemotactic factor for neutrophils is C5a, a peptide product of complement activation. Neutrophils contain abundant cytoplasmic granules, which contain toxic proteins. They are short-lived cells. They engulf microorganisms, destroy them but die quickly thereafter. Neutrophils are activated by antibodies, complements and cytokines.

PMN granules are of two kinds: primary (azurophilic) and secondary (specific).

- Primary azurophilic granules are characteristic of immature and very young neutrophils. They contain cationic proteins, defensins (low molecular weight proteins), proteases (elastase, cathepsin G, etc.), lysozyme and myeloperoxidase.
- Secondary granules are more characteristic of (specific for) mature granulocytes. They contain lysozyme and NADPH oxidase co-factors, lactoferrin and B-12 binding protein, the last two being characteristic of secondary granules.

Within the granules of neutrophils, there are numerous enzymes. They can induce two types of responses against invading organisms. They are oxygen-dependent mechanism and oxygen-independent mechanism.

Oxygen-independent mechanism There are two categories of proteins involved in it. These include general lysosomal proteases and glycolases that can disrupt the membrane functions of parasites, bacteria and viruses. Another type of protein, defensins, enters into the membrane of pathogens and disrupts membrane permeability of these cells. Defensins are also used by the CD8 Tc cells in cell-mediated immunity (CMI). The defensins and cathelicidins are innate immune factors present in airway surface liquid and make up part of the lung's natural defence. These peptides are produced by several different cell types including airway epithelial cells, macrophages and neutrophils. They attack the outer surface of the cell membrane surrounding the pathogen eventually punching lethal holes in it. Unlike eukaryotes, the phospholipids in the outer membrane of bacteria carry a surplus of negative charges and the positive charges on the defensins probably enable them to penetrate the bacterial membranes while sparing host membranes.

Oxygen-dependent mechanism Neutrophilic granules contain enzymes such as superoxide dismutase and myeloperoxidase which form toxic oxygen radicals (O_2^-, OH, H_2O_2). Also nitric oxide synthetase (NOS) produces NO, NO_2 and HNO_2. In addition, lipid peroxidases induce plasma membrane lipid oxidation.

PHAGOCYTOSIS

Phagocytosis is mediated by macrophages, polymorphonuclear leucocytes, dendritic cells and B cells.

Basically phagocytosis involves the ingestion and digestion of the following:

- microorganisms
- insoluble particles
- damaged or dead host cells
- cell debris
- activated clotting factors

It includes the following steps:

1. Activation
2. Chemotaxis
3. Attachment
4. Engulfment and phagosome formation
5. Destruction

Activation

Normally the phagocytes will be in the resting stage. These resting phagocytes are activated by inflammatory mediators such as bacterial products, complement proteins, pro-inflammatory cytokines and prostaglandins. This will lead the phagocytes to produce surface glycoprotein receptors that increase their ability to **adhere to surfaces and recognize microbes**. These glycoprotein molecules, known as **endocytic pattern-recognition receptors**, are so named because they **recognize and bind to pathogen-associated molecular patterns** (peptidoglycan, teichoic acids, lipopolysaccharide and mannose that are not found in human cells).

There are normally two important types of receptors in the phagocytic cells. They are mannose receptors and scavenger receptors. These receptors function to bind and ingest microbes. The mannose receptor is a macrophage lectin that binds terminal mannose and fucose residues of glycoproteins and glycolipids. These sugars are the part of molecules found on microbial cell walls. However, mammalian glycoproteins and glycolipids contain terminal sialic acid or *N*-acetylgalactosamine and thus cannot be recognized by this receptor. Therefore, the macrophage mannose receptor recognizes microbes and not host cells.

Scavenger receptors were originally defined as molecules that bind and mediate endocytosis of oxidized or acetylated low-density lipoprotein (LDL) particles that can no longer interact with the conventional LDL receptor.

Macrophage scavenger receptors bind a variety of microbes in addition to modified LDL particles (Figure 13.2). Macrophage integrins, notably Mac-1 (CD11b/CD18), may also bind microbes for phagocytosis.

The efficiency of phagocytosis is greatly enhanced when microbes are opsonized by specific proteins (opsonins) for which the phagocytes express high-affinity receptors.

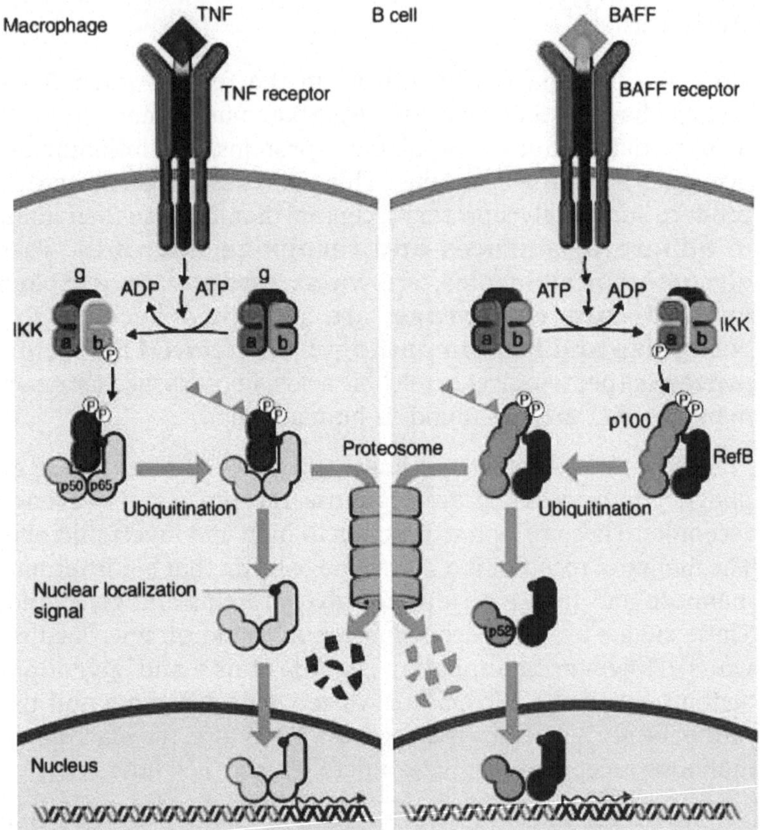

Figure 13.2 Activation of macrophage and B cell for the phagocytosis.

> The BAFF (B-cell activating factor belonging to the TNF family) is a member of TNF family. The expression of BAFF receptor appeared to be restricted to B cells.

The major opsonins are IgG antibodies, the C3b breakdown product of complement, and certain plasma lectins, notably MBL (mannan-binding lectin), all of which are recognized by specific receptors on leucocytes.

The activated macrophages exhibit increased metabolic and microbicidal activity by increasing their production of ATPs, lysosomal enzymes, lethal oxidants, etc.

Mac-1

In response to injury, monocytes migrate to the site of inflammation, where they differentiate into macrophages and participate in various biological processes. However, their fate during the resolution of acute inflammation is not fully understood. The inflammatory macrophages do not die locally by apoptosis; rather they migrate across the peritoneal mesothelium to the lymphatics, through which they further migrate to the lymph nodes and to the blood circulation. Macrophage efflux is enhanced considerably on cell activation and such accelerated macrophage migration is dependent specifically on integrin Mac-1 and can be blocked by addition of its antagonist. Mac-1 is involved specifically in the efflux of activated macrophages to the lymphatics, suggesting that Mac-1 may play an important role in the removal of local inflammatory macrophages and in their subsequent migration to the lymph nodes, a process that is critical to the development of the adaptive immunity.

Chemotaxis

Chemotaxis is the movement of phagocytes towards an increasing concentration of some chemotactic factors. It may be directly induced by a substance like C5a, produced as a result of complement activation or can be indirectly induced as a consequence of release of preformed mediators within mast cells by the action of C3a or C5a, e.g. eosinophil chemotactic factor or neutrophil chemotactic factor.

Leukotrienes, produced by the metabolism of mast cell arachidonic acid, are also chemotactic. Other chemotactic agents are chemokines (chemotactic cytokines such as interleukin-8 secreted by various cells), fibrin split products, kinins and phospholipids released by injured host cell and bacterial factors (bacterial proteins, capsules, cell wall fragments and endotoxin).

Attachment

At the site of infection, phagocytes can attach to bacteria through receptors for bacterial polysaccharides (scavenger receptor) or host proteins that act as opsonins (proteins which aid phagocytosis: fibronectin, complement and IgG antibody). This attachment triggers further activation of the phagocyte.

There are two types of attachments. They are:

Unenhanced attachment It is a general recognition of molecules irrespective of a microbe which are called pathogen-associated molecular patterns (PAMPs). It includes the components of common molecules such as peptidoglycan, teichoic acids, lipopolysaccharides, mannans and glucans common in microbial cell walls but not found on human cells by means of glycoproteins known as endocytic pattern-recognition receptors on the surface of the phagocytes.

Enhanced attachment It is the attachment of microbes to phagocytes by way of an antibody molecule or the complement proteins such as C3b and C4b produced during the complement pathways. Molecules such as IgG, C3b and C4b that promote enhanced attachment are called opsonins and the process is known as opsonization. Enhanced attachment is much more specific and efficient than unenhanced.

It is evident that organisms recognize invading microorganisms by common microbial patterns (i.e., molecular patterns). Surprisingly, it appears that essentially all eukaryotic organisms recognize and respond to the same microbial patterns. This strongly suggests that these innate mechanisms for self/non-self recognition are phylogenetically ancient.

Following are few microbial patterns:

- LPS, i.e., lipopolysaccharide (associated with the outer membrane of gram-negative bacteria)
- Mannose, fucose and other sugar residues (not just absence/presence but also factors like spacing between sugars on the cell surface)

- *Teichoic* acid (associated with the peptidoglycan cell wall of gram-positive bacteria)
- *N*-formyl peptides (all prokaryotic protein sequences begin with a formyl-methionine)

These common microbial patterns are recognized by the phagocytic receptor which have been termed pattern recognition molecules (PRMs) or pattern recognition receptors (PRRs).

The following are some of the pattern recognition molecules

- *Toll-like receptors* These are receptors which are
  - located on the membranes of certain host cells and
  - detect specific chemicals/structures associated with microorganisms and initiate an immune response, e.g. peptidoglycan, flagella, specific DNA sequences.
- *f-Met-Leu-Phe receptor* It binds to N-formyl peptides and when present attracts neutrophils.
- *Complement receptors* It is designated as CRs. It binds to complement components such as C3b and C4b which opsonize microorganisms as a consequence of the activation of the complement cascade.
- *Macrophage–Mannose receptor* It binds to mannose residues commonly present on the surface of microorganisms.
- *Scavenger receptors* There are at least 6 different scavenger receptors with different specificities that have been described (recognize certain anionic polymers and acetylated low-density lipoproteins).
- *CD14* It is the receptor present on the surface of phagocytes which allows for the recognition of LPS.

The interaction between a PRM and its microbial pattern leads to a rapid cascade of events. A good example is the interaction between CD14 and its ligand CD14L which cannot bind to LPS directly. A protein termed LBP (lipopolysaccharide-binding protein) must first bind to LPS. The LPS–LBP complex then binds to CD14 and the receptor–ligand complex is internalized. In addition, CD14 is associated with a protein known as toll-like receptor 4 (TLR-4).

The toll-like receptors (TLRs) are a family of pattern-recognition receptors that play a critical role in innate immune recognition. These receptors recognize PAMPs such as bacterial lipopolysaccharide (LPS), peptidoglycan and various viral products, and thus function as sensors of microbial infection. Upon recognition of PAMPs, TLRs induce inflammatory responses and a variety of antimicrobial effector responses. Members of the toll-like receptor (TLR) gene family convey signals stimulated by these factors, activating signal transduction pathways that result in transcriptional regulation and stimulate immune function. TLR2 is activated by bacterial lipoproteins, TLR4 is activated by LPS and TLR9 is activated by CpG DNA. Peptidoglycan recognition protein (PGRP) is activated by peptidoglycan (PGN). TLRs transduce their signals through MyD88 and the serine/threonine kinase IRAK (IL-1 receptor associated kinase). The IRAK family consists of two active kinases, IRAK and IRAK-4, and two inactive kinases, IRAK-2 and IRAK-M. IRAK-M expression is restricted to monocytes/macrophages, whereas other IRAKs are ubiquitous. The downstream signalling pathways used by these receptors are similar to those used by the IL-1 receptor, activating the IRAK through the MyD88 adaptor protein. Thus the signal created is passed through TRAF-6 and protein kinase cascades to activate NF-κB and c-Jun. This in turn activates the transcription of genes such as the proinflammatory cytokines IL-1 and IL-12.

Nuclear factor κB (NF-κB) It is a transcriptional regulator that plays a central part in responses to inflammatory signalling not only through toll-like receptors, but also through TNF receptors and the IL-1 receptor as well as in the diverse responses to other signals operating through the TNF receptor superfamily. It is also essential for responses to signalling through the variable antigen receptors of lymphocytes. NF-κB is a group of related homodimeric and heterodimeric transcription factors that are likely to activate distinct sets of target genes.

c-Jun is a component of the transcription factor AP-1, which binds and activates transcription at TRE/AP-1 elements. The *c-Jun* oncogene encodes a nuclear protein, p39, which interacts with the AP-1 *c-Fos* oncogene product and forms a transacting heterodimer. c-Jun plays an important role in the regulation of gene expression and signal transduction processes.

As a consequence of the CD14-LPB/LPS interaction at the level of the membrane, TLR-4 becomes activated. TLR-4 plays an important role in signal transduction, i.e., transfer of the signal received at the cell membrane eventually to DNA sequences located in the nucleus of the cell. After TLR-4 activates the transcription factor NF-κB, there will be transcriptional activation resulting in the synthesis of:

- ROIs or reactive oxygen intermediates and RNIs or reactive nitrogen intermediates—(highly toxic to microorganisms antimicrobial peptides, such as defensins)
- Cytokines (the small proteins which function as the chemical messengers of the immune response facilitating cell-to-cell communication)
- Chemokines (small proteins which function in the chemotaxis of leucocytes)
- Adhesion molecules (proteins which regulate the adhesive properties of leucocytes leading to alterations in leucocyte migration and trafficking)
- Acute-phase proteins (proteins synthesized largely in the liver and secreted rapidly following infection or tissue injury)

TLR ligation on specialized antigen presenting cells like dendritic cells (DCs) directly induces a differentiation program called DC maturation, which is characterized by the induction of co-stimulatory molecules on the cell surface. The co-stimulatory signal recognizes the antigenic peptides as foreign and is required (along with the TCR ligand–MHC/peptide complex) for the activation of T lymphocytes.

Engulfment and Phagosome Formation

Following attachment, polymerization and then depolymerization of actin filaments will take place and send pseudopods out to engulf the microbe. These will lead into an endocytic vesicle called a **phagosome**.

Destruction

Phagocytes contain membranous sacs called **lysosomes** which are in turn produced by the Golgi apparatus. They contain various digestive enzymes, microbicidal chemicals and toxic oxygen radicals. The lysosomes fuse with the phagosomes containing the ingested microbes to the **phagolysosome** (Figure 13.3) and the destruction of microbes occur.

There are **2 killing systems** found in neutrophils and macrophages. They are the **oxygen-dependent system** and the **oxygen-independent system** (Figure 13.4).

Oxygen-dependent System in Phagocytosis

The cytoplasmic membrane of phagocytes contains the respiratory burst oxidase which catalyses the reduction of oxygen by NADPH to form **superoxide anion (O^{2-})**. During oxidase activation, cytosolic oxidase proteins translocate to the phagosome or plasma membrane, where they assemble around a central membrane-bound component known as **flavocytochrome b**. This process is highly regulated, involving phosphorylation, translocation and multiple conformational changes. The superoxide anion can combine with water by way of the enzyme dismutase to form **hydrogen peroxide (H_2O_2)** and **hydroxyl (OH) radicals**. In the case of neutrophils, unlike macrophages, the hydrogen peroxide can then combine with chloride (Cl^{2-}) ions by the action of the enzyme **myeloperoxidase (MPO)** to form **hypochlorous acid (HOCl)**, and **singlet oxygen**. In addition, nitric oxide (NO) can combine with hydrogen peroxide to form **peroxynitrite radicals** (Table 13.4). These compounds are highly microbicidal because they are powerful **oxidizing agents** which oxidize most of the chemical groups found in proteins, enzymes, carbohydrates, DNA and lipids. Lipid oxidation will lead to the breakdown of cytoplasmic membranes. Neutrophils also contain catalase and glutathione (GS) which detoxify excess H_2O_2. GS in its reduced form (GSH), also recycles NADP to NADPH.

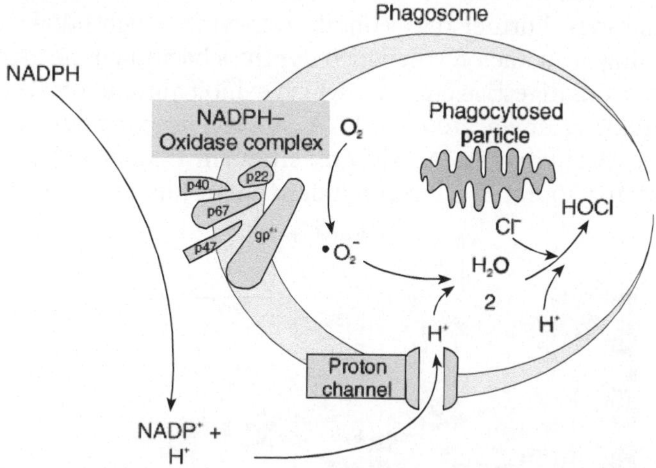

Figure 13.3 Mechanism of destruction of ingested particle by phagolysosome

Initially, sugars are rapidly oxidized within the phagosome with the production of the toxic oxygen products, viz. hydroxyl radicals. After the oxygen is consumed, there is a switch-over to fermentation leading to the production of lactic acid as an end product. This lowers the pH within the phagosome. Thus

Table 13.4 Oxygen-dependent reactions and enzymes involved

Reaction	Enzyme
$H_2O_2 + Cl^- \rightarrow OCl^- + H_2O$	Myeloperoxidase
$OCl^- + H_2O \rightarrow O_2 + Cl^- + H_2O$	
$2O_2 + 2H^+ \rightarrow O_2^- + H_2O_2$	Superoxide dismutase
$H_2O_2 \rightarrow H_2O + O_2$	Catalase

when lysosomes fuse with the phagosome, the pH is low enough for the acid hydrolases to effectively break down cellular proteins.

Oxygen-independent System in Phagocytosis

Some lysosomes contain cationic proteins that alter cytoplasmic membranes. Further they contain lysozyme which breaks down peptidoglycan, lactoferrin which deprives bacteria of needed iron and various digestive enzymes which exhibit antimicrobial activity by breaking down proteins, RNA, phosphate compounds, lipids and carbohydrates. As this mechanism occurs without the respiratory burst it is oxygen-independent killing (Table 13.5).

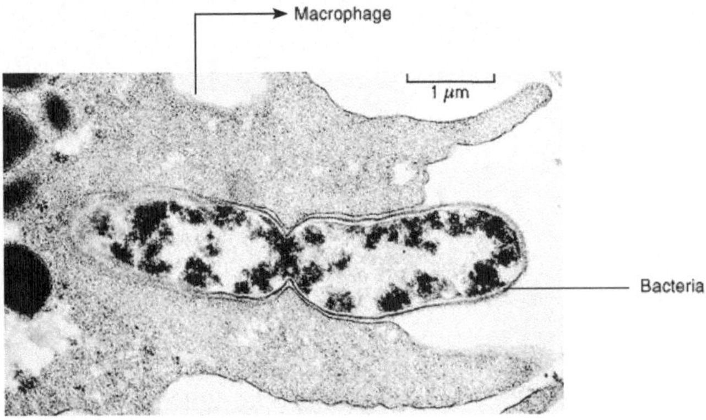

Figure 13.4 Electron microscopic view of phagocytosis

The complete mechanism of phagocytosis is shown in Figure 13.5.

Table 13.5 Molecules involved in oxygen-independent mechanism and its function

Effector molecule	Function
Cationic proteins	Damage to microbial membranes
Lysozyme	Splits mucopeptide in bacterial cell wall
Lactoferrin	Deprives proliferating bacteria of iron
Proteolytic and hydrolytic enzymes	Digestion of killed organisms

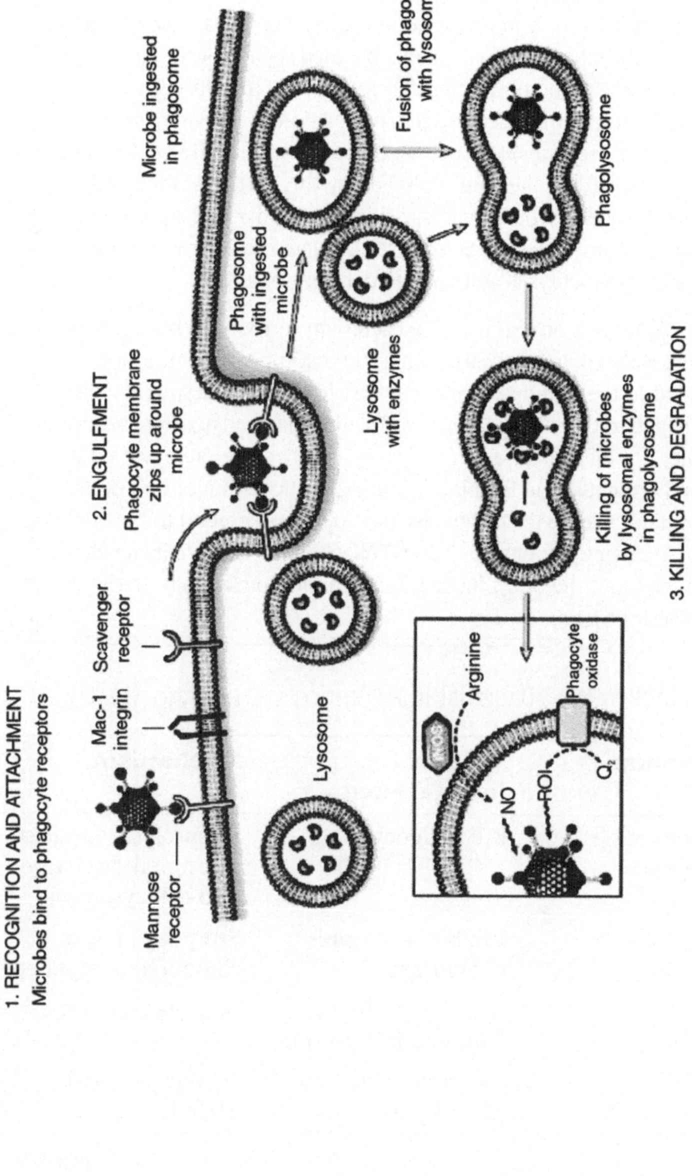

Figure 13.5 Complete mechanism of phagocytosis

Difficult Pathogens

Sometimes phagocytes have a difficult time with certain pathogens. Tuberculosis is an important example. A macrophage can usually engulf the tuberculosis bacterium, but then apparently the bacterium has a means for preventing the lysosomes from fusing with the phagosome. If the macrophage is not "activated" by paracrines from a specific immune response, the bacteria may remain alive for long periods within the macrophage. In this circumstance, other macrophages surround and wall off the infected macrophages, forming a type of chronic inflammation called a granuloma. Leprosy is another bacterium that is difficult for macrophages to destroy.

Anthrax is an example of a bacterium surrounded by a capsule that makes phagocytosis difficult. Anthrax spores from the lungs or a cut in the skin make their way first to lymph nodes, where they change to their "vegetative form" and begin dividing. But because they are difficult to destroy, they quickly become quite numerous and accumulate in the blood, causing septicemia. Not only do the bacteria release toxins but also macrophages respond to the crisis by releasing enough IL-1 and TNF-alpha to cause inflammation throughout the body. Indeed, this hyperinflammation by itself can quickly be fatal.

Table 13.6 Bacterial interference with phagocytosis

Bacterium	**Type of interference**	**Mechanism**
Streptococcus pyogenes	Kill phagocyte	Streptolysin induces lysosomal discharge into cell cytoplasm
	Inhibit neutrophil chemotaxis	Streptolysin is a chemotactic repellent
	Resist engulfment (unless Ab is present)	M protein on fimbriae
	Avoid detection by phagocytes	Hyaluronic acid capsule

(Contd.)

Table 13.6 (Continued)

Bacterium	**Type of interference**	**Mechanism**
Staphylococcus aureus	Kill phagocyte	Leucocidin induces lysosomal discharge into cytoplasm
	Inhibit opsonized phagocytosis	Protein A blocks Fc portion of Ab; Polysaccharide capsule in some strains
	Resist killing	Carotenoids, catalase, superoxide dismutase detoxify toxic oxygen radicals
	Inhibit engulfment	Cell-bound coagulase hides ligands for phagocytic contact
Bacillus anthracis	Kill phagocyte	Anthrax toxin EF
	Resist killing	Capsular polyglutamate
Streptococcus pneumoniae	Resist engulfment (unless Ab is present)	Capsular polysaccharide
Klebsiella pneumoniae	Resist engulfment	Polysaccharide capsule
Haemophilus influenzae	Resist engulfment	Polysaccharide capsule
Pseudomonas aeruginosa	Kill phagocyte	Exotoxin A kills macrophages; Cell-bound leukocidin
	Resist engulfment	Alginate slime and biofilm polymers
Salmonella typhi	Resist engulfment and killing	Vi (K) antigen (microcapsule)

(Contd.)

Table 13.6 (Continued)

Bacterium	Type of interference	Mechanism
Salmonella typhimurium	Survival inside phagocytes	Bacteria develop resistance to low pH, reactive forms of oxygen, and host"defensins" (cationic proteins)
Listeria monocytogenes	Escape from phagosome	Listeriolysin, phospholipase C lyse phagosome membrane
Clostridium perfringens	Inhibit phagocyte chemotaxis	ø toxin (toxin of *Clostridium*)
	Inhibit engulfment	Capsule
Yersinia pestis	Resist engulfment and/or killing	Protein capsule on cell surface
Yersinia enterocolitica	Kill phagocytes	Yop proteins injected directly into neutrophils
Mycobacteria	Resist killing and digestion	Cell wall components detoxify toxic oxygen radicals; prevent acidification of phagolysosome
Mycobacterium tuberculosis	Inhibit lysosomal fusion	Mycobacterial sulphatides modify lysosomes
Legionella pneumophila	Inhibit phagosome-lysosomal fusion	Unknown
Neisseria gonorrhoeae	Inhibit phagolysosome formation; possibly reduce respiratory burst	Involves outer membrane protein (porin) P.I.

(Contd.)

Table 13.6 (Continued)

Bacterium	**Type of interference**	**Mechanism**
Rickettsia	Escape from phagosome	Phospholipase A
Chlamydia	Inhibit lysosomal fusion	Bacterial substance modifies phagosome
Brucella abortus	Resist killing	Cell wall substance (LPS)
Treponema pallidum	Resist engulfment	Polysaccharide capsule material
Escherichia coli	Resist engulfment	O antigen (smooth strains); K antigen (acid polysaccharide)
	Resist killing	K antigen

OTHER INNATE IMMUNITY MECHANISMS

Apart from the above innate immunity mechanisms, there are two more important innate immunity mechanisms. They are:

1. Inflammation
2. Acute-phase response

Inflammation

Inflammation is the first response of the immune system to infection or irritation. It constitutes a complex network of molecular and cellular interactions directed to facilitate a return to physiological homeostasis and tissue repair. The response is composed of both local events and a systemic activation mediated by cytokines. Most of the body defence elements are located in the blood and inflammation is the means by which body defence cells and defence chemicals leave the blood and enter the tissue around the injured or infected site.

Cationic proteins

Cationic proteins are proteins that are present in lysosomes of neutrophils. They produce a range of effects during inflammation including the following:

- cause degranulation of mast cells with histamine release
- increase vascular permeability directly
- promote chemotaxis of neutrophils
- inhibit movement of neutrophils and eosinophils

Mechanism of inflammation First it starts with dilation of blood vessels (vasodilation) to bring more blood (hyperaemia). Along with the blood, the mediators of defence and healing will reach the area of inflammation. The endothelial cells that make up the wall of the smaller blood vessels will contract. This increases the space between the endothelial cells resulting in increased capillary permeability (Figure 13.6). Blood vessels become leaky allowing escape of fluids, proteins and cells (vasodilation) into tissue space causing oedema.

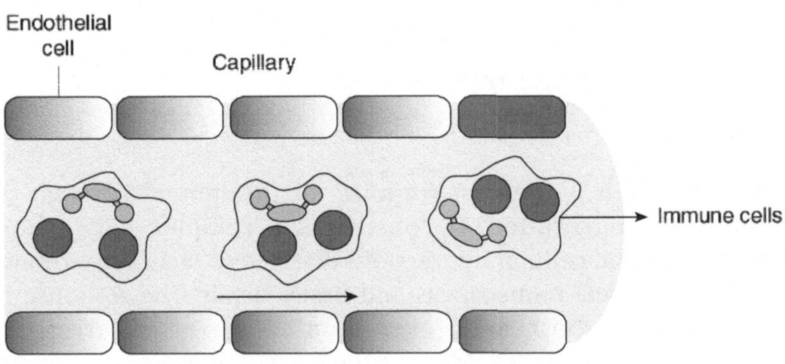

Figure 13.6 Blood capillary

Adhesion molecules are activated on the surface of the endothelial cells on the inner wall of the capillaries (Figure 13.7).

The molecules on the surface of leucocytes attach to these adhesion molecules allowing the leucocytes to flatten and squeeze through the space between the endothelial cells. Leucocytes cross the endothelial lining of the capillary and migrate interstitially and the whole process of it is called diapedesis or extravasation (Figure 13.8). Chemotactic factors drive this process, particularly during an inflammatory response when immune-related cells migrate and become concentrated at the site of inflammation.

Rolling adhesion

Rolling is a process in physiology where a white blood cell connects with endothelial cells via complementary surface adhesion molecules (selectins), rolling along it like a tumbleweed. As a result this can be the standstill of the white blood cell against the endothelium (referred to as tight binding), allowing for diapedesis to start.

Thus inflammation will bring all the cells involved in the immunity and will lead to further immune defence mechanisms, both innate and specific, at the area of infection.

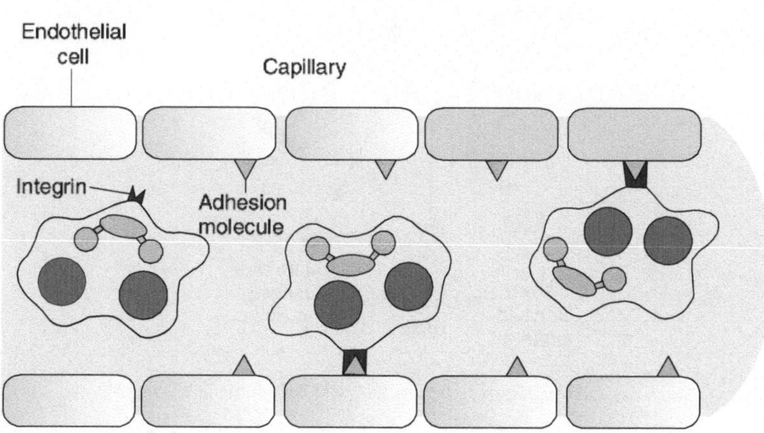

Figure 13.7 Rolling adhesion of the cells

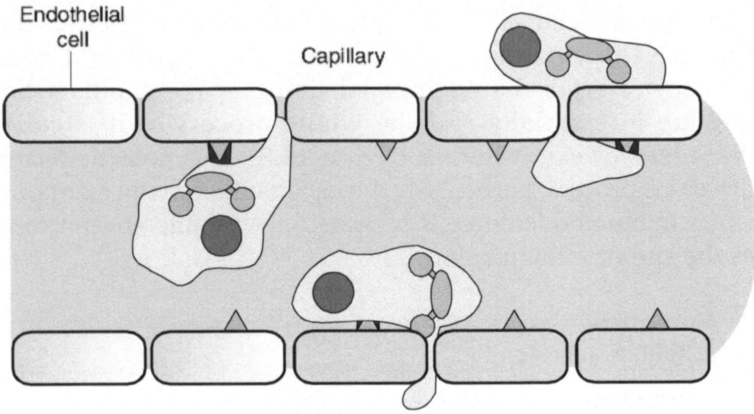

Figure 13.8 Diapedesis

Acute-phase Response

Acute-phase proteins are a class of proteins that are synthesized in the liver in response to inflammation. This response is called the acute-phase reaction (Figure 13.9).

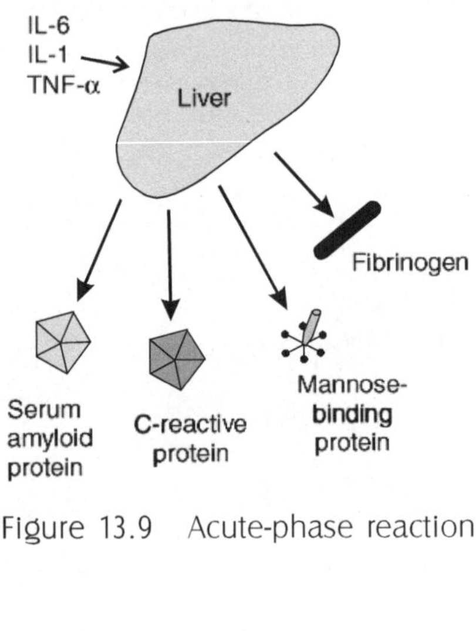

Figure 13.9 Acute-phase reaction

In response to injury, local inflammatory cells (neutrophils, granulocytes and macrophages) secrete a number of cytokines into the bloodstream, most notable of which are the interleukins IL-1, IL-6 and IL-8, and TNF-alpha.

The liver responds by producing a large number of acute-phase reactants, most notable of which are:

1. C-reactive protein
2. Mannan-binding lectin
3. Alpha 1-antitrypsin
4. Alpha 1-antichymotrypsin
5. Alpha 2-macroglobulin
6. Some coagulation factors (Fibrinogen, prothrombin, factor VIII, von Willebrand factor, plasminogen)
7. Complement factors
8. Ferritin
9. Serum amyloid P component

Two physiological responses in particular are regarded as being associated with acute inflammation. The first one is the alteration of the temperature set-point in the hypothalamus and the generation of fever. The second involves alterations in metabolism and gene regulation in the liver.

Three cytokines are released from the site of tissue injury. They are IL-1, TNF-α and IL-6. They are considered to regulate the febrile response, possibly as a protective mechanism. These cytokines mediate fever through the induction of PGE2 (prostaglandin E2). At the same time, IL-1 and IL-6 can act on the adrenal pituitary axis to generate adrenocorticotropic hormone (ACTH) and subsequently, induce the production of cortisol. This provides a negative feedback loop, since corticosteroids inhibit cytokine gene expression.

Prostaglandin E2

It is one of the prostaglandins, a group of hormone-like substances that participate in a wide range of body functions such as the contraction and relaxation of smooth muscle, the dilation and constriction of blood vessels, control of blood pressure and modulation of inflammation. Prostaglandin E2 (PG-E2) is released by blood vessel walls in response to infection or inflammation that acts on the brain (hypothalamus) to induce fever. The enzyme mPGES-1 is involved in the production of PG-E2 and is an important "switch" for activating the fever response.

It is important to consider the acute-phase response (and inflammation) as a dynamic homeostatic process that involves all of the major systems of the body, in addition to the immune, cardiovascular and central nervous systems. Normally, the acute-phase response lasts only a few days. However, in cases of chronic or recurring inflammation, an aberrant continuation of some aspects of the acute-phase response may contribute to the underlying tissue damage that accompanies the disease and may also lead to further complications, as for example, cardiovascular diseases or protein deposition diseases such as reactive amyloidosis.

The second important aspect of the acute-phase response is the altered biosynthetic profile of the liver. Under normal circumstances, the liver synthesizes a characteristic range of plasma proteins at steady-state concentrations. Many of these proteins have important functions, and higher plasma levels of these acute-phase proteins (APPs) are required during the acute-phase response following an inflammatory stimulus. Most of the acute-phase proteins are synthesized by hepatocytes and some are produced by other cell types, including monocytes, endothelial cells, fibroblasts and adipocytes.

C-reactive protein (CRP) The **major acute-phase protein** is the **C-reactive protein (CRP)** (Figure 13.10). CRP was first **discovered in 1930 by Tillet and Francis** in the serum of patients with pneumonia but it was not actually isolated until 1941. The name is derived from the ability of the C-reactive protein to react with **C-polysaccharide isolated from pneumococcal cell walls**. Early laboratory methods were only qualitative in nature until the late 1970s when significant advances in isolating CRP and measuring to the picogram range were made. Most clinical laboratories now use laser nephelometric assay because of its ease of use, speed and reproducibility. **CRP is synthesized by hepatocytes** and is classified as an acute-phase protein on the basis of its increase in plasma concentration during infection and inflammation.

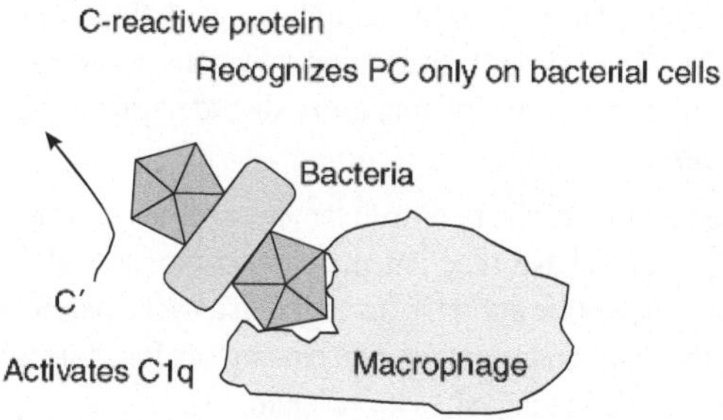

Figure 13.10 Action of C-reactive protein

Cytokines, particularly IL-6, induce CRP synthesis in the liver. The clearance rate of CRP is constant. Therefore, the level of CRP in the blood is regulated solely by synthesis. CRP acts as an opsonin for bacteria, parasites and immune complexes, activating the classical complement pathway.

The plasma levels of CRP in most healthy subjects is usually 1 mg/L with normal being defined as <10 mg/L. Plasma levels begin increasing within 4–6 hours after initial tissue injury and continue to increase several hundredfold within 24–48 hours. CRP remains elevated during the acute-phase response and returns to normal with restoration of tissue structure and function. The rise in CRP is exponential, doubling every 8 to 9 hours. The half-life is less than 24 hours. CRP is a direct and quantitative measure of the acute-phase reaction. Serial CRP measurements can be used as a diagnostic tool for infection, monitoring effect of treatment or early detection of relapse.

CRP

Tillet and Francis first discovered CRP in 1930 at Rockefeller University while examining the serum of adult patients diagnosed with acute pneumococcal pneumonia. They observed a precipitation reaction between CRP and the C-polysaccharide cell wall of the pneumococcal bacteria. This reaction later was determined to be the result of CRP binding to C-polysaccharide in the presence of calcium, forming CRP–ligand complexes. CRP's unique binding characteristics have led to the identification of elevated CRP levels in over 70 different infectious and non-infectious disorders associated with acute and chronic inflammatory disorders in adults, children and infants.

CRP is a member of the pentraxin family of proteins, which are non-specific, acute-phase reactant proteins composed of 5 identical 23-kDa polypeptide subunits arranged in a cyclic pentamer shape. Each of these subunits contains one binding site for a phosphocholine molecule and 2 binding sites for calcium.

These binding sites allow CRP to recognize and bind to a variety of biological substrates, including phosphocholine and phospholipid components of damaged cell walls and chromatin and nuclear antigens resulting in the formation of CRP–ligand complexes. CRP–ligand complexes can activate the complement system, thereby facilitating phagocytosis and the removal of materials released from damaged cells as well as potentially toxic materials from invading microorganisms. CRP–ligand complexes also bind directly to neutrophils, macrophages and other phagocytic cells, stimulating an inflammatory response and the release of cytokines.

Mannan-binding lectin Mannan-binding lectin (MBL) is a plasma protein (Figure 13.11) which, upon binding to microbial carbohydrate structures, elicits activation of the complement system. Mannan-binding lectin (MBL), a member of the collectin family, is known to have opsonic function. The mannan-binding lectin pathway is homologous to the classical complement pathway. The MBL pathway uses a protein similar to C1q of the classical pathway which binds to mannose residues as well as other sugars in a pattern that allows binding on multiple pathogens. MBL, produced by the liver, can initiate complement by binding to these pathogen surfaces.

MBL is a 6-headed molecule that forms a complex with MASP-I (Mannan-binding lectin-associated serine protease) and MASP-II, two protease zymogens. MASP-I and MASP-II are very similar to C1r and C1s of the classical pathway and are thought to have a common evolutionary ancestor.

When MBL binds to a pathogen's phospholipid bilayer, MASP-I and MASP-II are activated to cleave C4 and C2 into C4a, C4b, C2a and C2b. C4b and C2b combine on the pathogen's surface forming C3 convertase (C4bC2b), while C4a and C2a act as chemoattractants.

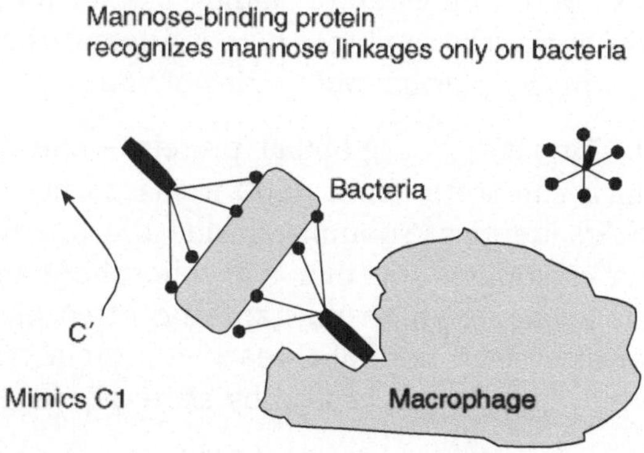

Figure 13.11 Action of mannose-binding lectin

Alpha-1 antitrypsin Alpha 1-antitrypsin or α,1-antitrypsin (α1AT) is a serine protease inhibitor (serpin). Serpins are a group of structurally related proteins, many of which inhibit peptidases)]. It protects tissue from enzymes of inflammatory cells, especially elastase. It is present in human blood at 1.5–3.5 gram/L.

Alpha-1 antichymotrypsin It is a glycoprotein found in alpha (1)-globulin region in human serum. It inhibits chymotrypsin-like proteinases *in vivo* and has cytotoxic killer-cell activity *in vitro*. The protein has a role as an acute-phase protein and is active in the control of immunological and inflammatory processes and as a tumour marker. It is a member of the serpin superfamily. Alpha 1 antichymotrypsin is an early-stage acute-phase plasma protein, which inhibits neutrophil proteinase cathepsin G and mast cell chymases and protects the lower respiratory tract from damage by proteolytic enzymes. It contains a reactive centre loop, which interacts with cognate proteinases, resulting in loop cleavage and a major conformational change.

Alpha-2 macroglobulin Alpha 2 macroglobulin is a large plasma protein found in the blood. It is produced by the liver and is a major component of the alpha-2 band in protein electrophoresis. Alpha 2-macroglobulin (alpha 2M) and related proteins share the function of binding host or foreign peptides and particles, thereby serving as humoral defence barriers against pathogens in the plasma and tissues of vertebrates.

Ferritin **Ferritin** is a globular protein found mainly in the liver, which can store about 4500 iron ions in a hollow shell made of 24 identical subunits. Inside the ferritin shell, iron ions form crystallites together with phosphate and hydroxide ions. It is an acute-phase reactant and often elevated in the course of disease. A normal C-reactive protein can be used to exclude elevated ferritin caused by acute-phase reactions.

Serum amyloid P component (SAP) Amyloid P component is a small, non-fibrillar glycoprotein found in normal serum and in all amyloid deposits. It has a pentagonal (pentaxin) structure. It is a 25-kDa pentameric protein first identified as the pentagonal constituent of *in vivo* pathological deposits called "amyloid". It acts as an acute-phase protein and modulates immunological responses in man, inhibits elastase and has been suggested as an indicator of liver disease.

POINTS TO REMEMBER

- The term, innate immunity, refers to the basic resistance to disease that a species possesses, i.e., the first line of defence against infection.
- There are three elements of innate immunity—anatomical barriers, secretory molecules, cellular components.
- Phagocytosis is mediated by macrophages, polymorphonuclear leucocytes, dendritic cells and B cells.
- Basically phagocytosis involves ingestion and digestion.
- The efficiency of phagocytosis is greatly enhanced when microbes are opsonized by specific proteins (opsonins) for which the phagocytes express high-affinity receptors.
- Inflammation is the first response of the immune system to infection or irritation.
- Acute-phase proteins are a class of proteins that are synthesized in the liver in response to inflammation. This response is called the acute-phase reaction.
- The major acute-phase protein is the C-reactive protein (CRP).

REVIEW QUESTIONS

1. Write short notes on:
   i. Difference between innate and adaptive immunity
   ii. Human b-defensins
   iii. B1 cells
   iv. Lysozyme
   v. Defensins
   vi. Lactic and fatty acids
   vii. Lactoferrin and transferrin
   viii. Fibronectin
   ix. C-reactive protein (CRP)

2. Write a detailed note on phagocytosis.
3. Write elaborately on the acute-phase response during an infection.
4. What is innate immunity? Explain in detail the innate immunity mechanisms of the human body.

14

HYPERSENSITIVITY

INTRODUCTION

Hypersensitivity refers to undesirable reactions produced by the normal immune system. The term hypersensitivity is used to describe immune responses which are damaging rather than helpful to the host. Gell and Coombs proposed a classification scheme which defined 4 types of hypersensitivity reactions. The first three are mediated by antibody and the fourth by T cells.

Schultz–Dale reaction

In the 1920s, Sir Henry Dale established that at least some of the phenomena associated with immediate hypersensitivity were caused by the chemical histamine. Dale sensitized guinea pigs against various antigens. He then observed that, when the muscles from the uterus were removed and exposed to the same antigen, histamine was released and the muscles underwent contraction, a phenomenon which is known as the Schultz–Dale reaction.

Thus hypersensitivity reactions can be divided into four types, type I, type II, type III and type IV, based on the mechanisms involved and time taken for the reaction. Many times, a particular clinical condition (disease) may involve more than one type of hypersensitivity reaction.

HYPERSENSITIVITY REACTIONS

- *Type I hypersensitivity* IgE-mediated anaphylactic reactions
  - Systemic anaphylaxis
  - Localized anaphylaxis
- *Type II hypersensitivity* Antibody-mediated cytotoxic reactions
  - Transfusion reactions
  - Newborn haemolytic disease
  - Autoimmune haemolytic anaemia
  - Drug-induced haemolytic anaemia
- *Type III hypersensitivity* Immune complex-mediated reactions
  - Localized reaction: Arthus reaction
  - Generalized reaction: Serum sickness
- *Type IV hypersensitivity* T-cell-mediated delayed-type reactions.

The general features of the four types of hypersensitivity reactions are listed in Table 14.1.

TYPE I HYPERSENSITIVITY

Type I hypersensitivity is also known as immediate or anaphylactic hypersensitivity. The reaction may involve:

- skin (urticaria and eczema),
- eyes (conjunctivitis),
- nasopharynx (rhinorrhoea, rhinitis),
- bronchopulmonary tissues (asthma) and
- gastrointestinal tract (gastroenteritis).

The reaction may cause a range of symptoms from minor inconvenience to death. The reaction usually takes 15–30 minutes from the time of exposure to the antigen, although

Table 14.1 Features of hypersensitivity reactions

Feature	**Type I**	**Type II**	**Type III**	**Type IV**
Other name	Immediate, IgE-mediated	Cytotoxic	Immune-complex	Delayed, cell-mediated,
Cell type responsible	B cells	B cells	B cells	T cells
Other cells involved	Basophils, mast cells	Red blood cells, white blood cells, platelets	Various host cells	Various host cells
Location of antigen	Soluble	Cell-bound	Soluble	Soluble or cell-bound
Type of antibody	IgE	IgG, IgM	IgG	None
Mediators	Histamine, serotonin, leucotrienes	Complement proteins, antibody-dependent cellular cytotoxicity (ADCC)	Complement proteins, neutrophil proteases	Cytokines
Timing of reaction	Immediate, or within 30 minutes	Hours to days	Hours to days	Hours to days, peaks at 48–96 hours
Skin reaction	Wheal and Flare	None	Arthus reaction (tissue death or injury)	Thickening or tissue death

(Contd.)

Table 14.1 (Continued)

Feature	**Type I**	**Type II**	**Type III**	**Type IV**
Other reactions	Vascular permeability, transudation, erythema, oedema, bronchospasm	Cell lysis, phagocytosis, inflammation, hypo-complementemia	Vaculitis, vasoconstriction, thombi, necrosis, hypo-complementemia	Erythema tissue thickening, granuloma formation
Target tissue	Vascular endothelium, bronchial smooth muscle	Blood or tissue cells	Vascular endothelium and epithelium cells	Modified self cells, infected cells, foreign tissue cells
Examples	Hay fever, insect venom, foods, drugs, vaccines	Transfusion reaction, haemolytic disease of newborns, rheumatic fever	Farmer's lung, serum sickness, malaria, vasculitis	Tuberculin reaction, contact dermatitis, tissue transplant rejection, poison ivy
Treatment	Avoidance, antihistamines, epinephrine, catecholamines, methyxanthine, cromolyn, hyposensitization	Immuno-suppressants, corticosteroids, cyclophosphamide, azathioprine, plasmapheresis	Immuno-suppressants, corticosteroids, cyclophosphamide, azathioprine	Anti-inflammatory corticosteroids, surgery

sometimes it may have a delayed onset (10–12 hours). Immediate hypersensitivity is mediated by IgE. The primary cellular component in this hypersensitivity is the mast cell or basophil. The reaction is amplified and/or modified by platelets, neutrophils and eosinophils. A biopsy of the reaction site demonstrates mainly mast cells and eosinophils.

History of type I hypersensitivity

Type I hypersensitivity reaction was first described in 1839 through experiments in which dogs that were repeatedly injected with egg albumin developed an immediate fatal shock. The term "anaphylaxis" was coined for this phenomenon in 1902, when Paul Portier and Charles Richet observed that dogs repeatedly immunized with extracts of sea anemone tentacles suffered a similar fate. Richet was awarded the 1913 Nobel Prize in Physiology or Medicine for his work on anaphylaxis.

Mechanism of Type I Hypersensitivity

Type I hypersensitivity (Figure 14.1) is mediated by IgE antibodies. The mechanism of reaction involves production of IgE, in response to certain substances called as allergens. It occurs in two phases.

Sensitization phase When the immune system of atopic individuals is first exposed to an allergen, B lymphocytes produce an excess of IgE antibodies. The high affinity FcεRI receptors are found on mast cells and basophils. Each cell has a high density of these receptors (40–250,000 per cell). IgE has very high affinity for its receptor on mast cells and basophils. IgE coats the surface of these cells by binding to receptors.

Activation phase Subsequent exposure to the same allergen crosslinks the cell-bound IgE. IgE cross-linking of mast cells stimulates the aggregation of the IgE–FcεRI resulting in the activation of a kinase-mediated signalling cascade, degranulation and the release of various pharmacologically active substances. Cross-linking of IgE Fc-receptor is important in mast cell triggering. Mast cell degranulation is preceded by increased Ca^{2+} influx, which is a crucial process. The methylation of membrane phospholipids leads to the formation of phosphatidylcholine (PC). PC increases membrane fluidity and facilitates formation of Ca^{2+} channels. Ca^{2+} activates an enzyme that cleaves PC to yield arachidonic acid. The arachidonic acid metabolism yields two important chemical mediators, viz. prostaglandins and leukotrienes. Ca^{2+} also promotes assembly of microtubules and contraction of microfilaments to move granules to plasma membrane. There is a transient increase in cAMP due to Ca^{2+} activation of membrane-bound adenylate cyclase. cAMP-dependent protein kinase then phosphorylates granule-membrane proteins leading to change in permeability to H_2O and Ca^{2+}. Thus the granules swell and facilitate membrane fusion. cAMP drop is required for degranulation (drugs that maintain high cAMP inhibit degranulation, e.g. epinephrine/ adrenalin). Fusion of granules with plasma membrane will lead to degranulation.

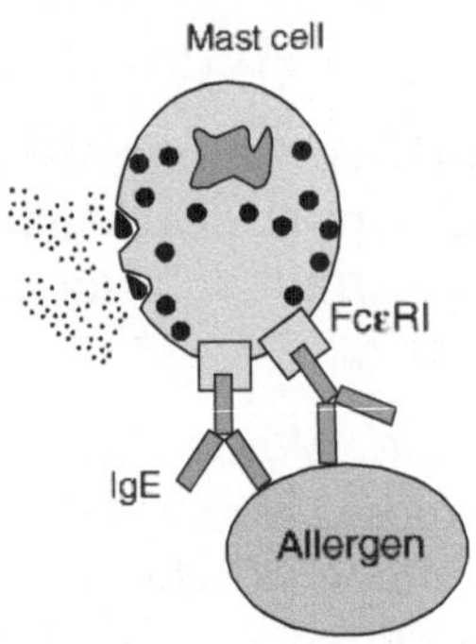

Figure 14.1 Type I hypersensitivity reaction

Mast cell degranulation releases many pharmacologically active substances that are responsible for the various diseases related to type I hypersensitivity (Table 14.2).

The most important among these compounds is histamine. It causes smooth muscle contraction, mucus release, vasodilatation and increased capillary permeability. The other substances are proteolytic enzymes, which break down tissue matrix proteins and TNF-α, which up-regulates adhesion molecule expression on vascular endothelial cells to promote leucocyte extravasation. Increased blood flow and fluid release at mucous membranes and into tissues are designed to wash away antigens or permit phagocytes and antibodies to eliminate it. Depending on the site of allergen contact, release of these mediators results in different clinical symptoms.

Mast cells are stimulated by antigen-binding to IgE-FcεRI complexes to synthesize another group of mediators, which cause prolonged symptoms (late-phase response) several hours after initial antigen-binding. These mediators include chemokines and platelet-activating factors that attract leucocytes, cytokines (including IL-4) that activate eosinophils and stimulate their synthesis in the bone marrow and slow-reacting substances of anaphylaxis (SRS-A) that promote blood flow, smooth muscle constriction and mucus secretion.

Table 14.2 Chemical mediators released in mast cell degranulation

Molecule	Effects
Primary mediators	
Histamine	Vascular permeability, smooth muscle contraction
Serotonin	Vascular permeability, smooth muscle contraction
ECF-A	Eosinophil chemotaxis
NCF	Neutrophil chemotaxis
Proteases	Mucous secretion, connective tissue degradation
Secondary mediators	
Leukotrienes	Vascular permeability, smooth muscle contraction
Prostaglandins	Vasodilation, smooth muscle contraction, platelet activation
Bradykinin	Vascular permeability, smooth muscle contraction
Cytokines	Numerous effects including activation of vascular endothelium, eosinophil recruitment and activation

Primary Mediators

Histamine Histamine (Figure 14.2) (hist—histidine residues, amine—because it is a vasoactive amine) causes several allergic symptoms. Histamine is 2-(4-imidazolyl)ethylamine and has the formula $C_5H_9N_3$. Histamine is synthesized by the decarboxylation of the amino acid histidine, a reaction catalysed by the enzyme L-histidine decarboxylase. It contributes to an inflammatory response and causes constriction of smooth muscle.

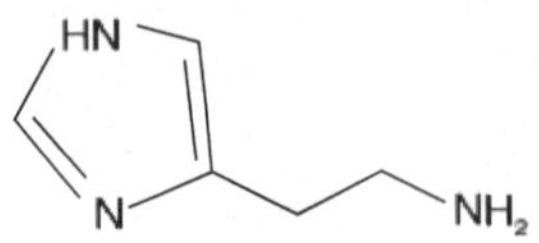

Figure 14.2 Structure of histamine

Histamine exerts its action by combining with specific cellular receptors located on cells. The four histamine receptors that have been discovered are designated as follows:

H_1 histamine receptor It is found on smooth muscle, endothelium and central nervous system tissue. It causes vasodilation, bronchoconstriction, smooth muscle activation and separation of endothelial cells (responsible for "urticaria" or "hives", also called "nettlerash"), and pain and itching due to insect stings. It is the primary receptor involved in allergic rhinitis symptoms.

H_2 histamine receptor It is located on parietal cells, which primarily regulate gastric acid secretion.

H_3 histamine receptor It is involved in decreased neurotransmitter release like histamine, acetylcholine, norepinephrine, serotonin, etc.

H_4 histamine receptor It has unknown physiological role. It is found primarily in the thymus, small intestine, spleen and colon. It is also found on basophils and in the bone marrow.

Histamine can cause inflammation directly as well as indirectly. Upon release of histamine by an antigen-activated mast cell, permeability of vessels near the site is increased. Thus, body fluids (including leucocytes, which participate in immune responses) enter the area causing swelling. This is accomplished due to histamine's ability to induce phosphorylation of an intercellular adhesion protein (called VE-cadherin) found on vascular endothelial cells.

Histamine's second type of allergic response is one of the major causes for asthma. In response to an allergen, histamine, along with other chemicals, causes the contraction of smooth muscle. Consequently, the muscles surrounding the airways constrict causing shortness of breath and possibly complete tracheal-closure, an obviously life-threatening condition.

Serotonin Serotonin (5-hydroxytryptamine, or 5-HT) is a monoamine neurotransmitter (Figure 14.3). It is also a vasoactive amine, with vasodilatory effects, released by several cells, particularly platelets.

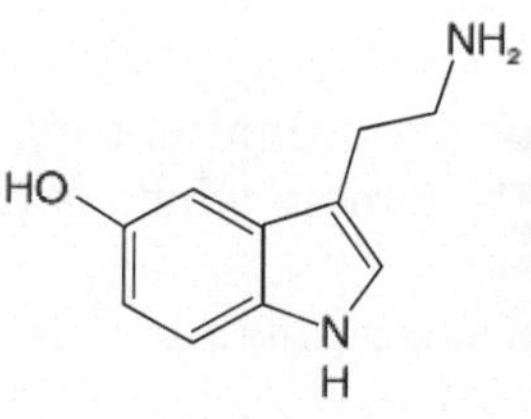

Figure 14.3 Structure of serotonin

Eosinophil chemotactic factor of anaphylaxis (ECF–A) It is the substance released from mast cells and basophils during anaphylaxis which attracts eosinophils. It is a tetrapeptide mediator of immediate hypersensitivity. ECF-A activity is associated with two acidic tetrapeptides (ala-gly-ser-glu and val-gly-ser-glu) and with less well characterized larger peptides, which are chemotactic for eosinophils and, to a lesser degree, for neutrophils.

Neutrophil chemotactic factor (NCF) It is a poorly characterized chemotactic factor. Its molecular weight is approximately 750,000 dalton. It attracts neutrophils but not eosinophils or monocytes and is released by basophils or mast cells in immediate hypersensitivity reactions.

Secondary Mediators

Prostaglandins They were first discovered and isolated from human semen in the 1930s by Ulf von Euler of Sweden. Thinking they had come from the prostate gland, he named them prostaglandins. It has since been determined that they exist and are synthesized in virtually every cell of the body.

Prostaglandins are like hormones in that they act as chemical messengers that but do not move to other sites but work right within the cells where they are synthesized.

Prostaglandins are unsaturated carboxylic acids, consisting of a 20 carbon skeleton that also contains a five-member ring. They are biochemically synthesized by the action of the enzyme phospholipase A2 which breaks down cell membrane components into arachidonic acid and then prostaglandin by cyclooxygenase.

They are highly proinflammatory, bronchospastic and vasodilatory. Unlike histamine which is produced in both mast

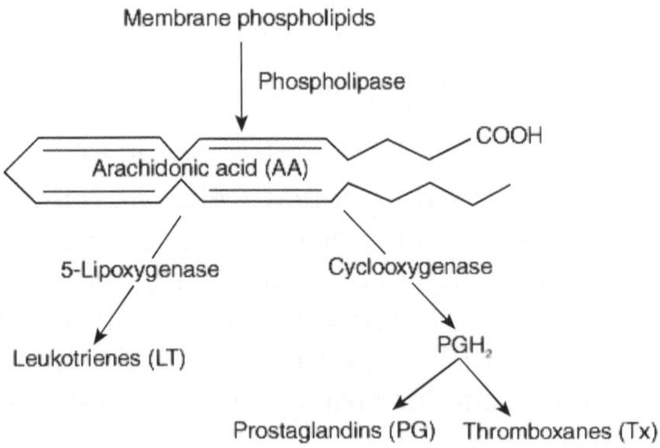

Figure 14.4 Biochemical synthesis of leukotrienes and prostaglandins

cells and basophils, prostaglandin D2 (PGD2) is only produced in the mast cells. PGD2 is a potent bronchoconstrictor, more powerful than histamine, though less so than the leukotrienes. Elevated PGD2 levels have been measured in secretions aspirated from the lungs of asthmatics and in nasal secretions from patients with nasal allergies (Figure 14.4).

Leukotrienes (SRS-A—slow reacting substance of anaphylaxis) Histamine directly contributes to inflammation but its release also begins a chain reaction or chemical cascade that results in the generation of leukotrienes. The name **leukotriene** comes from the words leucocyte and triene (a compound with three double bonds). Histamine activates the enzyme phospholipase A, which in turn releases arachidonic acid, a fatty acid, from the phospholipid membrane of the mast cell. The arachidonic acid is acted upon by an enzyme called 5-lipoxygenase and converted to an unstable intermediate chemical, viz. leukotriene A. It is immediately metabolized to form either leukotriene B4 or leukotriene C4, D4 or E4. These leukotrienes, especially leukotriene D4, are more than ten times potent than histamine. In addition to their constricting effect on bronchial muscle, the leukotrienes also act on blood vessels, causing them to become leaky and resulting in the swelling of the skin. Leukotriene B4 has a chemotactic effect on migrating neutrophils (Figure 14.4).

Bradykinin It is a physiologically and pharmacologically active peptide of the kinin group of proteins, consisting of nine amino acids. The amino acid sequence of bradykinin is: arg-pro-pro-gly-phe-ser-pro-phe-arg. Its empirical formula is $C_{50}H_{73}N_{15}O_{11}$. Bradykinin is a potent endothelium-dependent vasodilator, causes contraction of non-vascular smooth muscle and increases vascular permeability. It acts on endothelial cells to activate phospholipase A2. It is also spasmogenic for some smooth muscles and will cause pain. In some aspects, it has similar action as that of histamine and like histamine it is released from venules rather than arterioles.

Symptoms of Type I Hypersensitivity Reactions

Acute anaphylaxis The following are the symptoms of acute anaphylaxis

- Shock
- Very low blood pressure due to systemic arteriolar dilation
- Bradycardia and decreased respiratory rate

Examples Bee sting, penicillin allergy.

Acute anaphylaxis is treated with epinephrine.

Urticaria Urticaria also called nettle-rash or hives or wheals in common language, simply means itching with rash. Medically, urticaria (Figure 14.5) may be defined as skin eruption, which is allergic in origin and is characterized by profound itching, red circular or irregularly shaped eruptions on any part of the body.

These eruptions can remain on the body for variable period, anywhere between few seconds to even hours. They have the tendency to disappear and reappear. They tend to disappear without leaving behind any trace.

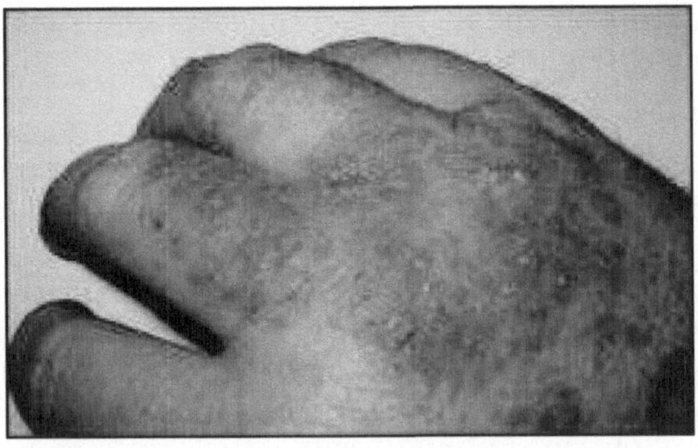

Figure 14.5 Urticaria

Urticaria is an allergic skin disorder. Characteristically the skin eruptions are erythematous, raised above the skin level, with intense itching and usually worsened by itching and with slight local warmth.

Urticaria may appear on any part of the skin. Angioedema is a condition when deep tissues are affected. The typical lesions may last for one minute to half an hour. Some may last even longer. Some patients may get the eruptions once in a while and some may have many times during the day. It may be restricted to a couple of spots in some patients, while some may have widespread rashes appearing for days or even months together.

Urticaria may be acute, subacute, chronic and recurring variants as far as the frequency and duration are concerned.

Under the microscope, a typical urticarial rash may exhibit perivascular, cellular infiltrate consisting of lymphocytes and eosinophils, is indicative of its allergic behaviour. There are findings related to oedema (swelling) and mucosal inflammation.

Allergic rhinitis (Hay fever) Rhinitis is defined as inflammation of the nasal membranes and is characterized by a symptom complex that consists of any combination of the following: sneezing, nasal congestion, nasal itching and rhinorrhoea. The eyes, ears, sinuses and throat can also be involved. Allergy is the most common cause of rhinitis.

Allergic asthma Allergic asthma is a chronic inflammatory disorder of the airways. Its symptoms are made worse by exposure to an allergen (e.g. dust, molds, pollen, animal dander) to which the patient has been sensitized.

The symptoms of allergic and non-allergic asthma are the same. They include coughing, wheezing, shortness of breath or rapid breathing and chest tightness.

Tests for Type I Hypersensitivity

Skin test This is the most commonly used allergy test. During this test, small amounts of allergens are introduced on or into the skin. Redness, itching and a raised wheal appear within 20 minutes if there is a positive reaction to an antigen (Figure 14.6). This test can identify specific allergens or classes of allergens. Because it involves introduction of possible allergens, it does carry some risks, including the rare but serious occurrence of a life-threatening anaphylactic reaction.

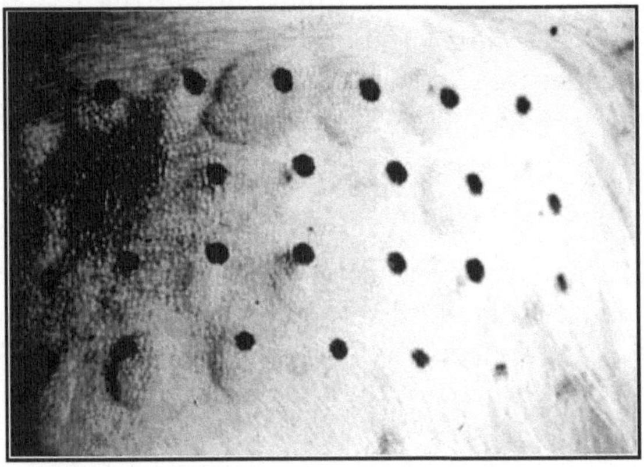

Figure 14.6 Skin test showing the wheal and flare reaction

Radioallergosorbent test (RAST) This test evaluates allergen-specific IgE levels in serum. Like the skin test, the RAST provides allergen-specific information. Because it is performed in the lab on serum only, there are no risks associated with this test. However, this test is very costly.

Differential leucocyte count A differential leucocyte count can be performed. Eosinophils are often elevated with allergic reactions. This test is non-specific and provides no information about specific allergenic substances to which the patient is allergic.

Enzyme-linked immunosorbent assay (ELISA) ELISA is very sensitive and can measure various immunoglobulins. Even though not so popular, it is very useful because of its ability to detect both immediate and delayed hypersensitivity reactions. It provides an indirect determination of allergen to which a person may be allergic to. Like RAST, it carries no direct risk to the patient.

Elimination-challenge diet The elimination-challenge diet is useful for detecting food allergies. This diet involves eating only hypoallergenic food for several weeks. The elimination diet is followed by a systematic reintroduction of possible triggers. Symptoms should be closely monitored.

Passive cutaneous anaphylaxis (PCA) It is a passively transferred local anaphylactic reaction used to study specific IgE antibodies. The skin of an animal is sensitized by an intradermal injection of serum from a sensitized animal. After a 24- to 72-hour latent period, the suspected allergen and Evans blue dye are injected intravenously. Reaction of the antigen with skin-fixed antibody causes the release of histamine, which increases vascular permeability, permits leakage of the albumin-bound dye, and produces a blue spot at the site of the intradermal injection.

Leucocyte histamine release test (LHRH) LHRH is a technique to evaluate the *in vitro* release of histamine from leucocytes (i.e., basophils) in response to exposure to an allergen. Initially, measurements of histamine release requires isolation of leucocytes from whole blood followed by the isolation of the released histamine. As such the technique is very difficult and time-consuming. Thus LHRT was primarily used as a research tool only. Recently, a special type of glass fibre has been developed that binds histamine with high affinity and selectivity. These glass fibres can be used as a "solid phase" to absorb the histamine that is released directly into the blood.

Bronchial challenge test Histamine or methacholine is used to perform this test when it is necessary to determine if the patient has hyper-responsive airways. Volatile chemicals are used to perform this test when the allergy is encountered in an occupational setting. If dust, ragweed or other common allergens are the suspected cause of the problem, this test is not medically necessary, since skin tests can be used in these situations.

Prausnitz–Kustner test or P–K test

The existence of a component in human serum which mediates hypersensitive reactions was demonstrated by Otto Prausnitz, a Polish bacteriologist and Heinz Kustner, a Polish gynaecologist, in 1921. Kustner had a strong allergy to fish. Prausnitz removed a sample of serum from his colleague and injected it under his own skin. The next day, Prausnitz injected fish extract in that same region. Hives immediately appeared, indicating that the serum contained components that mediated the allergy. For some time, the Prausnitz–Kustner test, or P–K test, remained a means of testing for allergens under circumstances in which a person could be tested for sensitivity. (It is no longer in use because of safety concerns.) In this test, a serum sample from the test subject was injected under the skin of a surrogate (usually a relative) and later followed with test allergens. The presence of a wheal and flare reaction (hives) (Figure 14.6) indicated sensitivity to the allergen. The serum component responsible for this sensitivity was later identified as the antibody IgE by K. Ishizaka and T. Ishizaka and S. G. O. Johansson in 1967. The target cells to which the IgE bound were later identified as mast cells and basophils.

TYPE II HYPERSENSITIVITY

It is also known as cytotoxic hypersensitivity or antibody-mediated cytotoxic reactions and may affect a variety of organs and tissues. It is caused by specific antibody-binding to cells or tissue antigens. The antibodies are of the IgM or IgG classes and cause cell destruction by Fc-dependent mechanisms either directly or by recruiting complement via the classical pathway. The antigens are normally endogenous, although exogenous chemicals which can attach to cell membranes can also lead to type II hypersensitivity (Figure 14.7).

The following are the three important mechanisms that are involved in type II hypersensitivity.

- Opsonization of the host cells whereby phagocytes stick to host cells by way of IgG, C3b or C4b and discharge their lysosomes.
- Activation of the classical complement pathway causing MAC lysis of the cells and
- ADCC (antibody-dependent cellular cytotoxicity) destruction of the host cells whereby NK cells attach to the Fc portion of the antibodies. The NK cells then release pore-forming proteins called perforins and proteolytic enzymes called granzymes. Granzymes pass through the pores and activate the enzymes that lead to apoptosis of the infected cell by means of destruction of its structural cytoskeleton proteins and by chromosomal degradation.

In practice, type II hypersensitivity reactions are usually seen in blood transfusion recipients and patients with certain autoimmune diseases. The classic ABO incompatibility reaction is a type II reaction, with IgM antibodies causing complement-mediated lysis of erythrocytes. In the Rhesus disease (or haemolytic disease of the newborn) the IgG antibodies cause destruction of foetal red blood cells by antibody-dependent cellular cytotoxicity (ADCC). The antibodies will be passively acquired by the host through the placenta. Drug-induced haemolytic anaemia, granulocytopenia and thrombocytopenia are some of the other examples of type II hypersensitivity. The reaction time is minutes to hours. Type II hypersensitivity is primarily mediated by antibodies of the IgM or IgG classes and complement. Phagocytes and K cells may also play a role in ADCC.

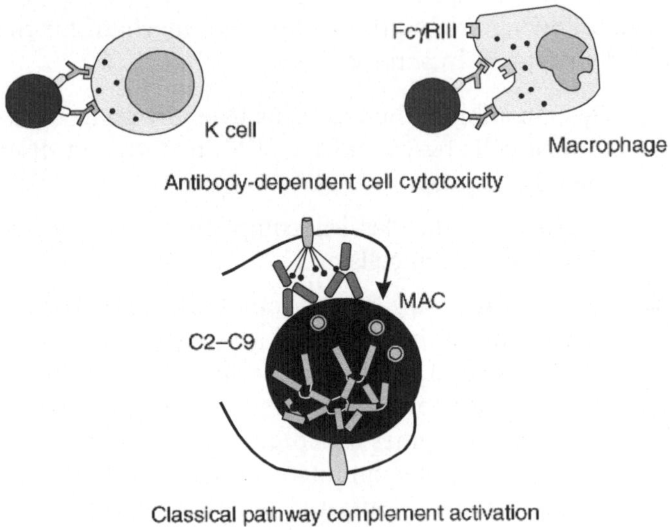

Figure 14.7 Various forms of type II hypersensitivity

Some Type II Hypersensitivity Reactions

Transfusion reaction—ABO system Transfusing a patient with the incorrect ABO group blood may have fatal consequences. Donor red cells may be destroyed by an antibody (IgM) in the recipient's plasma. The rapid intravascular haemolysis which occurs in ABO incompatible transfusions can precipitate severe disseminated coagulation (DIC), prolonged hypotension, acute uraemia and even death. It has also been recognized that a potent anti-A or anti-B in donor blood of group O may destroy the A or B red cells of a non-O recipient.

Newborn haemolytic disease (Erythroblastosis foetalis) It is a severe haemolytic disease of the newborn. It occurs in Rh– woman carrying Rh+ baby. Upon birth, when placenta separates, mother is exposed to Rh+ blood of the infant and elicits a humoral response leading to memory B cells. In second pregnancy, with another Rh+ child, some of the Rh+ antigens are recognized by the memory cells and IgG is produced. This crosses the placenta and binds foetal RBC. This will lead to the destruction of RBCs. Mild to severe anaemia may result and can even cause death. Bilirubin becomes elevated and can cause brain damage.

Protection is provided if Rhogam (anti-Rh AB) is given within 72 hrs of the first delivery. Antibody binds and clears any Rh+ antigen before humoral response is elicited.

ABO incompatibility between mother and foetus is not usually a problem since isohaemagglutinins are usually IgM and they do not cross the placenta.

Autoimmune haemolytic anaemia It is the production of antibodies to one's own RBCs that can lead to antibody-mediated complement lysis that may lead to anaemia. It is not clear in most cases how autoimmunity arises, but it is thought to arise by responses to infectious agents.

Drug-induced haemolytic anaemia Various kinds of drugs, especially antibiotics such as penicillin, cephalosporin, and streptomycin, can adsorb to RBC surfaces in a non-specific fashion. This sets up a complex analogous to a hapten–carrier complex and elicits an antibody response that results in lysis and anaemia.

Drug-induced immune haemolytic anaemia is an acquired form of haemolytic anaemia caused by interaction of certain drugs with the immune system. The result is the production of antibodies against the red blood cells and premature red blood cell destruction.

Drug-induced immune haemolytic anaemia occurs when certain drugs start an immune reaction against red blood cells. In some instances, the drugs interact with the red blood cell membrane, causing the cell to become antigenic. The immune system identifies these cells as not belonging to the body and thus antibodies form against the red blood cells. The antibodies attach to red blood cells and cause their premature destruction. This condition is rare in children.

Drugs that can cause immune haemolytic anaemia include the following:

- Penicillin and its derivatives
- Cephalosporins
- Levodopa

- Methyldopa
- Quinidine
- Some anti-inflammatory drugs

Goodpasture's syndrome This is a type II hypersensitivity disease that is characterized by autoantibodies directed against type IV collagen that is found in basement membrane. It commonly affects the renal glomerulus and the pulmonary alveoli, leading to glomerulonephritis and pulmonary haemorrhage. It is frequently preceded by an infection of the respiratory tract that could cause antigen exposure. It is also more common in smokers. Other organs that may be affected are the eye and the choroid region of the brain. Without treatment the prognosis is poor as it progresses to a rapid progressive renal failure.

TYPE III HYPERSENSITIVITY

Type III hypersensitivity (Figure 14.8) is also known as **immune complex hypersensitivity**. Type III hypersensitivity is mediated by immune complexes essentially of IgG and IgM antibodies with soluble antigens. Antibody–antigen complexes form in proper concentrations of antigen and antibody. Depending on the size of these complexes, they may be cleared efficiently by phagocytic cells or in circulation through the liver. Sometimes they deposit in various organs or tissues and are not cleared efficiently. These small complexes **lodge in the capillaries**, pass between the endothelial cells of blood vessels—especially those in the skin, joints and kidneys and become trapped on the surrounding basement membrane beneath these cells. **The antigen–antibody complexes then activate the classical complement pathway**. This may further cause

- **massive inflammation** due to complement protein C5a
- **influx of neutrophils** due to complement protein C5a resulting in neutrophils discharging their lysosomes and causing tissue destruction and further inflammation

- **MAC (membrane attack complex) lysis** of cells of surrounding tissue due to the membrane attack complex formed by complement pathway
- **aggregation of platelets,** resulting in more inflammation and the formation of microthrombi that block capillaries

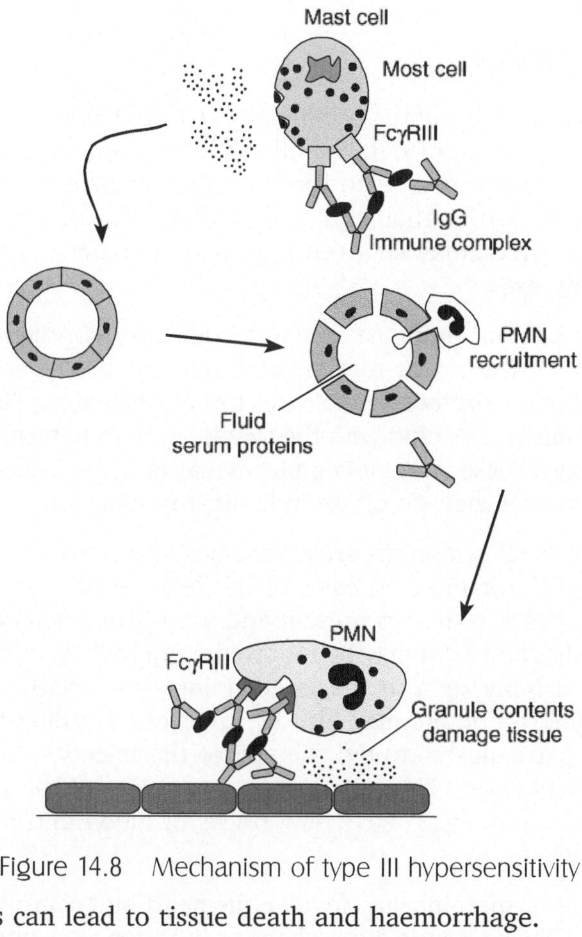

Figure 14.8 Mechanism of type III hypersensitivity

This can lead to tissue death and haemorrhage.

The reaction may be general (e.g. serum sickness) or may involve individual organs including the skin (e.g. systemic lupus erythematosus, Arthus reaction), kidneys (e.g. lupus nephritis), lungs (e.g. aspergillosis), blood vessels (e.g. polyarteritis nodosa), joints (e.g. rheumatoid arthritis) or other organs. This reaction may be the pathogenic mechanism by which many microorganisms cause diseases. The reaction may take 3–10 hours after exposure to the antigen.

Some examples of type III hypersensitivity include the following:

Serum Sickness

It is a systemic immune-complex-mediated disorder. Serum sickness is a type of delayed allergic response, appearing 4 to 10 days after exposure to some antibiotics or antiserum, the portion of serum that contains antibodies, such as gamma globulin, which may be given to provide immunization against some diseases.

The patient's immune system recognizes the proteins in the drug or antiserum as foreign proteins and produces its own antibodies to protect against the foreign proteins. The newly formed antibodies bind with the foreign protein to form immune complexes. These immune complexes may enter the walls of blood vessels where they set off an inflammatory reaction.

The usual symptoms are severe skin reactions, often on the palms of the hands and soles of the feet. Fever, sometimes as high as 104°F, is always present and usually appears before the skin rash. Joint pain may be reported in up to 50% of the cases. This is usually seen in the larger joints, but occasionally the finger and toe joints may also be involved. Swelling of lymph nodes, particularly around the site of the injection, is seen in 10–20% of cases. There may also be swelling of the head and neck. Urine analysis may show traces of blood and protein in the urine.

Other symptoms may involve the heart and central nervous system. These may include changes in vision and difficulty in movement. Breathing difficulty may also occur.

Traditionally, antitoxins were the most common cause of serum sickness, especially ones which are raised in horse.

Arthus Reaction

The model for immune complex diseases is Arthus reaction named after its discoverer in 1903. When antigen is injected intradermally into a hypersensitized animal, a localized cutaneous inflammatory reaction occurs that progresses to a haemorrhagic necrotic lesion. Histological examination of tissue biopsies of the site show an influx of PMNs. Staining with appropriate antibodies reveals immune complexes (usually IgG-Ag) and C3b.

When antigen is injected, the antigen–antibody complexes form and complement is activated. Complement components lead to mast cell degranulation releasing pharmacologically active substances that increase inflammation. Meanwhile, complement provides chemotaxis for neutrophils. Neutrophils are activated and lytic enzymes are released causing tissue damage. More number of neutrophils, macrophages and platelets invade the site and vessel blockage occurs. Vessels rupture leading to haemorrhage and more tissue is killed.

Nicolas Maurice Arthus (1862–1945)

Arthus studied in Paris and was conferred Doctor of Medicine in 1886. He made numerous contributions to the understanding of the immune and allergic mechanisms of serum. His important work on anaphylaxis and immunity was written while Arthus was the Professor of Physiology at the University of Lausanne. Arthus studied subcutaneous injections of horse serum into rabbits. After the fourth injection he noted that absorption was slow and a local oedematous reaction occurred. After the fifth it became purulent and after the seventh gangrenous. Intravenous injection, however, produced characteristic anaphylaxis, with rapid respiration, hypotension and incoagulability of the blood and if a high-enough dose was given, respiratory and cardiac death occurred.

Glomerulonephritis

Glomerulonephritis is the inflammation of the renal glomeruli. It is initiated by immunopathological mechanisms such as deposition of circulating antigen–antibody complexes in the glomerular basement membrane or binding of antibodies to antigens expressed in the glomerulus. The antibodies can activate complement and phagocytes. The resulting inflammatory response can lead to renal failure.

Antibodies deposited in tissues recruit neutrophils and macrophages. These bind to the antibodies or attached complement proteins by Fc and complement receptors. These leucocytes are activated and their products induce tissue injury.

Glomerulonephritis is a clinical manifestation of systemic lupus erythematosus (SLE). It can also occur after a streptococcal infection or after chronic and persistent parasitic infections.

Rheumatoid Arthritis

It is characterized by inflammation of the synovium associated with destruction of the joint cartilage and bone. Its morphology is similar to a localized immune response. The site of immune response is involved with $CD4^+$ T cells, activated B lymphocytes, plasma cells and macrophages and various other inflammatory cell types are also found. In severe cases, numerous cytokines, IL-1, IL-8, TNF and IFN (gamma) have been detected in the synovial fluid.

Cytokines are believed to activate resident synovial cells to produce proteolytic enzymes such as collagenase. These mediate the destruction of the cartilage ligaments and tendons of the joints. Increased osteoclast activity in the joints which may be related to the production of TNF family cytokine ligand by activated T cells, causes bone destruction.

The diagnostic test used for rheumatoid arthritis is the presence of these autoantibodies, called rheumatoid factors. Even though activated B cells and plasma cells are often present in rheumatoid arthritis cases, the specificities of the antibodies produced by these cells or their roles in causing joint lesions are unknown.

TYPE IV HYPERSENSITIVITY

It is also called as delayed type hypersensitivity (DTH). Type IV hypersensitivity (Figure 14.9) is involved in the pathogenesis of many autoimmune and infectious diseases (tuberculosis, leprosy, blastomycosis, histoplasmosis, toxoplasmosis, leishmaniasis, etc.) and granulomas due to infections and foreign antigens. Another form of delayed hypersensitivity is contact dermatitis (poison ivy, chemicals, heavy metals, etc.) in which the lesions are more papular.

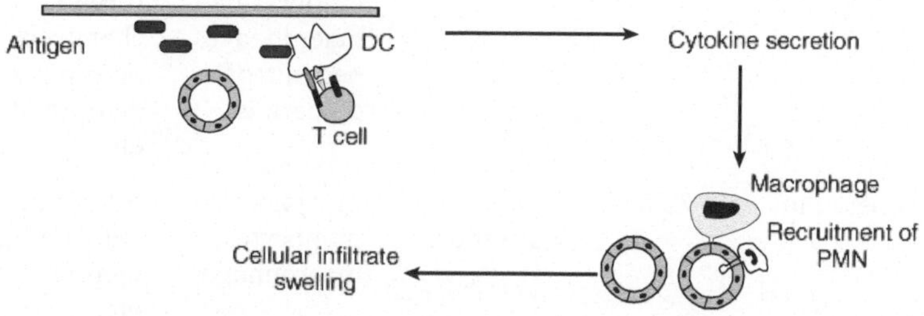

Figure 14.9 Mechanism of type IV hypersensitivity

Delayed type hypersensitivity results when an antigen-presenting cell, typically a tissue dendritic cell which presents the processed antigen in the form of peptide fragments along with class II MHC to antigen-specific T_H1 cells present in the tissue. The resulting activation of the T cell produces cytokines such as chemokines for macrophages, other T cells and to a lesser extent, neutrophils as well as TNF-β and IFN-γ. The consequences are a cellular infiltrate in which mononuclear cells (T cells and macrophages) tend to predominate. It is usually maximal in 48–72 hours. Type IV reactions absolutely require the presence of "natural" IgM antibody for initiation. Due to the nature and kinetics of the reaction, it is still believed that activation of memory T_H1 cells is primarily responsible for propagating the reponse, but initiation may require IgM and probably also complement.

Type IV hypersensitivity can be classified into three categories (Table 14.3) depending on the time of onset and clinical and histological presentation. They are:

- Contact hypersensitivity
- Tuberculin hypersensitivity
- Granuloma hypersensitivity

Table 14.3 Three categories of type IV hypersensitivity and its features

Type	**Reaction time**	**Clinical appearance**	**Histology**	**Antigen and site**
Contact	48–72 hr	Eczema	Lymphocytes, followed by macrophages, oedema of epidermis	Epidermal (organic chemicals, poison ivy, heavy metals, etc.)
Tuberculin	48–72 hr	Local induration	Lymphocytes, monocytes, macrophages	Intradermal (tuberculin, lepromin, etc.)
Granuloma	21–28 days	Hardening	Macrophages, epitheloid and giant cells, fibrosis	Persistent antigen or presence of foreign body (tuberculosis, leprosy, etc.)

Tuberculin Hypersensitivity

The classic form of type IV hypersensitivity is induced by injecting an antigen preparation of *Mycobacterium tuberculosis* intradermally. If the host has been previously exposed to the bacterium, swelling and induration will result. Approximately after 4 hours of injection of antigen, neutrophils rapidly accumulate around the post-capillary venules at the injection site.

The neutrophil infiltrate rapidly subsides and by about 12 hours the injection site becomes infiltrated with T cells and blood monocytes and some basophils, also organized in perivenular distribution. The endothelial cells lining these venules swell, show increased biosynthetic organelles and become leaky to plasma macromolecules.

Fibrinogen escapes from the blood vessels into the surrounding tissues, where it is converted into fibrin. The deposition of fibrin and to a lesser extent accumulation of T cells and monocytes within the extravascular tissue space around the injection site cause the tissue to swell and become indurated. Induration, the hallmark of DTH, is usually detectable by about 18 hours and maximal by 24 hours in mice, 48 hours in guinea pigs, and 48–72 hours in humans. The inflammation then subsides.

The Mantoux test (or Mantoux screening test, tuberculin sensitivity test, Pirquet test, or PPD test for purified protein derivative) is a diagnostic tool for tuberculosis. A standard dose of 10 tuberculin units (0.2 ml) is injected intradermally (into the skin) and read, 48–72 hours later. A person who has been exposed to the bacteria is expected to mount an immune response in the skin containing the bacterial proteins.

Charles Mantoux

Charles Mantoux (1877–1947) was a French physician, the developer of the eponymous serological test for tuberculosis.

He graduated from the University of Paris where he studied under Broca. For health reasons he relocated to Cannes but continued to work in Paris during the long vacation periods granted to patients in sanatoriums.

In 1908 he presented his first study of intradermal injections to the French Academy of Sciences and published this work in 1910, and in the following years the intradermal test replaced the subcutaneous test (Pirquet test).

The reaction is read by measuring the diameter of induration (palpable raised hardened area) across the forearm (perpendicular to the long axis) in millimeters. No induration should be recorded as "0 mm". Erythema (redness) should not be measured.

Von Pirquet

Clemens Peter Freiherr von Pirquet (May 12, 1874–February 28, 1929) was an Austrian scientist and pediatrician, best known for his contributions to the fields of bacteriology and immunology.

After graduating in medicine in 1900 he started practising at the Childrens's Clinic in Vienna.

In 1906, he noticed that patients who had previously received injections of horse serum or smallpox vaccine had quicker, more severe reactions to a second injection. He, along with Bela Schick, coined the word allergy (from the Greek *allos* meaning "other" and *ergon* meaning "reaction") to describe this hypersensitivity reaction.

Soon after, the observation with smallpox led Pirquet to realize that tuberculin, which Robert Koch isolated from the bacteria that cause tuberculosis in 1890, might lead to a similar type of reaction. Mantoux expanded upon Pirquet's ideas and the Mantoux test, in which tuberculin is injected under the skin, became a diagnostic test for tuberculosis in 1907.

Contact Hypersensitivity

Contact hypersensitivity (CHS) is another form of T-cell-mediated immunity that is characterized as delayed type hypersensitivity.

The mechanism of CHS is well understood. There are three critical events that must occur in generating a reaction. They are sensitization, trafficking and elicitation.

In the sensitization phase, a naive subject is exposed to hapten, usually through the skin but sometimes through inhalation or ingestion. Usually no symptoms of exposure are evident. The hapten binds covalently to any cell-associated protein or extracellular protein. The chemically modified proteins can then be presented by antigen-presenting cells (APC) to T cells which recognize the modified protein as foreign.

Once the APC has pinocytosed or phagocytosed the protein, it will traffic back into the draining lymph node where it will present the antigen to reactive T cells and expand the clonal population. These memory T cell clones that are expanded in the lymph node then can generate the elicitation phase of the response.

In the elicitation phase, local APCs present the hapten-protein to transiting memory T cells. The T cells then recruit more inflammatory cells to the antigen deposition site. In skin sensitization, the local Langerhans' cells have been shown to play an important role by presenting modified self antigen in the context of MHC class II. However, it has also been demonstrated that CHS reactions can occur in an MHC class I restricted manner. One possible mechanism that can explain these results is that keratinocytes can also act as APCs by presenting hapten-modified self proteins in the context of MHC class I molecules. Another group has shown that Langerhans' cells present hapten-peptide in both the MHC class I and MHC class II molecules. It is clear that both MHC systems appear to be involved in the reaction. There is experimental evidence that $CD4^+$ T cells regulate the magnitude and duration of the $CD8^+$ T-cell response. In fact, some of the current evidence suggests that the $CD8^+$ cells may play a more important role in CHS.

The molecular mechanisms involved in CHS are also becoming clearer. The antigen-presenting cell whether it is a Langerhans' cells or another form of APC must not only recognize the T-cell receptor and $CD8^+$ or $CD4^+$ molecule but also co-stimulatory molecules. The B7/CD28 ligand pair has been shown to be important in regulating the T-cell response to contact sensitizers.

Allergic contact dermatitis As the name implies, it is the result of direct contact with a contact allergen. One of the most common causes of this form are buttons and rivets in jeans, which contain the metal salt, nickel. Allergic contact dermatitis

(Figure 14.10) is considered a delayed-response immune reaction, because elicitation of an allergic reaction typically takes 48 to 72 hours to occur.

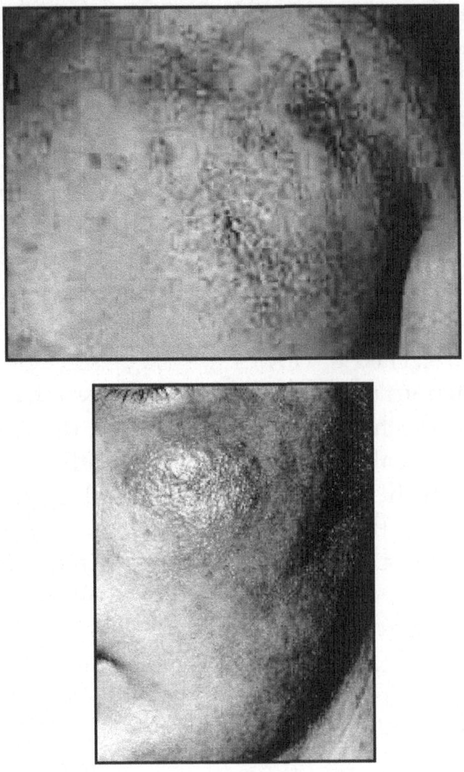

Figure 14.10 Contact dermatitis

Patch test This test is performed in cases of contact dermatitis (eczema) where allergy is suspected.

The allergens are prepared in appropriate concentrations in white soft paraffin (e.g. vaseline) and are then spread onto discs, 1 cm diameter (Figure 14.11).

The discs (which are made of a special metal, cannot themselves provoke a reaction) are placed on the skin, usually on the back, and are kept in place by hypoallergenic tape.

The skin is coded appropriately and the patient is asked to keep the skin dry. The patches are left in place for 48 hours.

After 48 hours, the discs are removed, the skin is examined and any redness or swelling is noted. The skin is re-examined after 48 hours for any remaining local redness or swelling (Figure 14.12).

Any professional, interpreting skin, blood or patch tests must first interpret the results in the light of the patient's history. No test should be read in isolation.

Figure 14.11 Patch test

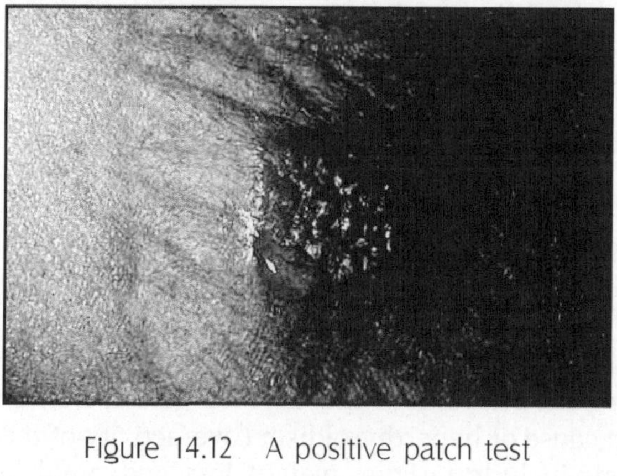

Figure 14.12 A positive patch test

Granuloma Hypersensitivity

The granulomatous reactions develop over a period of weeks. The granulomas are formed by the aggregation and proliferation of macrophages and may persist for weeks. This reaction is, in terms of its clinical consequences, by far the most serious type of delayed type hypersensitivity response. The position is complicated because these different types of reactions may overlap or occur sequentially following a single antigenic challenge.

The delayed-type hypersensitivity reactions are probably important for host defence against intracellular parasites such as tuberculosis and certain viruses and are prevalent in certain diseases such as sarcoidosis, Wegener's granulomatosis and polymyositis. In some diseases such as chronic granulomatous disease of childhood, granuloma formation can lead to obstruction of vital parts such as the oesophagus or ureters.

TYPE V—STIMULATORY HYPERSENSITIVITY

This is an additional type of hypersensitivity that is sometimes used as a distinction from type II.

Instead of binding to cell-surface components and causing destruction to it, the antibodies recognize and bind to the cell-surface receptors, which either prevents the intended ligand binding with the receptor or mimics the effects of the ligand, thus impairing cell signalling.

Some clinical examples:

- Graves' disease
- Myasthenia gravis

Graves–Basedow Disease

Graves–Basedow disease is a form of thyroiditis, an autoimmune disorder that stimulates the thyroid gland, being the most common cause of hyperthyroidism (overactivity of the thyroid). Symptoms include fatigue, weight loss and rapid heart beat.

Since antibodies similar to those stimulating the thyroid also affect the eye, eye symptoms are also commonly reported. Treatment is with medication that reduces the production of thyroid hormone (thyroxin) or with radioactive iodine, if refractory.

Myasthenia Gravis

Myasthenia gravis (MG, Latin: "grave muscle weakness") is a neuromuscular disease leading to fluctuating weakness and fatiguability. It is one of the best known autoimmune disorders and the antigens and disease mechanisms have well been identified. Weakness is caused by circulating antibodies that block acetylcholine receptors at the post-synaptic neuromuscular junction, inhibiting the stimulative effect of the neurotransmitter acetylcholine. Myasthenia is treated with immunosuppression and cholinesterase inhibitors.

Signs and symptoms The hallmark of myasthenia gravis is muscle weakness that increases during periods of activity and improves after periods of rest. Certain muscles such as those that control eye and eyelid movement, facial expression, chewing, talking, and swallowing are often, but not always, involved in the disorder. The muscles that control breathing and neck and limb movements may also be affected.

POINTS TO REMEMBER

- The term hypersensitivity is used to describe immune responses which are damaging rather than helpful to the host.
- Hypersensitivity reactions can be divided into four types: type I, type II, type III and type IV, based on the mechanisms involved and time taken for the reaction.
- Type I hypersensitivity: IgE-mediated anaphylactic reactions
- Type II hypersensitivity: Antibody-mediated cytotoxic reactions

- Type III hypersensitivity: Immune complex-mediated reactions.
- Type IV hypersensitivity: T-cell-mediated delayed-type reactions.
- Type V stimulatory hypersensitivity is an additional type of hypersensitivity that is sometimes used as a distinction from Type II. Instead of binding to cell-surface components and causing destruction to it, the antibodies recognize and bind to the cell-surface receptors.

REVIEW QUESTIONS

1. Write short notes on:
   - i. Histamine
   - ii. Serotonin
   - iii. Eosinophil chemotactic factor of anaphylaxis (ECF-A)
   - iv. Neutrophil chemotactic factor (NCF)
   - v. Prostaglandins
   - vi. Leukotrienes
   - vii. Bradykinin
   - viii. Urticaria
   - ix. Allergic rhinitis (Hay fever)
   - x. Schultz–Dale reaction
   - xi. Prausnitz–Kustner test or P–K test
   - xii. Passive cutaneous anaphylaxis (PCA)
   - xiii. ADCC
   - xiv. Erythroblastosis foetalis
   - xv. Serum sickness
   - xvi. Arthus reaction
   - xvii. Contact hypersensitivity

xviii. Tuberculin hypersensitivity

xix. Granuloma hypersensitivity

2. Write a detailed note on each

   i. Type I hypersensitivity

   ii. Type II hypersensitivity

   iii. Type III hypersensitivity

   iv. Type IV hypersensitivity

15

IMMUNE RESPONSE

INTRODUCTION

Whenever the foreign particles (including microorganisms) enter into the body, the immune system will recognize it and will elicit a reaction towards it. This is called immune response. There are two types of immune responses antibody-mediated immunity or humoral immunity and cell-mediated immunity.

Antibody-mediated immunity (AMI) or humoral immunity is a response towards the foreign particles in which the antigen specific antibodies are produced by the B cells. It refers to antibody production and all the accessory processes that accompany it—activation and cytokine production, germinal centre formation and isotype switching, affinity maturation and memory cell generation. It also refers to the effector functions of the antibody which include pathogen and toxin neutralization, classical complement activation and opsonin promotion of phagocytosis and pathogen elimination.

Cell-mediated immunity (CMI) is the response of the immune system towards the foreign antigen in which the antigen-specific cells will neutralize the foreign particles.

Cytokines regulate both the initiation and the maintenance of immune responses against foreign antigens (Ag). Moreover, they control the types of immune responses generated and therefore the effector mechanisms that ultimately mediate resistance.

$CD4^+$ T helper (T_H) cells are central to the regulation of immune responses. These cells can be subdivided into distinct subsets based on the cytokines they produce.

T_H1 cells They produce gamma interferon (IFN-γ) and predominantly induce cell-mediated immune responses. They are responsible for the virus-neutralizing antibody responses of the immunoglobulin G2a (IgG2a) isotype. The production of IFN-γ by T_H cells is often induced by interleukin-12 (IL-12) which is secreted by antigen-presenting cells (APC).

T_H2 cells They secrete IL-4 and stimulate B-cell proliferation and differentiation to produce predominantly IgG1 and IgE antibodies. The balance between T_H1 and T_H2 cells plays a major role in immunity and pathogenesis for several infectious diseases.

ANTIBODY-MEDIATED IMMUNITY (AMI) OR HUMORAL IMMUNITY

Humoral immunity refers to antibody production, and all the accessory processes that accompany it. Humoral immunity is the aspect of immunity that is mediated by secreted antibodies, produced in the cells of the B-lymphocyte (B cells) lineage. Secreted antibodies bind to antigens on the surfaces of invading microbes, which leads them to destruction.

Primary and Secondary Response

When an antigen is exposed to the immune system for the first time, there is a lag of several days before the specific antibody becomes detectable. The first-formed antibody is IgM. After a brief period of time, the IgM antibody level will start to decline. These are the main characteristics of the primary response.

If the immune system is re-exposed to the same antigen, there is a far more rapid appearance of antibody and in greater amount. It is of IgG class and remains detectable for months or even years. These are the features of the secondary response.

Induction of Primary Responses

Induction of a primary immune response (Figure 15.1) begins when an antigen enters the body and comes into contact with macrophages or any other antigen-presenting cells (APCs), which include B cells, monocytes, dendritic cells, Langerhans' cells and endothelial cells. Antigens are internalized by endocytosis, "processed" by the APC and then "presented" to immunocompetent lymphocytes to initiate the early steps of the immunological response.

The antigen-presenting cells will attach the processed antigen to the MHC class II molecule. The antigen–class II MHC complex is presented to a T helper (T_H2) cell which is able to recognize processed antigen associated with the class II MHC molecule on the membrane of the macrophage. This interaction, together with stimulation by interleukin 1 (IL-1), produced by the macrophage, will activate the T_H2 cell.

Activation of the T_H2 cell causes that cell to begin to produce interleukin 2 (IL-2), and to express a membrane receptor for IL-2. The secreted IL-2 will attach to the receptor of IL-2 and autostimulate the proliferation of the T_H2 cells. Stimulated T_H2 cells produce a variety of lymphokines including IL-2, IL-4, IL-6, and gamma interferon which mediate various aspects of the immune response.

IL-2 binds to IL-2 receptors on other T cells (which have bound to the Ag) and stimulates their proliferation, while IL-4 causes B cells to proliferate and differentiate into antibody-secreting plasma cells and memory B cells. IL-4 activates only B cells in the vicinity which have bound the antigen and not others, so as to sustain the specificity of the immune response.

B-cell activation is initiated by the cross-linking of BCR (B-cell receptor). Antigen must have at least two identical epitopes situated so that they can cross-link B cell-surface Ig to activate B cells. Aggregation of BCR–IgαIgβ complexes promotes phosphorylation of ITAMs (immunoreceptor tyrosine-based activation motifs) by cytoplasmic tyrosine kinases Fyn, Blk and Lyn, to initiate a second messenger cascade. Once antigen is bound, it is internalized with its BCR. The antigen can then be degraded (processed), combined with class II MHC, and presented to effector T_H2 cells, which must bind to specific antigen in order to provide help.

Effector T_H2 cells bind B cells via CAMs (cell adhesion molecules) and then specific peptide on class II MHC with TCR. T cell CD40 ligand (CD40L) binds B cell CD40 and sends the co-stimulatory signal for B-cell activation. The T_H2 cell reorganizes its membrane molecules and Golgi complex to concentrate CD40L-binding and cytokine secretion. This helps the specific B cell to maintain the antigen specificity of the response.

CD40 ligand

It is expressed on the surface of activated $CD4^+$ T cells, basophils and mast cells. Binding of CD40L to its receptor, CD40, on the surface of B cells stimulates B-cell proliferation, adhesion and differentiation. CD40L is also known as CDC154 or gp39 or TRAP.

Role of Complement in Humoral Immunity

Complement (classical pathway) especially by the Fc region of IgM and IgG leads eventually to death of bacteria by the terminal complement components which punch holes in the cell wall, leading to an osmotic death. Complement components also facilitate phagocytosis by cells possessing a receptor for C3b, e.g. polymorphs.

Role of Cytokines

IL-4 is a cytokine that stimulates the proliferation of activated B cells, T cells, and differentiation of $CD4^+$ T cells into T_H2 cells.

IL-5 is an interleukin produced by T_H2 cells and mast cells. Its functions are to stimulate B-cell growth and increase immunoglobulin secretion.

IL-6 plays critical roles in the immune response and is a potent B-cell differentiation factor inducing antibody-forming plasma cells.

When a B cell is bound to the antigen and simultaneously stimulated by IL-4 produced by a nearby T_H2 cell, the B cell is stimulated to grow and divide to form a clone of identical B cells, each capable of producing identical antibody molecules. The activated B cells further differentiate into plasma cells which synthesize and secrete large amounts of antibody.

Some B cells, however, do not differentiate into plasma cells. Instead, these cells undergo secondary DNA rearrangements that place the constant region of the IgG, IgA or IgE genes in conjunction with the VDJ genes. T-helper-cell cytokines direct isotype switching in activated B cells. Although T_H1 cells promote cellular immunity and do not initiate antibody formation, they can induce isotype switching to certain isotypes. Depending on which lymphokine is in highest concentration near a given B cell, it can follow several differentiation pathways. T_H1 cells secrete mainly IL-2 and IFN-γ, while T_H2 produce mainly IL-4, IL-5 and IL-6. IL-2 stimulates production of J chain. IFN-γ stimulates switching to IgG2a. IL-4 stimulates switching to IgE or IgG_1, while IL-5 stimulates switching to IgA. IL-6 promotes antibody secretion. T_H cells are thought to produce certain cytokines in response to their location and antigen density on the APC. For example, T_H2 cells in mucosal tissues produce cytokines that promote isotype switching to α chain. Each antibody isotype has specific effector functions in humoral immunity.

This "class switch" establishes the phenotype of these newly differentiated B cells. These cells remain as long-lived "memory cells". Upon subsequent encounter with antigen, these cells respond very quickly to produce large amounts of IgG, IgA or IgE antibody, generating the "secondary response".

Induction of a Secondary Immune Response

On re-exposure to same antigens (secondary exposure to antigen), there is an accelerated immunological response, known as the secondary or memory response or anamnestic response (Figure 15.1). Larger amounts of antibodies are formed in just 1 to 2 days. This is due to the activities of specific memory B cells or memory T cells which were formed during the primary immune response. These memory cells, when stimulated by homologous antigen, rapidly divide and differentiate into effector cells. Stimulating memory cells to rapidly produce very high (effective) levels of persistent circulating antibodies is the basis for giving "booster" type of vaccinations to humans and other animals.

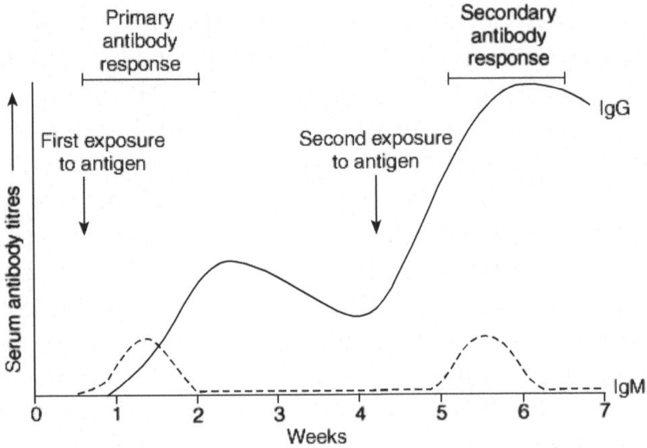

Figure 15.1 Graph showing the primary and secondary immune response

Activation of B Lymphocytes by T-independent Antigens

T-independent antigens are usually large carbohydrate and lipid molecules with multiple repeating subunits. B lymphocytes mount an antibody response to T-independent antigens without the requirement of interaction with T4 lymphocytes. Bacterial LPS from the gram-negative cell wall and capsular polysaccharides are examples of T-independent antigens. The resulting antibody molecules are generally of the IgM isotype and do not give rise to a memory response.

> **B cells and germinal centres**
>
> Most activated B cells go to the B-cell areas of the lymph node, called primary follicles. They contain mainly follicular dendritic cells (FDC) and IgM^+IgD^+ B cells (naive resting B cells). B cell blasts (dividing cells) form secondary follicles (germinal centres) within a few days of antigenic stimulation. Germinal centres are sites of intensive B-cell proliferation and are surrounded by the T_H2 cells that activate the B cells and migrate with them. Within a week after initial antigen contact, many germinal centres are seen.

CELL-MEDIATED IMMUNITY (CMI)

Cell-mediated immunity (CMI) is the immune response in which the cells are involved in the elimination of foreign particles. In contrast to humoral immunity, the CMI response cannot be transferred (passively) from animal to animal by antibodies or serum but can be transferred by lymphocytes removed from the blood.

Cell-mediated immunity involves the following:

- Phagocytosis and killing of intracellular pathogens
- Direct cell killing by cytotoxic T cells
- Direct cell killing by NK and K cells

Thus the CMI includes the activation of macrophages and NK cells, the production of antigen-specific cytotoxic T lymphocytes and the release of various cytokines in response to an antigen.

Cell-mediated immunity aids in

- removing virus-infected cells
- defending against fungi, protozoans and cancers
- defending against intracellular bacteria
- transplant rejection

Mechanism of Cell-mediated Immunity

The important mechanism in the cell-mediated immune response is the activation of various subsets of T lymphocytes to develop into effector T cells. **Three classes** of **effector cells** are specialized to deal with three different kinds of antigen.

- **CD8 CTL** kill infected cells displaying cytosolic pathogen peptides on class I MHC.
- **CD4 T_H1** cells activate macrophages with persistent vesicular pathogens whose peptides are displayed on class II MHC. They also activate B cells to produce opsonizing antibodies.
- **Natural killer (NK) cells** exhibit cytotoxic activity against a number of tumour cell lines, particularly against virus-infected or virus-transformed cells.

T cells that generate CMI are present in the lymphoid tissues, blood and lymph. Due to constant recirculation between blood and lymph nodes through lymphatics and back to the blood, a T cell circulates once in about 24 hours. These T cells possess receptors for the specific antigen with which they can react. Recognition of T cell by antigen occurs only when the antigen is associated with the MHC molecule. The T cells have receptors (TCR) complementary to the complexed MHC determinant and the antigenic determinant. T_H1 cells and T_H2 cells recognize antigen that is associated with MHC II (that is displayed by macrophages and other APCs). CTLs recognize antigen on cells complexed with MHC I (that is displayed by altered self cells).

APC and effector cells. TNF-α and TNF-β act with IFN-γ to activate macrophages and to directly kill some target cells.

Tapasin

It is a transmembrane protein that is located within the endoplasmic reticulum (ER). Its primary function is to assist in the binding of the MHC I protein with its antigen peptide fragment. It acts as an accessory molecule, binding to both the TAP and the MHC I protein within the lumen of the ER. More specifically, it binds to the TAP1 molecule of the TAP complex, and the β2m (beta-2 microglobulin) region of the MHC I protein. Initially, the alpha chains of the MHC I protein are held in place within the ER by the protein calnexin. The β2m region then binds to the alpha chains. It is not known whether or not tapasin plays a role in assisting in the binding of the β2m region to the alpha chains, but once the β2m region becomes associated with the MHC I protein, the MHC complex disassociates from calnexin and binds its β2m region with tapasin, which is also bound to TAP1.

Once this binding occurs, chaperone molecules, Erp57 and calreticulin, that assist in the MHC I peptide formation bind to the alpha chains of the MHC I molecule. As soon as this complex of TAP, tapasin, MHC I, Erp57, and calreticulin is formed, the TAP protein channel opens and allows peptide fragments that have been cleaved by a proteasome external to the ER to enter the ER.

Once the peptide fragments enter the ER, tapasin assists in the binding of the peptide fragments to the MHC I complex. The MHC I complex is then complete and ready to be exported to the outer cell membrane so that it can present its peptide fragment to immunoglobulins.

The activated cytotoxic T lymphocytes kill the target cell by three possible ways. They are:

1. CTLs may release a substance known as perforin in the space between the CTL and its target. In the presence of calcium ions, perforin polymerizes forming channels

in the target cell's membrane. These channels may cause the target cell to lyse.

2. CTLs may also release various enzymes (importantly granzymes) that pass through the polyperforin channels, causing target cell damage. Granzymes are serine proteases which, once inside the cell, proceed to cleave the precursors of caspases. The caspases are proteases that destroy the protein structural scaffolding of the cell (the cytoskeleton), degrade the cell's nucleoprotein and activate enzymes that degrade DNA.
3. CTLs may release lymphokines that interact with specific receptors on the target cell surface, causing internal responses that lead to destruction of the target cell. CTLs express on their surface the death activator designated Fas ligand (FasL) (Figure 15.3). Most potential CTL targets (cells that have to be destroyed) express a receptor for FasL designated Fas (also called CD95). When cytotoxic T cells recognize (bind to) their target, they produce more FasL at their surface. This binds with the Fas on the surface of the target cell leading to its death by apoptosis (Figure 15.4).

The CTLs act to eliminate endogenous antigens.

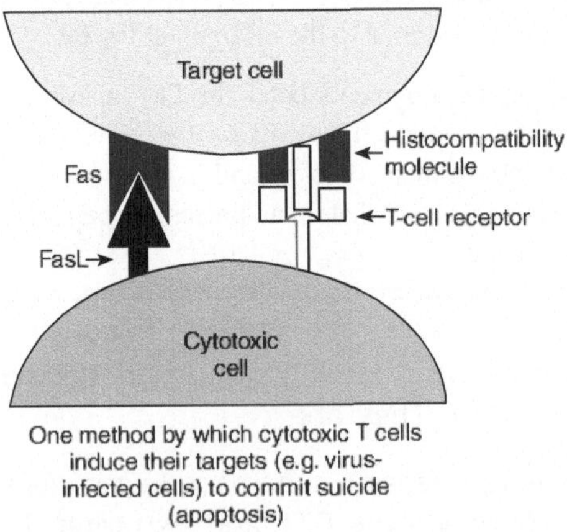

Figure 15.3 Binding of CTL to Fas receptor by the Fas ligand to mediate the target cell apoptosis

the requirement of interaction with T4 lymphocytes. Bacterial LPS from the gram-negative cell wall and capsular polysaccharides are examples of T-independent antigens. The resulting antibody molecules are generally of the IgM isotype and do not give rise to a memory response.

B cells and germinal centres

Most activated B cells go to the B-cell areas of the lymph node, called primary follicles. They contain mainly follicular dendritic cells (FDC) and IgM^+IgD^+ B cells (naive resting B cells). B cell blasts (dividing cells) form secondary follicles (germinal centres) within a few days of antigenic stimulation. Germinal centres are sites of intensive B-cell proliferation and are surrounded by the T_H2 cells that activate the B cells and migrate with them. Within a week after initial antigen contact, many germinal centres are seen.

CELL-MEDIATED IMMUNITY (CMI)

Cell-mediated immunity (CMI) is the immune response in which the cells are involved in the elimination of foreign particles. In contrast to humoral immunity, the CMI response cannot be transferred (passively) from animal to animal by antibodies or serum but can be transferred by lymphocytes removed from the blood.

Cell-mediated immunity involves the following:

- Phagocytosis and killing of intracellular pathogens
- Direct cell killing by cytotoxic T cells
- Direct cell killing by NK and K cells

Thus the CMI includes the activation of macrophages and NK cells, the production of antigen-specific cytotoxic T lymphocytes and the release of various cytokines in response to an antigen.

Cell-mediated immunity aids in

- removing virus-infected cells
- defending against fungi, protozoans and cancers
- defending against intracellular bacteria
- transplant rejection

Mechanism of Cell-mediated Immunity

The important mechanism in the cell-mediated immune response is the activation of various subsets of T lymphocytes to develop into effector T cells. **Three classes** of **effector cells** are specialized to deal with three different kinds of antigen.

- **CD8 CTL** kill infected cells displaying cytosolic pathogen peptides on class I MHC.
- **CD4 T_H1** cells activate macrophages with persistent vesicular pathogens whose peptides are displayed on class II MHC. They also activate B cells to produce opsonizing antibodies.
- **Natural killer (NK) cells** exhibit cytotoxic activity against a number of tumour cell lines, particularly against virus-infected or virus-transformed cells.

T cells that generate CMI are present in the lymphoid tissues, blood and lymph. Due to constant recirculation between blood and lymph nodes through lymphatics and back to the blood, a T cell circulates once in about 24 hours. These T cells possess receptors for the specific antigen with which they can react. Recognition of T cell by antigen occurs only when the antigen is associated with the MHC molecule. The T cells have receptors (TCR) complementary to the complexed MHC determinant and the antigenic determinant. T_H1 cells and T_H2 cells recognize antigen that is associated with MHC II (that is displayed by macrophages and other APCs). CTLs recognize antigen on cells complexed with MHC I (that is displayed by altered self cells).

During a primary CMI response, antigen is presented to the precursor cytotoxic T lymphocytes ($CD8^+$) in association with MHC class I proteins. All nucleated cells express MHC I on their surfaces, so virtually any cell in the animal expressing a new (non-self) antigen on its surface will activate the cytotoxic T lymphocytes.

Activation of T_H Cells

T_H cells ($CD4^+$) reacting with antigen may produce a variety of lymphokines. Interleukin-2 (IL-2) stimulates T-cell activation and IL-4 stimulates B cells.

As already discussed T_H1 cells recognize foreign antigen on the surface of the cells, that is presented along with MHC II molecule. Mainly, T_H1 cells produce IL-2, gamma IFN and lymphotoxin. This results in macrophage activation and the delayed type hypersensitivity (DTH) reaction, very important in the activation of cytotoxic T lymphocytes.

The lymphokines produced by T_H cells stimulate B cells and cytotoxic T lymphocytes, inducing them to proliferate and mature into effector cells. Gamma interferon activates macrophages and NK cells to exhibit their full cytolytic activity.

IL-12 and its role in CMI

The cytokine interleukin-12 (IL-12), which is produced during the immune response to a variety of stimuli, is critical for the development of cell-mediated immunity. Among its various functions, IL-12 is a potent inducer of interferon-gamma production by T lymphocytes and natural killer (NK) cells and is a cofactor in the mitogenic stimulation of these cell types. However, inappropriate cell-mediated immune responses, often accompanied by over-expression of IL-12, have been identified as having a key role in the development of a variety of autoimmune disorders, such as rheumatoid arthritis, multiple sclerosis, graft-versus-host disease, diabetes mellitus, systemic lupus erythematosus and Crohn's disease. Reduction of IL-12 levels, for example, by administration of antibodies to IL-12, has been effective in controlling autoimmune diseases in animal models.

Cytotoxic T Lymphocytes (CTLs)

Cytotoxic T lymphocytes (CTLs) are the body's major defence against viruses, intracellular bacteria and cancers. It is involved in the destruction of infected cells and tumour cells. Cytotoxic T cells (CTLs) mediate antigen-specific, MHC-restricted cytotoxicity. These CTLs are effector cells derived from T8 lymphocytes during cell-mediated immunity. They kill cells bearing foreign antigens on their surface in association with MHC I molecule. CTLs can kill cells that are harbouring intracellular pathogens (bacteria or viruses) as long as the infected cell is displaying a microbial antigen on its surface. They also kill tumour cells and are involved in the rejection of transplanted cells. These cells recognize Ag–MHC I complexes on target cells, contact them, and release the contents of granules directly into the target cell membrane and lead to the lysis of the cell (Figure 15.2).

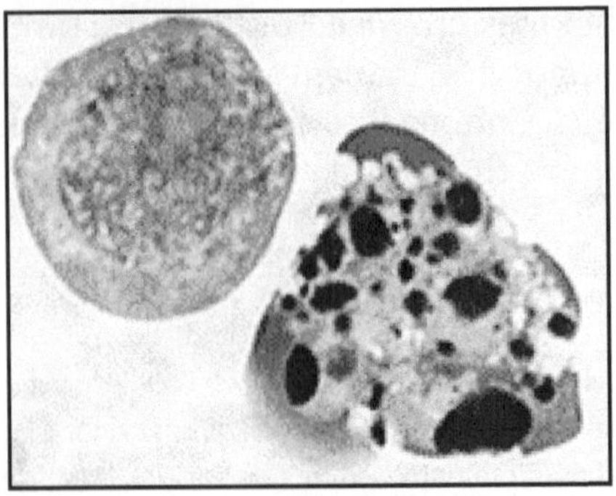

Figure 15.2 CTL causing destruction of target cell

CTLs also regulate immune responses by releasing IFN-γ, TNF-α, and TNF-β. IFN-γ inhibits viral replication, activates IL-1 and TAP (the transporter associated with antigen processing is a member of the ATP-binding cassette transporter family that specializes in delivering cytosolic peptides to class I molecules in the endoplasmic reticulum) expression by infected cells to promote antigen presentation and recruits and activates macrophages as

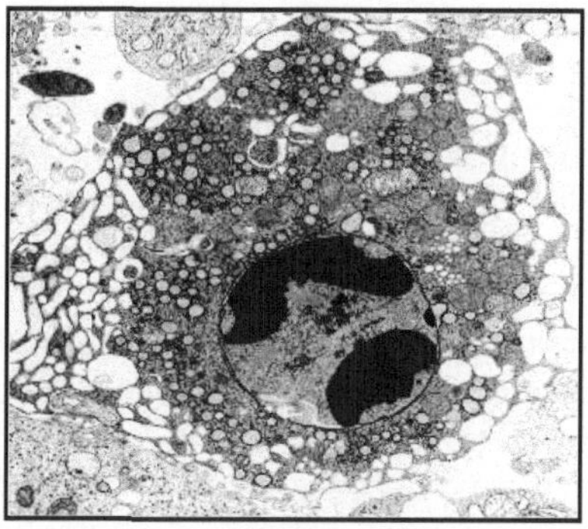

Figure 15.4 An apoptosized cell showing massive condensation of chromatin, condensed cytoplasm with intact organelles, and peripheral blebbing

Apoptosis

The term "apoptosis" describes the biological processes leading to controlled cellular death and was first introduced in a seminal publication by Kerr, Wyllie and Currie. Apoptosis is of Greek origin meaning "falling off". The term is used in an analogy to the apparent suicide of leaves resulting in the very visible colour changes associated with the autumn/fall and that eventually lead to the leaves falling from the trees. Similarly, cells go through a preordained sequence of events resulting in death and removal from the body.

Process of apoptosis by granzyme

Granzyme B and perforin, proteins released by an effector cell (cytotoxic T cell), can induce apoptosis in target cells by forming transmembrane pores and through cleavage of effector caspases such as caspase-3. In addition, caspase-independent mechanisms of granzyme B-mediated apoptosis have been suggested. Caspase-activated DNase (CAD) is activated through the cleavage of its associated inhibitor ICAD by caspase-3. CAD is then able to interact with components such as topoisomerase II (Topo II) to condense chromatin, leading to DNA fragmentation and ultimately apoptosis.

Caspases

The name "caspase" is derived from cysteine-aspartic acid-protease, an enzyme that contains predominantly these two amino acids (cysteine and aspartic acid) and is a proteolytic enzyme. Caspases are essential in cells for apoptosis. Caspases are a group of cysteine proteases, enzymes with a crucial cysteine residue that can cleave other proteins after an aspartic acid residue, a specificity which is unusual among proteases.

Macrophages in Cell-mediated Immunity

During a cell-mediated immune response, macrophages play their usual role in the presentation of antigen to T helper cells and in producing cytokines that are involved in the initiation of immune reactions. In addition, macrophages play a role in the expression of CMI. The TCRs and CD4 molecules on the T_H1 cell interact with the MHC II molecule with bound peptide epitope on the macrophage. Co-stimulatory molecules such as CD40L on the T_H1 cell then bind to CD40 on a macrophage. This triggers the T_H1 cells to secrete the cytokine interferon-gamma (IFN-γ). IFN-γ subsequently binds to IFN-γ receptors on the macrophage causing its activation. IFN-γ causes the local macrophage population to develop an increased number of lysosomes and also increased secretion of microbicidal products.

Macrophage involvement in CMI may be part of the pathology of certain diseases. When there is difficulty in elimination of an intracellular parasite (e.g. the tuberculosis bacillus) the chronic CMI response to local antigens leads to the accumulation of densely packed macrophages. This leads to the release of fibrinogenic factors and stimulate the formation of granulation and fibrosis. The resulting structure, called a granuloma, actually represents an attempt by the host to isolate a persistent infection.

Natural Killer Cells (NK cells) and CMI

Another class of cytotoxic lymphocytes distinct from cytotoxic T lymphocytes may be stimulated during the cell-mediated immune response. They are called as natural killer cells or NK cells. NK cells are found in blood and lymphoid tissues, especially the spleen. They do not bear T-cell or B-cell CD markers. NK cells are characterized by the presence of CD56 and/or CD16. Like CTLs, they are able to recognize and kill cells that are displaying an altered antigen on their surfaces. However, they do not display TCR and they are not MHC-restricted.

NK cells are present in an animal in the absence of antigenic stimulation and it is because of this that they are referred to as "natural" killers. They might also be considered part of the constitutive defences; however, NK cells become activated in a CMI response by T-cell lymphokines, including interleukin-2 and gamma interferon.

Some NK cells have surface receptors (CD16) for the Fc portion of IgG. They bind to target cells by receptors for the Fc portion of antibody that has reacted with antigen on the target cell. This type of CMI is called antibody-dependent cell-mediated cytotoxicity or ADCC. NK cells may also have receptors for the C3 component of complement and therefore recognize cells that are coated with C3 as targets. ADCC is thought to be an important defence against a variety of parasitic infections caused by protozoa and helminths.

To summarize the immune response, the antigens are processed by the APCs and are presented to T helper cells along with the MHC molecules. They in turn will stimulate either the T cell or B cell or both to get further activated. Some of the T cells and B cells will turn into memory cells and thus can be readily activated upon second exposure of the same antigen (Figure 15.5).

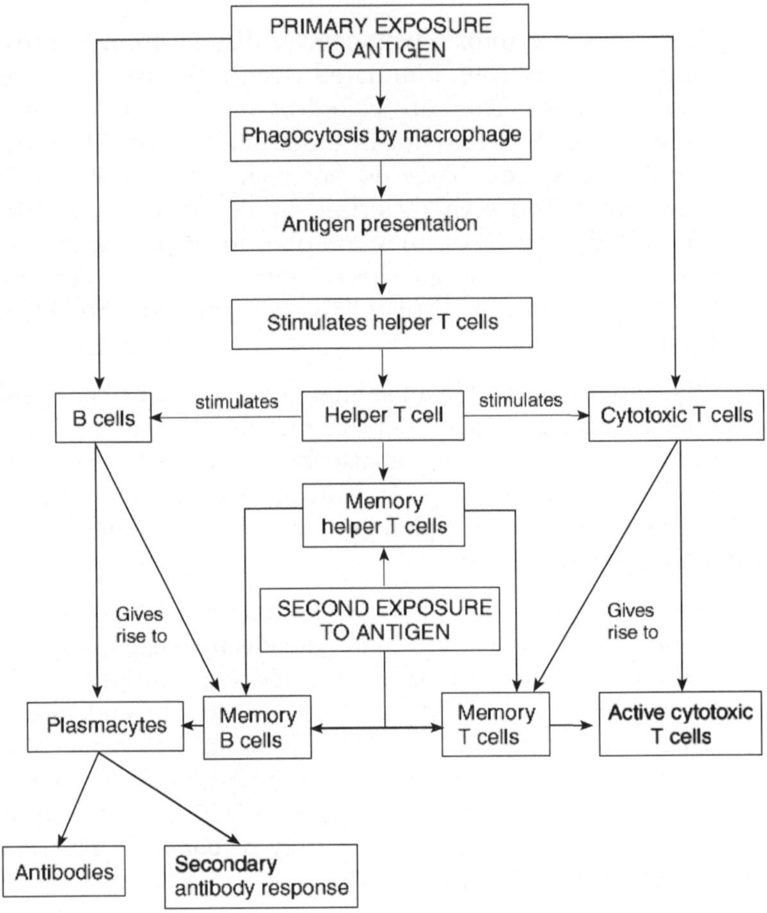

Figure 15.5 A brief outline of immune response

CLONAL SELECTION THEORY

The clonal selection theory has become a widely accepted model for how the immune system responds to infection and how certain types of B and T lymphocytes are selected for destruction of specific antigens invading the body.

In 1954, immunologist Niels Jerne put forward a theory which stated that there is already a vast array of lymphocytes in the body prior to any infection. The entrance of an antigen into the body results in only one type of lymphocyte to match it and produce a corresponding antibody to destroy it.

This selection of only one type of lymphocyte results in it being cloned or reproduced by the body extensively to ensure that there are enough antibodies produced to inhibit and prevent infection.

Australian immunologist Sir Frank Macfarlane Burnet worked on this model and was the first to name it "clonal selection theory". Burnet explained immunological memory as the cloning of two types of lymphocytes. One clone acts immediately to combat infection whilst the other is long-lasting, remaining in the immune system for a long time, which results in immunity to that antigen.

CD56

CD56 [neural cell adhesion molecule (NCAM), a natural killer cell-marker] is part of a family of cell-surface glycoproteins that plays a role in embryogenesis and contact-mediated interactions between neural cells. CD56 is expressed in a variety of normal and abnormal tissues including skin, small cell carcinoma, neuroblastoma, neurons, astrocytes, Schwann cells, NK cells and a subset of activated T-cell lymphomas.

Sir Frank MacFarlane Burnet (1899–1985) was an Australian biologist. He was awarded the Nobel Prize in Medicine in 1960, along with Peter Brian Medawar, for the discovery of acquired immunological tolerance. He was also awarded "Australian of the Year" at this time.

POINTS TO REMEMBER

- There are two kinds of immune response—humoral immunity and cell-mediated immunity.
- Antibody-mediated immunity (AMI) or humoral immunity is a response towards the foreign particles in which the antigen-specific antibodies are produced by the B cells.
- Cell-mediated immunity (CMI) is the response of immune system towards the foreign antigen in which the antigen-specific cells will neutralize the foreign particles.
- Cytokines regulate both the initiation and the maintenance of immune responses.
- Cell-mediated immunity involves phagocytosis and killing of intracellular pathogens, direct cell killing by cytotoxic T cells and direct cell killing by NK and K cells.

REVIEW QUESTIONS

1. Write short notes on
    i. T_H1 cells
    ii. T_H2 cells
    iii. Memory cells
    iv. Plasma cells
    v. Anamnestic response
    vi. Cytotoxic T lymphocytes (CTLs)
2. Write a detailed note on antibody-mediated immunity.
3. Write in detail about cell-mediated immunity.

16

VACCINES

INTRODUCTION

A vaccine is any preparation of killed or weakened microorganism that is given to a person orally or injected in order to prevent disease. Edward Jenner demonstrated that a person inoculated (Figure 16.1) into the skin with cowpox was protected against small pox, and he thus developed the principles of vaccination in 1796.

Figure 16.1 Edward Jenner's inoculation knives

Edward Jenner was a doctor in Berkeley, Gloucestershire. In 1796 he carried out his famous experiment on eight-year-old James Phipps (Figure 16.2). Jenner injected pus taken from a cowpox pustule on the hand of milkmaid Sarah Nelmes and injected it into an incision on the boy's arm. He was testing his theory, drawn from the folklore of the countryside, that milkmaids who suffered the mild disease of cowpox never contracted smallpox.

Jenner subsequently proved that having been inoculated with cowpox, Phipps developed immunity to smallpox. He submitted a paper to the Royal Society in 1797 describing his experiment but was told that his ideas were too revolutionary and that he needed more proof. Thus, Jenner experimented on several other children, including his own 11-month-old son. In 1798, the results were finally published.

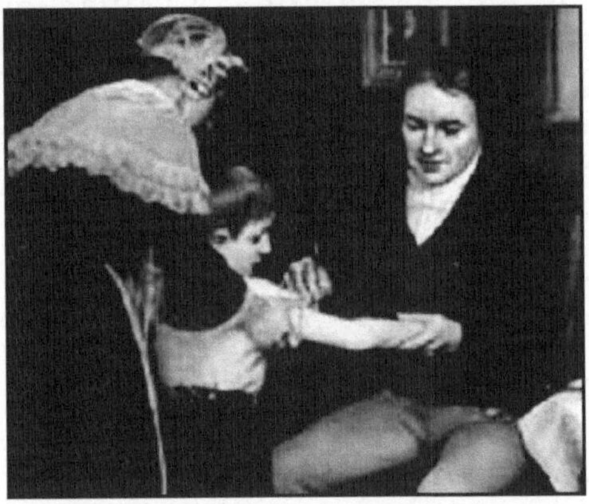

Figure 16.2 Historic photograph of the inoculation of the first vaccine by Edward Jenner to eight-year-old James Phipps

In 1881, Louis Pasteur honoured Jenner by christening the process "vaccination" and the substance used to vaccinate was called a "vaccine."

The principle of vaccination is to induce a primary immune response in the vaccinated subject (development of memory T and B cells) so that, following the exposure of a pathogen, a rapid secondary immune response is generated leading to the accelerated elimination of the organism and protection from clinical disease. Success depends on the generation of memory T and B cells and the presence of neutralizing antibody in the serum.

CONTRIBUTIONS OF LOUIS PASTEUR

Louis Pasteur was born in 1822 in Dole, France. He was the son of a tanner. As a young man, Pasteur studied at the Ecôle Normale in Paris. In 1843, he became a research chemist. He developed such a reputation that in 1854 aged just 32, he became the Dean of the Faculty of Science at the University of Lille.

Interestingly, Pasteur's (Figure 16.3) studies on chicken cholera led him to the discovery of vaccines. Cholera was a serious problem for farmers. Chicken cholera would spread rapidly and wipe out the entire flock in as little as 3 days. The disease spread by contaminated food or animal excrements. Pasteur had identified the cholera bacillus and was growing it in pure culture. When injected, chicken invariably died in 48 hours.

Figure 16.3 Historic photograph of Louis Pasteur seen working in his laboratory

During the heat of the summer, Pasteur returned to Paris leaving the cholera cultures used for infection stored on the shelves of the Arbois laboratory (Figure 16.4). Upon return, Pasteur's collaborators were disappointed to find that the stored cultures no longer killed injected chickens, nor even made them sick. The group set to work to make new fresh cultures of the bacillus and tested on new birds and also the old treated birds. The results were astonishing: The previously injected birds were unaffected by the bacillus, while the new birds all died. Pasteur understood the concept of attenuation and manufactured attenuated cultures of chicken cholera and could routinely prevent cholera in the vaccinated chickens.

Figure 16.4 Louis Pasteur's Laboratory which is still preserved

With this success, Pasteur decided to extend the discovery to prevent anthrax in sheep. By various techniques involving oxidation and aging, anthrax vaccines were prepared by him. This could prevent the anthrax in laboratory trials. Pasteur's reports on preventing sheep anthrax were so exciting to some and unbelievable to many, that he was challenged by the well-known veterinarian Rossignol. Thus Pasteur was instructed to demonstrate his finding in public (Figure 16.5). This took place at Pouilly le Fort, a farm in the town of *Melun*, south of Paris.

25 sheep were controls. Another 25 were vaccinated by Pasteur. Then all animals received a lethal dose of anthrax. Two days after final inoculation (May 5th, 1882), all the 25 control sheep were dead and all the 25 vaccinated sheep were alive and healthy. The fame of Pasteur and these experiments spread throughout France, Europe and beyond. Supported by the successes with anthrax and fowl cholera diseases, Pasteur identified and isolated over the next 2–3 years the microbes for many other diseases including swine erysipelas, childbirth fever and pneumonia.

Figure 16.5 Photograph depicting Louis Pasteur demonstrating his anthrax vaccine before the public

The most famous success of Pasteur's research was the development of a vaccine against rabies or hydrophobia as it is known in humans. Pasteur and his colleague, Roux realized that conquest of rabies would be recognized as a great achievement to the world of science and to the public at large.

Pasteur and Roux initially attempted to transfer infection by injecting healthy dogs with saliva from rabid animals. The results were variable and unpredictable. Later, recognizing that the active agent was in the spinal cord and brain, and because they were unable to detect a specific rabies microorganism, Pasteur and Roux applied extracts of rabid spinal cord directly to the brain of dogs. With this technique they could produce rabies in the test animals in a few days.

The next goal was to develop a vaccine that would provide protection to the subject before the rabies agent moved from the bite site to the spinal cord and then to the brain. This was achieved by injecting into test animals suspensions of spinal cord of rabid rabbits that were attenuated by air-drying over a 12-day period in the now-famous Roux bottle (Figure 16.6). A strip of spinal cord was suspended from a hanger in the centre of the bottle containing a hole at the top of the bottle and one on the lower side. Air entered from the bottom opening, passed over a drying agent and exited from the top. The longer the cord was dried, the less potent was the tissue in producing rabies.

Figure 16.6 The Roux bottle with the spinal cord hanging for drying

The treatment plan used to develop immunity to rabies was to inject under the skin of a dog the least potent preparation of minced spinal cord, followed by a stronger and stronger extract every day for the next 12 days. At the end, the animal was completely resistant to bites of rabid dogs and failed to develop rabies if the most potent extracts were applied directly to the brain.

On July 6, 1886, 9-year-old Joseph Meister and his mother approached Pasteur's laboratory. Two days earlier the young boy had been bitten repeatedly by a rabid dog. As the boy was almost confirmed of his death, Pasteur decided to test his vaccine with that young boy. Pasteur and his colleagues injected young Meister 13 times over 10 days with rabid rabbit's nerve tissue that had been dried in a bottle with potash. The boy recovered.

A few months later a second victim turned up. He was a young shepherd, named Jean-Baptiste Jupille of Villers-Farlay, also bitten by a mad dog. He too recovered. Pasteur became a hero and a legend. The Pasteur Institute funded by public and governmental subscriptions was built in Paris initially to treat victims of rabies who were coming to Pasteur's laboratory in increasing numbers.

Rabies was the last major research of Pasteur. Pasteur died in 1895. His funeral was attended by thousands of people. His remains, initially interred in the Cathedral of Notre Dame, were transferred to a permanent crypt in the Pasteur Institute, Paris.

Joseph Meister, the first person publicly to receive the rabies vaccine, returned to the Pasteur Institute as an employee where he served for many years as gate keeper. Tragically, when Meister, then 64 years old, was unable to prevent the Wehrmacht from entering Pasteur's crypt in 1940, he went home, took out his World War I service revolver and shot himself.

PROPERTIES OF A GOOD VACCINE

The following are the important properties of a good vaccine:

1. Ability to elicit the appropriate immune response for the particular pathogen. For example cell-mediated immunity should be developed for tuberculosis and viral diseases whereas humoral immunity for most of the bacterial diseases.
2. Long-term protection
3. Safety
4. Stability
5. Inexpensiveness

TYPES OF VACCINES

The following are the different types of vaccines:

1. Live vaccines
2. Killed whole organism vaccines
3. Subunit vaccines—purified or recombinant antigen
4. Toxoids
5. Recombinant virus vaccines
6. Synthetic peptide vaccines
7. Anti-idiotype antibodies
8. DNA vaccines

LIVE VACCINES

The live vaccines are also called as attenuated vaccines or sometimes as live attenuated vaccines. Attenuation is the process of reduction of virulence of an organism *in vitro*. Live vaccines are prepared from attenuated strains that are almost or completely devoid of pathogenicity but are capable of inducing a protective immune response. They multiply in the human host and provide continuous antigenic stimulation over a period of time.

The virulence of the microorganism can be artificially reduced under any adverse conditions such as:

1. By culturing in reduced temperature
2. By culturing in unnatural host
3. By repeated subculture
4. By inoculating the organism in the animal through unusual route.

Louis Pasteur discovered the method for the attenuation of virulent microorganisms.

Replication of the vaccine strain in the host reproduces many of the features of wild type infection, **without causing clinical disease**. Most successful viral vaccines belong to this group.

The immune response is usually good—when the virus replicates in the host cells, both antibody as well as cell-mediated immune responses are generated and immunity is generally long-lived. Often, only a single dose is needed to induce long-term immunity.

Contraindication to Live Vaccination

Live vaccines should not be administered to immunocompromised persons. Such patients include people who have cancers or are being treated with anti-cancer agents, transplant patients, patients taking steroids, etc.

Pregnant patients should not be given live vaccines as there is a risk of harm to the baby. Live vaccines are affected by circulating antibodies and therefore not usually effective under one year of age.

Potential Drawbacks of Live Vaccines

The potential drawbacks of live vaccine are:

- the danger of reversion to virulence and
- the possibility of causing extensive disease in immunocompromised individuals.

Use of a Related Microorganism from Another Animal

A good example was the use of cowpox to prevent smallpox. In this case, the microorganism may not cause any disease in human but however, due to some antigenic similarity it can produce the protective immunity against the similar disease in humans.

Advantages of Live Vaccine

- One initial dose is usually sufficient, rarely additional booster doses may be required
- Confers local immunity

- Provides more rapid protection
- Produces a wider spectrum of protection
- Less likely to cause allergic reactions or post-vaccination lumps
- Less susceptible to passive antibodies
- Less expensive

Disadvantages of Live Vaccines

- Potential to mutate to a virulent form
- Could exacerbate disease in immunosuppressed animals
- Some risk of causing abortion or transient infertility

Examples of Live Vaccines

B.C.G. vaccine (Bacillus Calmette–Guerin) The vaccine is a live attenuated strain of bovine *Mycobacterium tuberculosis*. The vaccine was developed by Calmette and Guerin in 1921. They noted that a glycerine–bile–potato mixture grew bacilli that seemed less virulent. The principle is to give controlled infection with a live attenuated bacillus, which will induce hypersensitivity and immunity. Subsequent exposure to infection will be contained locally without general dissemination.

Léon Charles Albert Calmette (July 12, 1863–October 29,1933) was a French physician, bacteriologist and immunologist and an important officer of the Pasteur Institute. He discovered the Bacillus Calmette–Guerin, an attenuated form of *Mycobacterium* sp. used in the BCG vaccine against tuberculosis. He also developed the first antivenom for snake venom, the Calmette's serum.

The present recommendation is to administer BCG at birth, preferably within 4 weeks of life. It is given intradermally, 0.1 ml at the insertion of deltoid muscle on the left side. One expects a local reaction at the site of vaccination over a period of 4 to 8 weeks leaving a puckered scar and also a mild enlargement of anterior axillary lymph nodes on the same side. A positive tuberculin test 6–8 weeks after vaccination signifies a successful vaccination. BCG vaccination can safely be given at any age after a prior negative tuberculin test.

Typhoid vaccine This is a live attenuated strain of *S. typhi* Ty21a that was developed in the early 1970s by chemical mutagenesis. Protection is markedly influenced by the number of doses and their spacing. When the vaccine is given in three doses two days apart, protective immunity is achieved seven days after the last dose. In disease-endemic areas a booster dose is recommended every three years. Travellers from non-endemic to disease-endemic regions are recommended a booster on a yearly basis. There are currently no field trial data to document the efficacy of this vaccine in children less than three years of age.

Measles vaccine Measles is a virus infection which was one of the most common diseases of modern man, but is now being eradicated by immunization. The cause of measles is the rubeola virus.

It is a live attenuated virus grown in chick embryo fibroblasts. It was first introduced in the 1960s. Live attenuated (dried) measles virus vaccine, is prepared in avian leucosis-free chick embryo fibroblast cultures from the Edmondston strain of attenuated measles virus, obtained from Dr. J.F. Enders and given an additional 69 passages in chick embryo cell cultures. The diluent for reconstitution for the first dose vial is water for injection (WFI). The diluent for reconstitution for the 10 dose vial is phosphate buffered saline with 40 to 70 ppm Tween 80.

Its extensive use has led to the virtual eradication of measles in the developed countries. In developed countries, the vaccine is administered to all children in the second year of life (at about 15 months). However, in developing countries, where measles is still widespread, children tend to become infected early (in the first year), which frequently results in severe disease. It is therefore important to administer the vaccine as early as possible (between six months and a year). However if the vaccine is administered too early, there is a poor intake rate due to the interference by maternal antibodies. For this reason, when vaccine is administered before the age of one year, a booster dose is recommended at 15 months.

Mumps vaccine Mumps is a relatively mild short-term viral infection of the salivary glands that usually occurs during childhood. Typically, mumps is characterized by a painful swelling of both cheek areas. Paramyxovirus causes mumps.

Live attenuated virus was developed in the 1960s. At least 10 strains of the mumps virus are in use throughout the world for live attenuated vaccine. The first vaccine strain to be developed, and that most often used, is the Jeryl Lynn strain which was named after the child from whom the virus was isolated. It was developed in USA by passaging it seven times in embryonated hen's eggs and ten times in chick embryo cell cultures.

Live mumps vaccine is available in monovalent vaccine and in combination with other vaccines. It is administered together with measles and rubella at 15 months in the MMR vaccine.

Rubella vaccine Rubella, also called German measles, is caused by a virus that is spread from person to person when an infected person coughs or sneezes. Rubella is also spread by direct contact with the nasal or throat secretions of an infected person. If a pregnant woman gets rubella during the first 3 months of pregnancy, her baby is at risk of having serious birth defects or dying.

The currently licensed rubella vaccines in international use are based on the live attenuated RA 27/3 strain of the virus. Other attenuated vaccine strains are available in China and Japan. 27/3 vaccines are propagated in human diploid cells and have proven to be safe and efficacious. Rubella vaccines are commercially available in a monovalent form, a bivalent combination with measles vaccine or mumps vaccine, or as trivalent measles–mumps–rubella vaccine (MMR).

MMR vaccine The first combined measles–mumps–rubella vaccine was licensed in 1971. MMR is a live attenuated vaccine. Also available, but not generally used today, are individual vaccines for each disease, as well as some combination vaccines against two of the diseases. Only the combination vaccine MMR is generally used today.

Two doses of the MMR vaccine have been recommended since 1989, given as follows:

- The first dose is given between 12 and 15 months of age.
- The second dose is given between 4 and 6 years of age.
- The second dose may be given at any age provided it is given at least 28 days after the first dose.

Polio vaccine The **live attenuated oral polio vaccine** by **Sabin** has been adopted in most parts of the world. Its chief advantages are low cost, induction of mucosal immunity and the possibility that, in poorly immunized communities, vaccine strains might replace circulating wild strains and improve herd immunity. Against this is the risk of reversion to virulence (especially of types 2 and 3) and the fact that the vaccine is sensitive to storage under adverse conditions.

Albert Bruce Sabin (1906–1993) developed the "live" polio vaccine, beating rival Jonas Salk's "killed" virus vaccine. Sabin's vaccine is widely used and has saved many from the paralysis associated with polio.

Sabin was born in Bialystock, Poland (then part of Russia) and emigrated with his family to the United States in 1921. He attended New York University, received his medical degree in 1931, and began research on the virus that causes polio. Known at the time as infantile paralysis, polio was a source of much fear because of its ability to cause paralysis and death, especially in infants and young children.

By 1936, Sabin and his colleagues were able to grow the polio virus in human tissue cultures outside the body. In 1941, Sabin established that the human polio virus enters the body via the digestive tract and not the nose as was then thought. World War II (1939–1945) interrupted Sabin's polio research. While in the Army, he studied several diseases affecting American troops, such as sandfly fever, dengue fever, toxoplasmosis and encephalitis lethargica. After the war, Sabin returned to polio research.

By 1954, Sabin had developed a vaccine that gave protection against polio using a live virus rather than the killed virus used by Salk. Sabin believed that an attenuated (weakened and harmless) live virus would provide more rapid and long-lasting protection than the Salk method.

The **inactivated Salk vaccine (killed vaccine)** is recommended for children who are **immunosuppressed.**

Yellow fever vaccine The 17D strain is a live attenuated vaccine developed in 1937. Yellow fever vaccine is a live, attenuated virus preparation made from the 17D yellow fever virus strain. Historically, the 17D vaccine has been considered to be one of the safest and most effective live virus vaccines ever developed. The virus is grown in chick embryos inoculated with a seed virus of a fixed passage level. The 17D yellow fever vaccine virus family is the foundation for both the 17D-204 lineage and 17DD lineage.

It is a highly effective vaccine which is administered to residents in the tropics and travellers to endemic areas. A single dose induces protective immunity to travellers and booster doses every 10 years, is recommended for residents in endemic areas.

Max Theiler, discoverer of yellow fever vaccine

Max Theiler was born on January 30 1899, in Pretoria, South Africa. In 1922 he joined the Department of Tropical Medicine at the Harvard Medical School, Boston, Massachusetts, first as an assistant, then was appointed instructor. In 1930, he joined the staff of the International Health Division of the Rockefeller Foundation, becoming Director of Laboratories of the Rockefeller Foundation's Division of Medicine and Public Health, New York in 1951.

His early work at Harvard dealt with amoebic dysentery and rat bite fever. He also worked on the problem of yellow fever, a subject in which he had become interested whilst still in London. This was to become his major interest. By 1927, he and his colleagues had proved that the cause of yellow fever was not a bacterium but a filterable virus. He also demonstrated that the disease could be readily transmitted to mice. Previously, laboratory work on this topic had been done using monkeys as experimental animals; the use of mice enabled the cost of such research to be greatly reduced. In 1930, when he joined the Rockefeller Foundation, it was engaged in a broad attack on the problem of yellow fever. Here, Theiler and his colleagues worked on vaccines against the disease and eventually developed a safe, standardized vaccine, 17D, one advantage of which was its ready adaptability to mass production.

KILLED WHOLE ORGANISM VACCINE

A killed vaccine is a vaccine that is produced by growing the organism and then killing or inactivating it with heat and/or chemicals. These are not live vaccines, they cannot replicate and therefore cannot cause the disease it is being used for.

The killed vaccines are used when safe live vaccines are not available, either because attenuated strains have not been developed or else because reversion to wild type occurs too readily.

Preparation of Killed Vaccine

The organism is propagated in bulk, *in vitro* and inactivated with either *beta*-propiolactone or formaldehyde. These vaccines are not infectious and are therefore relatively safe. However, they are usually of lower immunogenicity and multiple doses may be needed to induce immunity. In addition, they are usually expensive to prepare.

Advantages of Killed Vaccines

- Available for a wide variety of diseases
- No risk of reverting to virulent form
- Little risk of causing abortion
- More stable in storage
- Excellent stimulant of passive antibodies in colostrums

Disadvantages of Killed Vaccines

- Can cause allergic reactions and post-vaccination lumps
- Two initial doses required at least 10 days apart
- Slower onset of immunity
- May not produce as strong or as long-lasting an immunity
- Produce a narrower spectrum of protection
- More expensive than the live vaccine

Examples of Killed Vaccines

Inactivated polio vaccine IPV (Salk vaccine) Poliomyelitis is caused by polio virus, an enterovirus and there are 3 types, type 1, 2 and 3. IPV was developed by Salk.

Inactivated polio vaccine needs to be administered parenterally and it stimulates development of circulating antibodies. Though individual immunity is adequate, it does not provide herd immunity and it is not cost-effective in developing countries.

Dr. Jonas Salk (October 28, 1914–June 23, 1995) had begun his medical research career studying immunology. In 1947, while at the University of Pittsburgh, he began his research on polio virus. His research was greatly helped in 1949, when a method of growing poliovirus in cell culture, instead of having to use monkeys for research was discovered. Salk needed to find a way to process the viruses so that they were less infectious, before using them in a vaccine. In 1952, Salk was the first to develop a successful vaccine using a mixture of the three types of virus, grown in monkey kidney cultures. He developed a process using formalin, a chemical that inactivated the whole virus.

Rabies vaccine A good but expensive killed virus vaccine (human diploid cell vaccine, HDCV) grown in human fibroblasts is available for safe use in man. The human diploid cell rabies vaccine (HDCV) was discovered in 1967. Human diploid cell rabies vaccines are made using the attenuated Pitman–Moore L503 strain of the virus.

The unusually long incubation period of the virus permits the effective use of active immunization with vaccine post-exposure. When used, vaccine has dramatically cut the rabies death rate.

SUBUNIT VACCINES

Subcellular Fractions

When protective immunity is known to be directed against only one or two proteins or of polysaccharide coat of an organism, it may be possible to use a purified preparation of these proteins or polysaccharides as vaccines. The organism is grown in bulk and inactivated, and then the protein or polysaccharide of interest is purified and concentrated from the culture suspension. These vaccines are safe and fewer local reactions occur at the injection site. Examples of subcellular fractions are influenza vaccine which contains purified haemagglutinins from the viruses currently in circulation around the world and the Vi polysaccharide vaccine of *Salmonella typhi*.

Recombinant Proteins

Immunogenic proteins of virulent organisms may be synthesized artificially by introducing the gene coding for the protein into an expression vector, such as *E. coli* or yeasts. The protein of interest can be extracted from lysates of the expression vector, then concentrated and purified for use as a vaccine. The only example of such a vaccine in current use is the hepatitis B vaccine. The gene encoding a protein expressed on the surface of the hepatitis B virus, called hepatitis B surface antigen or HBsAg, can now be expressed in *E. coli* cells which provides the material for an effective vaccine.

TOXOIDS

In some diseases (diphtheria and tetanus are notorious examples), it is not the growth of the bacterium that is dangerous, but the protein toxin that is liberated by it. Treating the toxin with, for example, formaldehyde, denatures the protein so that it is no longer dangerous, but retains some epitopes on the molecule that will elicit protective antibodies. The inactivated toxin is called as toxoid.

Triple Vaccine—DPT (Diphtheria, Pertussis, Tetanus)

DPT is called as triple vaccine as it can prevent three diseases, viz. diphtheria, pertussis (whooping cough) and tetanus.

In 1888, Emile Roux discovered the diphtheria toxin secreted by *Corynebacterium diphtheriae*, causative agent of diphtheria. In 1890, the work of Emil von Behring and Shibasaburo Kitasato on antibodies to diphtheria toxins made it possible to envision their use in treating the disease. In 1891, for the first time a patient was cured of diphtheria in Berlin when injected with an antitoxic serum produced in sheep. Behring won the Nobel Prize in 1901 for the development of these antitoxic serums. In 1894, Roux and Martin immunized horses to enable large-scale serum therapy. In 1897, Ehrlich established a standardized diphtheria toxin. These passive serum therapies would soon lead to active immunization.

In 1923, Alexander Glenny and Barbara Hopkins demonstrated that formalin was capable of reducing the toxicity of the diphtheria toxin. This discovery came after several containers of diphtheria toxin, which were too large to be placed in the autoclave, were cleaned with formalin. The residual formalin present in the casks detoxified the toxin. When these toxins were tested on guinea pigs, even doses 1000 times the normal toxin dose was not enough to kill guinea pigs. The same year, Gaston Ramon, working at the Pasteur Institute in France, described a method for chemically detoxifying the toxin secreted by *Corynebacterium diphtheriae* through the use of formaldehyde and incubation at 37°C for 15 to 20 days. Though no longer toxic, the diphtheria toxin retained its specific immunological potential. Vaccines obtained by this type of chemical treatment are known as toxoids or formalin-toxoids.

For the production of tetanus toxoid, *Clostridium tetani* culture is grown in a peptone-based medium and detoxified with formaldehyde. The detoxified material is then purified by serial ammonium sulphate fractionation, followed by sterile filtration. The toxoid is then diluted with physiological saline solution (0.85%) and thimerosal (a mercury derivative) is added to a final concentration of 1:10,000.

The pertussis vaccine protects against whooping cough. The first inactivated whole cell vaccine against pertussis was developed by Madsen in the year 1923. Protection against diphtheria is nearly 90%. Whole-cell pertussis vaccine was first used on a large scale by Madsen, during a 1925 epidemic in the Faroe Islands. Madsen reported that manifestation of pertussis among vaccinated cases was less severe than among unvaccinated cases.

The latest version of the pertussis vaccine was released in the fall of 1996. This vaccine is called the "acellular" pertussis vaccine (or aP) and is purer than the old "whole cell" pertussis vaccine. The old pertussis vaccine still contained a killed form of the whole pertussis bacteria. Because individual bacteria are sometimes called cells, the old pertussis vaccine was called the whole-cell vaccine. On the other hand, the new pertussis vaccine takes advantage of recent advances in protein chemistry and protein purification. Because the whole killed pertussis bacteria are no longer present, the new pertussis vaccine is called the "acellular" vaccine.

The old pertussis vaccine, called the "whole-cell" vaccine, had a high rate of mild and severe side effects. Mild side effects such as pain and tenderness where the shot was given, fever, fretfulness and drowsiness occurred in as many as one-half to one-third of children who received the vaccine. Severe side effects such as persistent, inconsolable crying occurred in one of every 100 doses, fever greater than 105 degrees occurred in one of every 330 doses, and seizures with fever occurred in one of every 1,750 doses.

The new "acellular" pertussis vaccine, the one that has been in use in the United States since 1996, has a rate of both mild and severe side effects that is at least 10-fold less than that found for the old pertussis vaccine.

Tetanus toxoid, for intramuscular or subcutaneous use, is a sterile solution of toxoid in isotonic sodium chloride solution.

The vaccine is clear or slightly turbid in appearance. For the preparation of tetanus toxoid, the bacteria are cultured under an anaerobic condition at 35°C for 4 days. Then the culture is filtered using a filter paper. To the thus obtained filtrate, formalin is added and incubated so that the final concentration of the formalin becomes 0.4 v/v in the presence of 0.005 to 0.25M lysine for 24 to 32 days at a pH of 6.0 to 8.0 and a temperature of 30 to 45°C.

Each 0.5 mL dose is formulated to contain 4 Lf (flocculation units) of tetanus toxoid and passes the guinea pig potency test. The residual formaldehyde content, by assay, is less than 0.02%.

Lf (Limit of flocculation)

Lf (Limit of flocculation), flocculation unit, is the amount of toxin or toxoid that shows the fastest flocculation after mixture with I antitoxin unit of antitetanus serum.

RECOMBINANT VIRUS VACCINES OR LIVE RECOMBINANT VACCINES

Using genetic engineering, it is possible to introduce a gene coding for an immunogenic protein from one organism into the genome of another, such as vaccinia virus (Figure 16.7). The organism expressing a foreign gene is called a recombinant. Following injection into the subject, the recombinant organism will replicate and express sufficient amounts of the foreign protein to induce a specific immune response to the protein. This is called live recombinant vaccine.

The most advanced and popular vector used is vaccinia virus. One big attraction is that the virus can induce strong cytotoxic T lymphocyte (CTL) immune responses, although some are better at doing this than others.

Vaccinia virus is a large dsDNA virus that replicates entirely in the cytoplasm. If foreign gene is inserted into the vaccinia virus genome under the control of vaccinia virus promoter, it will be expressed independently of host regulatory and enzymatic functions. The important features of a vector vaccine are the delivery and expression of a cloned gene encoding antigens that elicit neutralizing antibodies against pathological agents. The vaccinia virus genome lacks unique restriction endonuclease sites, therefore it is not possible to insert additional DNA into viral genome and hence genes for specific antigen must be introduced into virus genome by *in vivo* homologous recombination.

DNA sequence for a specific antigen is inserted into the plasmid vector along with a vaccinia virus promoter and vaccinia thymidine kinase sequences (tk). The plasmid is used to transform tk^- animal cell culture that has previously been infected with wild type vaccinia virus that is tk^+. Recombination between DNA sequences that flank promoter and antigen gene on plasmid and homologous sequences on viral genome results in the incorporation of cloned gene into viral DNA. Absence of *tk* activity in host cells and disruption of *tk* gene in recombinant virus renders host cells resistant to toxic effects of bromodeoxyuridine (BrdU). This selection enriches the cells that carry recombinant vaccinia virus. The final selection is made using a DNA hybridization probe for antigen gene.

The genes of several viruses can be inserted, so the potential exists for producing polyvalent live vaccines. HBsAg, rabies, HSV and other viruses have been expressed in vaccinia.

Hybrid virus vaccines are stable and stimulate both cellular and humoral immunity. They are relatively cheap and simple to produce. Being live vaccines, smaller quantities are required for immunization. As yet, there are no accepted laboratory markers of attenuation or virulence of vaccinia virus for man. Alterations in the genome of vaccinia virus during the selection of recombinant may alter the virulence of the virus. The use of vaccinia also carries the risk of adverse reactions associated with the vaccine and the virus may spread to susceptible contacts. At present, efforts are being made to attenuate vaccinia virus further and the possibility of using other recombinant vectors is being explored, such as attenuated poliovirus and adenovirus.

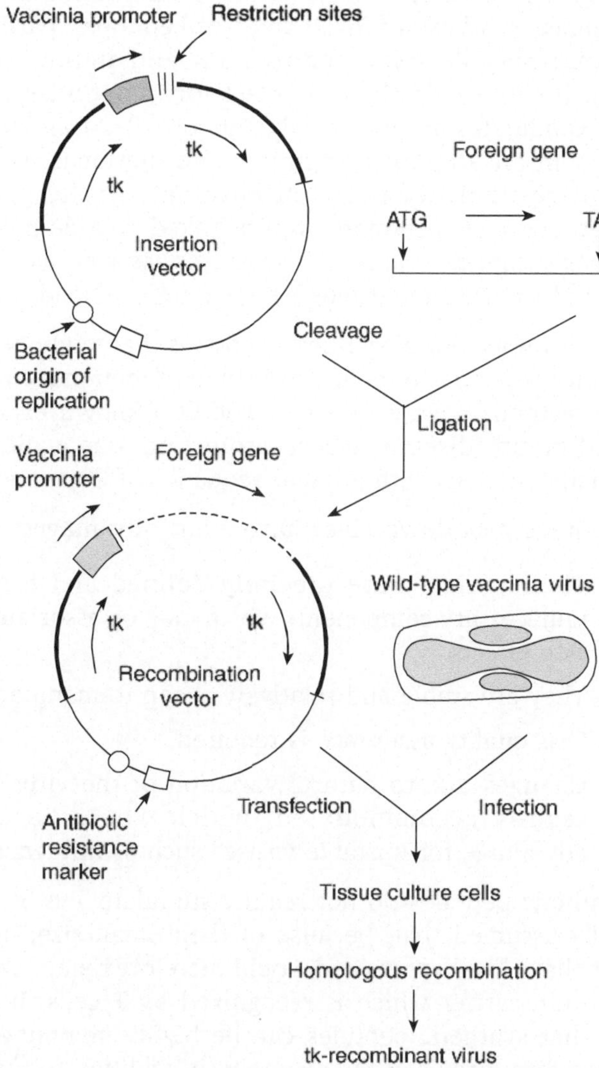

Figure 16.7 Methodology of preparation of recombinant vaccine using vaccinia virus

SYNTHETIC PEPTIDE VACCINE

The "peptide" vaccine concept is based on the identification and chemical synthesis of B-cell and T-cell epitopes, particularly those that may be immunodominant and induce specific immune functions. In their simplest form, peptides used as vaccine candidates are linear polymers of ~8–24 amino acids, however, in an attempt to mimic conformational structures (particularly those of the viral envelope) cyclic peptides, branched peptides, peptomers (cross-linked peptide polymers) and other complex multimeric structures, as well as peptides conjugated to other molecules have been developed.

The development of synthetic peptides that might be useful as vaccines depends on the identification of immunogenic sites. Several methods have been used. The best known example is foot-and-mouth disease, where protection was achieved by immunizing animals with a linear sequence of 20 amino acids.

Synthetic peptide vaccines have many advantages:

- Their antigens are precisely defined and free from unnecessary components which may be associated with side effects.
- They are stable and relatively cheap to manufacture.
- Less quality assurance is required.
- Changes due to natural variation of the virus can be readily accommodated, which would be a great advantage for unstable viruses such as influenza.

Synthetic peptides do not readily stimulate T cells. It was generally assumed that, because of their small size, peptides would behave like haptens and would therefore require coupling to a protein carrier which is recognized by T cells. It is now known that synthetic peptides can be highly immunogenic in their free form provided they contain, in addition to the B-cell epitope, T-cell epitopes recognized by T helper cells. Such T-cell epitopes can be provided by carrier protein molecules, foreign antigens or within the synthetic peptide molecule itself.

The synthetic peptide vaccine has had an impact on three major fronts:

1. Bacteria—synthetic peptide vaccines for bacterial diseases have focussed primarily on trying to neutralize toxins. For example, synthetic peptide vaccines have been developed for both the diphtheria and cholera toxins.
2. Viruses—currently, invariant regions (amino acid sequence highly conserved) are the focus of synthetic peptide vaccine design for viruses. Synthetic peptide vaccines for HIV proteins and glycoproteins are currently under development.
3. Parasites—scientists hope to achieve stage-specific anti-malarial immunity by constructing synthetic peptide vaccines for key malarial sporozoite epitopes.

ANTI-IDIOTYPIC VACCINE

The unique and characteristic site on an antibody that recognizes a specific antigen is the variable region. It can itself act as an antigen. More precisely, the variable region contains a number of antigen-like segments and these are known collectively as an idiotype. Like any other antigen, an idiotype can trigger complementary antibody. This secondary antibody is known as an anti-idiotype. An anti-idiotype, in turn, can trigger an anti-anti-idiotype. Like a series of mirrored reflections, the process can go on and on.

The ability of anti-idiotype antibodies to mimic foreign antigens has led to their development as vaccines to induce immunity against viruses, bacteria and protozoa in experimental animals. Anti-idiotypes have many potential uses as viral vaccines, particularly when the antigen is difficult to grow, or is hazardous. They have been used to induce immunity against a wide range of viruses, including HBV, rabies, Newcastle disease virus, reoviruses and polioviruses.

The concept of the idiotype is being put to practical use today in the development of experimental antigen-free vaccines.

DNA VACCINE

DNA vaccines are usually circular plasmids that include a gene encoding the target antigen (or antigens) under the transcriptional control of a promoter region active in human cells. The coding region of the inserted gene is followed by transcription termination and polyadenylation sequences. To permit selection of plasmid-containing bacteria during the production process, the plasmid also contains an antibiotic resistance gene with a bacterial origin of replication. DNA is generally less costly to produce than peptide or protein vaccines and is chemically stable under a variety of conditions. DNA vaccines are generally administered intramuscularly, using either a needle and syringe or a needle-free injector.

In contrast to conventional vaccines, DNA vaccines elicit cell-mediated, as well as antibody-mediated immune responses.

Cell-mediated Response

- The plasmid is taken up by an antigen-presenting cell (APC) like a dendritic cell.
- The gene(s) encoding the various components are transcribed and translated.
- The protein products are degraded into peptides.
- These are exposed at the cell surface nestled in class I histocompatibility molecules where they serve as a powerful stimulant for the development of cell-mediated immunity.

Antibody-mediated Response

- The plasmid is taken up by muscle cells.
- The proteins synthesized are released and can be engulfed by antigen-presenting cells (including B cells).

- In this case, the proteins are degraded in the class II pathway and presented to helper T cells.
- These secrete lymphokines that aid B cells to produce antibodies.

Conjugate vaccines

The bacteria that cause some diseases, such as pneumococcal pneumonia and certain types of meningitis, have special outer coats. These coats disguise antigens so that the immature immune systems of infants and younger children are unable to recognize these harmful bacteria. In a conjugate vaccine, proteins or toxins from a second type of organism, one that an immature immune system can recognize, are linked to the outer coats of the disease-causing bacteria. This enables a young immune system to respond and defend against the disease agent.

Currently, conjugate vaccines are available to protect against a type of bacterial meningitis caused by *Haemophilus influenzae* type B (HIB). Meningitis, an inflammation of the fluid-filled membranes that protect the brain and spinal cord, can be fatal or it can cause severe, permanent long disabilities such as deafness and mental retardation. Since HIB vaccines have been in widespread use in the United States, HIB meningitis has nearly disappeared among babies and young children.

DNA vaccines were first tested in human beings with HIV infection and subsequently in uninfected people as preventive vaccines against HIV and malaria. While immune responses to DNA alone have been relatively weak in humans, combination with adjuvants or with recombinant viral vectors in prime-boost approaches have resulted in appreciable HIV-specific CD8 responses and have induced protective responses in primate models. So far, most of the work on DNA vaccines has been done in mice where they have proved to be able to protect them against tuberculosis, SARS, smallpox and other intracellular pathogens.

CONSTITUENTS OF A VACCINE

Vaccines contain a number of substances which can be divided into the following groups:

1. Microorganisms, either bacteria or viruses, thought to be causing certain infectious diseases and which the vaccine is supposed to prevent. These are whole-cell proteins or just the broken-cell protein envelopes and are called antigens.
2. Chemical substances which are supposed to enhance the immune response to the vaccine, called adjuvants.
3. Chemical substances which act as preservatives and tissue fixatives, which are supposed to halt any further chemical reactions and putrefaction (decomposition or multiplication) of the live or attenuated (or killed) biological constituents of the vaccine.

All these constituents of vaccines are toxic and their toxicity may vary, as a rule, from one batch of vaccine to another.

ADJUVANTS

Certain substances, when administered along with a specific antigen, will enhance the immune response to that antigen. Such compounds are routinely included in inactivated or purified antigen vaccines. The substances are called as adjuvants.

Adjuvants in Common Use

Aluminium salts Aluminum salts (alum) have been useful for some vaccines including hepatitis B, diphtheria, polio, rabies and influenza, but may not be useful for others, especially if stimulation of cell-mediated immunity is required for protection. Reports indicate that alum failed to improve the effectiveness of whooping cough and typhoid vaccines and provided only a slight effect with adenovirus vaccines. Problems with alum include induction of granulomas at the injection site and lot-to-lot variation of alum preparations.

- The advantages of this adjuvant are safe and effective compound to be used in human vaccines.
- It promotes a good antibody response, but poor cell-mediated immunity.

Liposomes and immuno-stimulating complexes (ISCOMS)
The properties of ISCOMS are as follows:

- It is an alternative vaccine vehicle.
- The antigen is presented in an accessible, multimeric, physically well-defined complex.
- It is composed of adjuvant (Quil A) and antigen held in a cage-like structure.
- Adjuvant is held to the antigen by lipids.
- Can stimulate CMI.
- Mean diameter is 35 nm.

The immuno-stimulating complex (ISCOM) (Figure16.8) is a vaccine formulation which combines a multimeric presentation of antigen with a built-in adjuvant. It has a cage-like structure composed of Quillaja saponins, cholesterol, phospholipids and protein. Typically, ISCOMs have icosahedral symmetry, are 30–40 nm in diameter and are composed of 12 nm ring-like subunits.

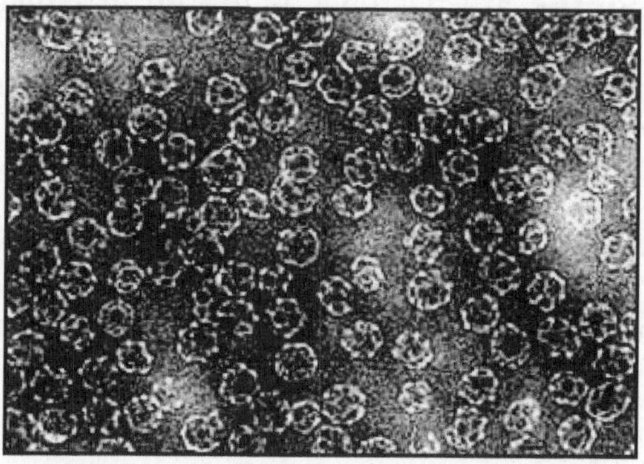

Figure 16.8 Immuno-stimulating complex (ISCOM)

The procedure for preparation of ISCOMs comprises solubilization of amphipathic proteins in preferably non-ionic detergents, addition of *Quillaja* saponins, cholesterol and phosphatidylcholine. In the presence of amphipathic proteins, ISCOM particles are formed on removal of the detergent. If no protein (antigen) is present in the mixture, ISCOM MATRIX is formed. The unique components of the ISCOM MATRIX are the Quillaja glycosides exhibiting a unique affinity to cholesterol facilitating the stability of the complex.

ISCOM-borne antigen induces an enhanced, cell-mediated immune response, delayed-type hypersensitivity reaction and cytotoxic T lymphocyte response under MHC class I restriction. Increased expression of MHC class II molecules has also been scored in primary as well as in recall immunization with ISCOMs.

Experimental ISCOM vaccine formulations have induced protective immunity to a number of microorganisms encompassing viruses including retroviruses, parasites and bacteria in several species, including primates.

The immuno-stimulating complex (ISCOM) is documented as a **strong adjuvant** and delivery system for parenteral immunization. Its effectiveness for mucosal immunization has also been proven with various incorporated antigens.

Liposomes are particles made up of concentric lipid membranes containing phospholipids and other lipids in a bilayer configuration separated by aqueous compartments. They have been used parenterally in people as carriers of biologically active substances and considered safe.

Complete and incomplete Freund's adjuvant It is an emulsion of oil and water. Even though it induces a good cell-mediated immune response, it is too toxic for man.

Muramyl dipeptide Muramyl dipeptide (MDP) has been found to be the minimal unit of the mycobacterial cell wall complex that generates the adjuvant activity. MDPs are potent pyrogens and their action is not completely understood. Hence they are not acceptable for use in humans.

Cytokines IL-2, IL-12 and interferon-gamma are also added as adjuvants.

Possible Modes of Action

The adjuvants trap antigen in the tissues, thus allowing maximal exposure to dendritic cells and specific T and B lymphocytes. Further, some adjuvants activate antigen-presenting cells to secrete cytokines that enhance the recruitment of antigen-specific T and B cells to the site of inoculation.

IMMUNIZATION IN INDIA

In India, prior to 1974, there was no organized immunization programme. **Expanded programme on immunization (EPI)** was introduced in 1974 by the WHO and adopted by India in 1978 without measles vaccination. Due to poor performance and results of EPI, Universal Immunization programme was introduced in 1985 globally and also in India. Now all our efforts are directed towards protection against six common communicable diseases, namely poliomyelitis, diphtheria, pertussis, tetanus, tuberculosis and measles. Simultaneously, efforts have been made to contain other equally dreaded diseases like typhoid fever, hepatitis B virus infection, pneumococcal infection, meningococcal infection and Japanese B encephalitis infection with vaccines.

The following is the immunization schedule followed in India:

Immunization Schedules

Mandatory

No.	**Age**	**Vaccination**
1	0–4 weeks	BCG/OPV 0 dose
2	6 weeks	DPT/OPV I dose
3	10 weeks	DPT/OPV II dose
4	14 weeks	DPT/OPV III dose
5	9 months	Measles vaccine/OPV IV dose
6	15–18 months	MMR
7	18–24 months	DPT/OPV I booster
8	5 years	DT/OPV II booster
9	10 years	Tetanus toxoid

BCG—Bacillus Calmette–Guerin
OPV—Oral Polio Vaccine
DPT—Diphtheria, Pertussis, Tetanus
MMR—Measles, Mumps, Rubella

Recommended

No.	**Age**	**Vaccination**
1.	0–1 week	HBV I dose
2.	6 weeks	HBV II dose
3.	6 months	HBV III dose

Optional

No.	**Age**	**Vaccination**
1.	12 years	Parenteral typhi M
2.	6 years	Oral typhoid vaccine
3.	2 months	HIB I dose
4.	4 months	HIB II dose
5.	6 months	HIB III dose
6.	18 months	Booster of HIB

HBV—Hepatitis B vaccine
HIB—*H. influenzae* B vaccine

Herd Immunity

If enough people in a community are immunized against certain diseases, then it is more difficult for that disease to get passed between those who are not immunized. This is known as herd immunity.

Herd immunity does not apply to all diseases because they are not all passed on from person to person. For example, tetanus can only be caught from spores in the ground.

PASSIVE IMMUNIZATION

Artificially acquired passive immunity is a short-term immunization by an injection of antibodies, such as gamma globulin, that are not produced by the recipient's cells. Naturally acquired passive immunity occurs during pregnancy, in which certain antibodies are passed from the maternal into the foetal bloodstream.

In passive immunization, antibodies against a specific infectious organism are given directly to a person. Passive immunization is used for people whose immune system does not respond adequately to an infection or for people who acquire an infection before they can be vaccinated (for example, after exposure to the rabies virus). Passive immunization can also be used to prevent disease when exposure is likely and the person does not have time to get or complete a vaccination series. An example of this is the use of gamma globulin (an antibody preparation) to help prevent hepatitis in people who travel to certain parts of the world. Passive immunization lasts for only a few days or weeks, until the body eliminates the injected antibodies.

Immunological tolerance for foreign antigens can be induced experimentally by creating conditions of high-zone tolerance, i.e., by injecting large amounts of a foreign antigen into the host organism, or low-zone tolerance, i.e., injecting small amounts of foreign antigen over long periods of time.

Active immunity is the development of antibodies in response to stimulation by an antigen, whereas in passive immunity, preformed antibodies are administered for protection from a disease.

Tetanus Antitoxin

Tetanus antitoxin is an antiserum that has been produced by actively immunizing an animal (e.g. a horse) with tetanus toxoid. This is normally administered under those conditions when a human

- may have been exposed to spores of the tetanus bacillus (e.g. in a dirty punctured wound) and
- has never or not for a long time (10 years) been actively immunized with tetanus toxoid.

The physician, fearing that disease symptoms may occur before the patient is able to mount an active immune response, will inject tetanus antitoxin in order to provide immediate protection. The protection is short-lived, lasting only until the last of the injected antibodies have been catabolized.

Antivenoms

These antisera (raised in horses or sheep against snake venom) provide immediate protection to people bitten by a snake.

Immune Globulin (IG)

Horse and sheep proteins are foreign to the human patient and will, in due course, elicit an active immune response. This may lead to an allergic reaction such as

- systemic anaphylaxis or
- serum sickness

To avoid such problems, humans are often used as the source of passive antibodies. Some immunoglobulin (Ig) is prepared from the gamma globulin fraction of pooled plasma from the outdated blood of several thousand blood donors on the assumption that this large pool will contain good levels of antibodies against many common diseases such as

- hepatitis A ("infectious hepatitis")
- measles
- rubella

Ig is also used to provide protection to men with **X-linked agammaglobulinemia**, who are unable to manufacture antibodies because of a mutation in their single X chromosome gene for Bruton's tyrosine kinase.

Some preparations of immune globulin are harvested from selected **individual** donors who have either recently recovered from the disease or who have been deliberately and intensively immunized against it. These are used to provide immediate protection against such diseases as

- rabies
- tetanus
- varicella (chickenpox)
- smallpox or complications arising from giving the smallpox vaccine (vaccinia immune globulin or VIG)
- **Rh immune globulin** (**RhIg**) or **Rhogam** is used to prevent Rh-negative mothers from becoming sensitized to the Rh antigen of their newborn child.

Advantages of human immune globulin

- The preparation contains fewer irrelevant serum proteins and of those that remain, being human proteins.
- They are far less immunogenic.
- They are catabolized more slowly than horse proteins.

However, care must be (and is) taken to ensure that the preparations are not contaminated with human pathogens such as HIV or Hepatitis viruses.

Non-antigen-specific effects of human immune globulin Intravenous injections of IG have helped patients with autoimmune disorders like

- immune haemolytic anaemia
- immune thrombocytopenic purpura
- myasthenia gravis

The therapeutic effect seems to have nothing to do with the antigen specificities (e.g. antitetanus) of the antibodies in the preparation.

Instead it is the C-region of the antibody molecules that provide the protection.

Animal studies suggest that it does so by binding to a class of receptors on macrophages, which inhibits them from phagocytosing antibody-coated cells. Some example are

- antibody-coated red cells in immune haemolytic anaemia
- antibody-coated platelets in idiopathic thrombocytopenic purpura

The spleen is packed with macrophages and it is here that most of red blood cell and platelet destruction occurs in these diseases (and explains why removal of the spleen so often helps the patient).

POINTS TO REMEMBER

- A vaccine is any preparation of killed or weakened microorganism that is given to a person by mouth or injection in order to prevent disease.
- The principle of vaccination is to induce a primary immune response in the vaccinated subject (development of memory T and B cells) so that, following the exposure of a pathogen, a rapid secondary immune response is generated.

- The live vaccines are also called as attenuated vaccines or sometimes as live attenuated vaccines.
- A killed vaccine is a vaccine that is produced by growing the organism and then killing or inactivating it with heat and/or chemicals.
- The inactivated toxin is called as toxoid.
- DNA vaccines are usually circular plasmids that include a gene encoding the target antigen which is under the transcriptional control of a promoter region active in human cells.
- Chemical substances which are supposed to enhance the immune response to the vaccine are called adjuvants.
- In passive immunization, antibodies against a specific infectious organism are given directly to a person.

REVIEW QUESTIONS

1. Write short notes on:
   i. Edward Jenner
   ii. Louis Pasteur
   iii. Live vaccines
   iv. Killed whole organism vaccines
   v. Subunit vaccines—purified or recombinant antigen
   vi. Toxoids
   vii. Recombinant virus vaccines
   viii. Synthetic peptide vaccine
   ix. Anti-idiotype antibodies
   x. DNA vaccines
   xi. Adjuvants used in vaccines
2. Write a detailed note on the immunization schedule followed in India.

17

TRANSPLANT IMMUNOLOGY

INTRODUCTION

Transplantation is the introduction or replacement of biological material such as organs, tissues, cells and fluids into a living being.

The living being from which the organ is obtained is called a donor and the living being receiving the organ transplant is called a recipient. We can distinguish 3 critical relationships between the donor and the recipient.

- Syngeneic transplants—usually the donor and recipient will be same individual (autologous) or from genetically identical individuals, i.e., grafts between identical twins or isogenic strains of experimental animals.
- Allogeneic transplants—from one individual to another of the same species.
- Xenogeneic transplants—between individuals of different species.

Normally the syngeneic transplant will be accepted by the recipient and thus never involves any immunity. However, the acceptance of allogeneic and xenogeneic transplants involves lot of immunological parameters. Often allografts and xenografts will be rejected by the recipient.

The acceptance or the rejection of a graft is determined by two important tissue antigens. They are:

1. HLA
2. Blood group antigens

The type of antigens present in an individual is determined by the genetic constituents of that particular individual. Thus the important problem of graft rejection is due to the genetic difference between the donor and recipient. No two *Homo sapiens* will be genetically identical (except for identical twins). Thus allograft rejection is quite common in humans.

Tissue compatibility is mainly determined by the major histocompatibility complex (MHC). They are responsible for the production of different kinds of HLA antigens. HLA antigens are controlled by a series of genes on chromosome 6, referred to as the human major histocompatibility complex (MHC). These genes have been classified into major categories. HLA-A and HLA-B are examples of class I genes and HLA-DR and HLA-DQ represent class II genes. These genes are highly polymorphic. For instance, HLA-A has more than 20 alleles and HLA-B has more than 50 alleles. Therefore, it is highly unlikely that two unrelated persons have the same HLA type. Both HLA class I and class II alloantigens can induce transplant immunity at humoral (antibody) and cellular (T lymphocyte) immune levels. The compatibility of a transplant is determined by the closeness of the HLA antigen patterns of the donor and recipient.

Next important set of antigens are the blood group antigens.

BLOOD GROUPS

ABO blood grouping system was the first discovered by Karl Landsteiner in 1900 and is the most important in assuring safe blood transfusions.

Table 17.1 shows the four ABO phenotypes ("blood groups") present in the human population and the genotypes that give rise to them.

Table 17.1 Antigens and antibodies present in different blood groups and their possible genotypes

Blood group	Antigen on RBCs	Antibody in serum	Genotypes
A	A	Anti-B	*AA* or *AO*
B	B	Anti-A	*BB* or *BO*
AB	A and B	Neither A or B	*AB*
O	Neither	Anti-A and anti-B	*OO*

When red blood cells carrying one or both antigens are exposed to the corresponding antibodies, they agglutinate, that is, clump together. People usually have antibodies against those red cell antigens that they lack.

Karl Landsteiner (June 14, 1868–June 26, 1943), was an Austrian biologist and physician. He is noted for his development in 1901 of the modern system of classification of blood groups from his identification of the presence of agglutinins in the blood. In 1930 he received the Nobel Prize in Physiology or Medicine. With Alexander S. Wiener, he identified the Rh factor in 1937. He was posthumously awarded the Lasker Award in 1946.

The ABO antigen system is a ubiquitous antigen system. The A, B or O (H antigen) is not limited only to a individual's red cells. The ABO group of an individual is expressed on diverse cells and tissues. In certain individuals who possess a specific gene, the ABO antigen is also expressed in saliva and other body fluids. They are called as secretors.

ABO antigens are carbohydrate structures present on most, perhaps all, cells and in most people on soluble molecules in serum and secretions. Majority of humans have a basic structure known as the H antigen (sometimes also referred to as O antigen). The acceptor polysaccharide for this carbohydrate can be one of several structures but the antigenic epitope is composed of the terminal 2 sugars.

The A, B or O blood group of an individual involves the interaction between enzymes which are the product of two genes:

- *H gene* It is present on chromosome 19. It codes for a fucosyl transferase that synthesizes the H antigen.
- *ABO gene* It is located on chromosome 9.

The ABO antigen expressed by the individual cells are carbohydrate antigens. These are constructed of oligosaccharide chains constructed in a stepwise manner with each sugar being added to the growing chain by a specific enzyme (glycosyl transferase).

These oligosaccharides are attached to different types of precursor chains such as polypeptides or ceramide which act as transmembrane anchors.

The successful creation of the final ABO blood group is however the final stage of a complex series of enzymatic actions with approximately 100 glycosyl transferases being involved in the creation of the carbohydrate oligosaccharides to which the H, A or B antigens are added.

The building block for the construction of the A, B or H antigen is a precursor oligosaccharide which can vary from a few sugar molecules in a simple linear chain to more complex structures. The oligosaccharide is constructed in a sequential manner by the enzymatic addition of sugars to the terminal end of the oligosaccharide chain; this is then attached to a membrane anchor which may be a glycoprotein, glycolipid, etc. The molecule is attached to the last three sugars on the chain, viz. galactose, *N*-acetyl glucose and a terminal galactose.

The synthesis of an ABO antigen is a sequential 2-step process involving the following.

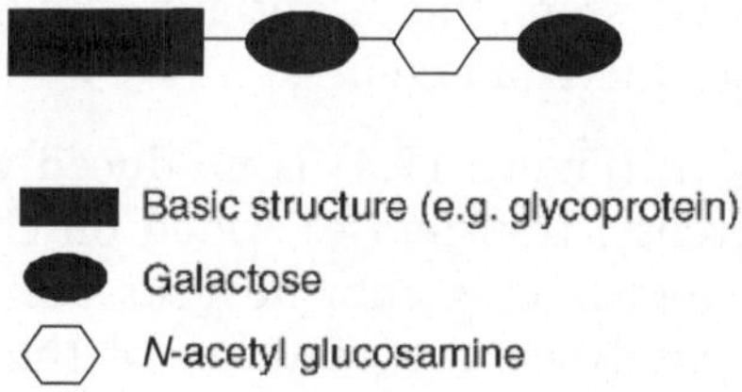

Figure 17.1 H antigen precursor structure

In the first step the H antigen (Figure 17.1) is synthesized. The synthesis of the H antigen is controlled by the *H* gene on chromosome 19. This gene codes for the synthesis of an enzyme called fucosyl transferase (FUT 1). This enzyme recognizes the terminal galactose of the precursor oligosaccharide and catalyses the addition of fucose to the terminal galactose. This enzymatic addition creates the H antigen (Figures 17.2 and 17.3) which is the antigen expressed on group O individuals and is used as the precursor for A and B group antigens.

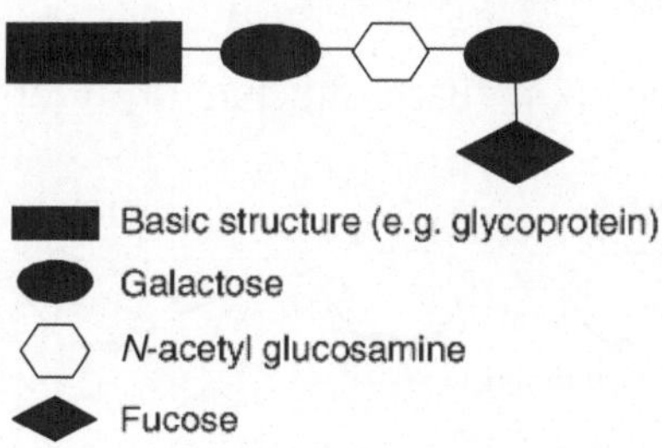

Figure 17.2 Basic structure of H antigen

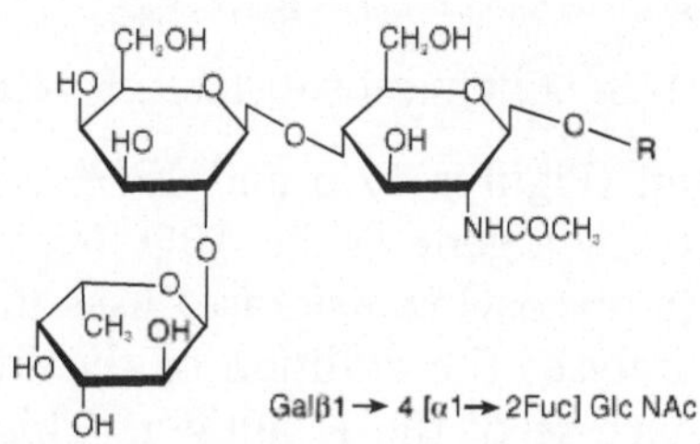

Figure 17.3 Chemical structure of H antigen

In the second step, the A antigen and/or B antigen is synthesized from the H antigen.

The A antigen (Figure 17.4) is produced via the action of an enzyme (glycosyl transferase) coded on chromosome 9 at the ABO gene locus. This enzyme uses the H antigen as a substrate and catalyses the addition of the sugar *N*-acetyl galactosamine to the terminal galactose to produce the A antigen. The A enzyme can only produce the A antigen (Figure 17.5) if a H antigen is already present.

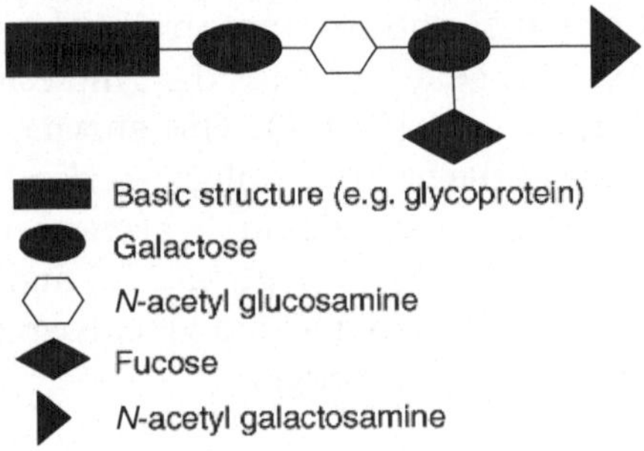

Figure 17.4 Basic structure of "A" antigen

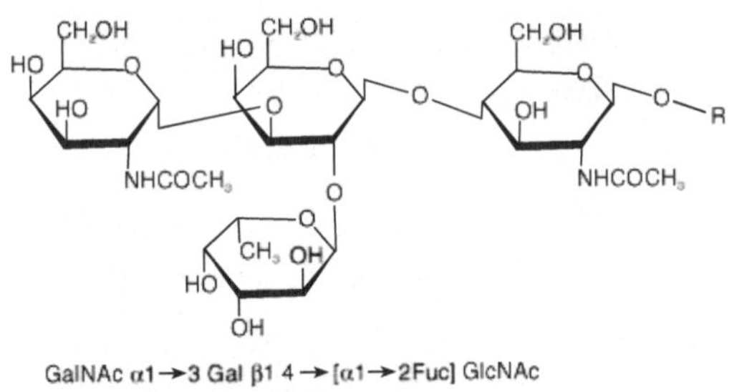

Figure 17.5 Chemical structure of "A" antigen

The B antigen (Figures 17.6 and 17.7) is produced by an enzyme coded for by a gene at the ABO locus on chromosome 9. The enzyme (a glycosyl transferase) uses the H antigen as a substrate and catalyses the addition of the sugar galactose to the terminal galactose of the H antigen. This produces the B antigen. The B enzyme can only produce the B antigen if a H antigen is present.

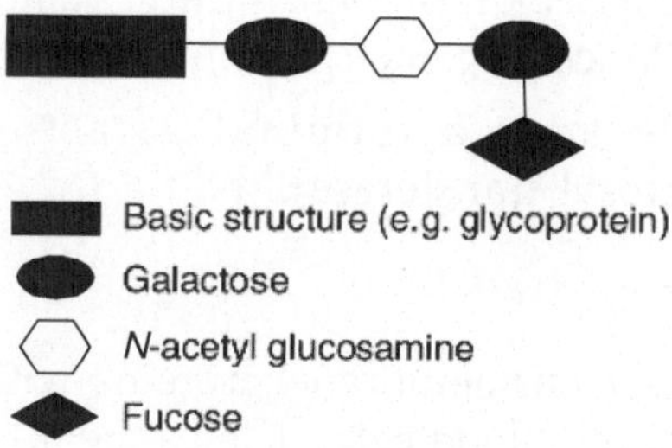

Figure 17.6 Basic structure of "B" antigen

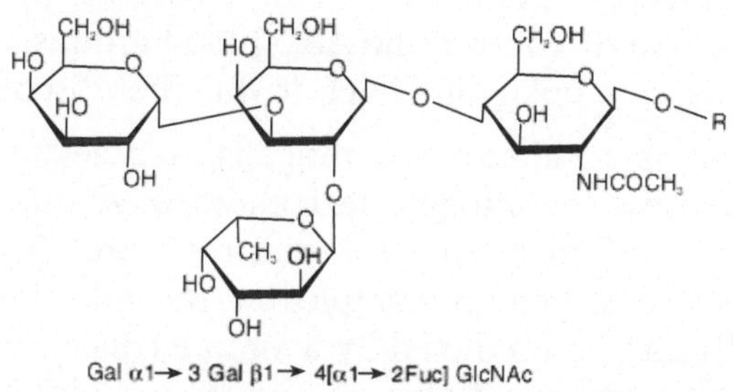

Figure 17.7 Chemical structure of "B" antigen

Genetics of the ABO System

The ability to produce H antigen is controlled by the H and Se loci. Both genes are tightly linked and both encode alpha 1, 2 L-fucosyl transferases.

The *H* gene product (FUT1) is expressed in haemopoietic cells and vascular endothelial cells and the Se (FUT2) in other cells, including secretory cells.

Very rarely in some individuals the ability to produce this substance is missing (genotype h/h). Such individuals do not express H (or A, B) on their red cells. They are called as Bombay blood group individuals. They will express H (and A, B if produced) in their body fluids if they are Se/se or Se/Se.

The H substance is modified by the products of the ABO locus.

- O allele is a null, i.e., cannot modify the H antigen.
- A allele adds a terminal *N*-acetyl galactosamine [alpha 1, 3 D-*N*-acetyl galactosaminyl transferase].
- B allele adds a terminal galactose [alpha 1, 3 D-galactosyl transferase].

The Rh System

Rh antigens are transmembrane proteins with loops exposed at the surface of red blood cells. They appear to be used for the transport of carbon dioxide and/or ammonia across the plasma membrane. Rh blood types were discovered in 1940 by Karl Landsteiner and Alexander Wiener. This was 40 years after Landsteiner had discovered the ABO blood groups. It is named after the rhesus monkey in which it was first discovered.

There are a number of Rh antigens. Red cells that are "Rh positive" express the antigen designated D. About 15% of the population have no D antigens and thus are "Rh negative". Despite its actual genetic complexity, the inheritance of this trait usually can be predicted by a simple conceptual model in which there are two alleles, D and d. Individuals who are homozygous dominant (DD) or heterozygous (Dd) are Rh+. Those who are homozygous recessive (dd) are Rh–.

The major importance of the Rh system in human health is to avoid the danger of Rh D incompatibility between mother and foetus.

During birth, there is often a leakage of the baby's red blood cells into the mother's circulation. If the baby is Rh positive (having inherited the trait from its father) and the mother Rh negative, these red cells will cause her to develop antibodies against the D antigen. The antibodies, usually of the IgG class, do not cause any problems for that child, but can cross the placenta and attack the red cells of a subsequent Rh positive foetus. This destroys the red cells producing anaemia and jaundice. The disease, called erythroblastosis foetalis or haemolytic disease of the newborn may be so severe as to kill the foetus or even the newborn infant. It is an example of antibody-mediated cytotoxicity disorder.

Although certain other red cell antigens (in addition to Rh) sometimes cause problems for a foetus, an ABO incompatibility does not. Why is Rh incompatibility so dangerous when ABO incompatibility is not?

It turns out that most anti-A or anti-B antibodies are of the IgM class and these do not cross the placenta. In fact, an Rh negative/type O mother carrying an Rh positive/type A, B, or AB foetus is resistant to sensitization to the Rh antigen. Presumably her anti-A and anti-B antibodies destroy any foetal cells that enter her blood before they can elicit anti-Rh antibodies in her.

This phenomenon has led to an extremely effective preventive measure to avoid Rh sensitization. Shortly after each birth of an Rh positive baby, the mother is given an injection of anti-Rh antibodies. The preparation is called Rh immune globulin (RhIg) or Rhogam. These passively acquired antibodies destroy any foetal cells that got into her circulation before they can elicit an active immune response in her.

Dr. Alexander S. Wiener (1907–1976), a lifelong resident of New York City, was recognized internationally for his contributions to science. He was an outstanding leader in the fields of Forensic Medicine, Serology and Immunogenetics. His pioneer work led to the discovery of the Rh factor in 1937, along with Dr. Karl Landsteiner and subsequently to the development of exchange transfusion methods that saved the lives of countless infants with haemolytic disease of the newborn. He received the Lasker Award for his achievements.

At first, Dr. Wiener and Dr. Landsteiner did not understand the full significance of their discovery. They were engaged in identifying a range of factors, such as the M factor, which proved to have much less significance. Dr. Wiener named the Rh factor after the Rhesus monkeys used as test subjects. However, by the time he and Dr. Landsteiner published in 1940, Dr. Wiener was able to demonstrate the role of Rh sensitization as a cause of intragroup haemolytic reactions, thus increasing the safety of blood transfusions.

HISTORY OF DISCOVERY OF BLOOD GROUPS

In 1628, William Harvey demonstrated blood circulation. The first transfusion from dog to dog was completed in England in 1665 by Richard Lower. Excited by the work of Harvey and others, Lower went on to transfuse animal blood into a human in 1667 (Figure 17.8). This transfusion was not used as fluid replacement but as an attempt to alter the mental characteristics of the patient. The result, not surprisingly, was unimpressive. It could have been fatal if the methods used had been adequate to permit the flow of larger amounts of blood.

Figure 17.8 Photograph published during the 17th century depicting the transfusion of animal's blood to human

During this time, a French physician named Jean Baptiste Denis transfused sheep's blood into a 15-year-old boy. The boy was allegedly "cured" and Denis went on to perform several more transfusions. At one point, he transfused a man who later died, and the widow sued Denis. After a considerable controversy, the Paris Society of Physicians in 1678 declared all transfusions illegal. Other countries, such as England, soon followed suit, and transfusion was not practised for some fifty years.

In 1818, a French physician named James Blundell wrote that species lines should not be crossed when transfusing blood. He performed the first human-to-human transfusion on a mother who had lost large quantities of blood during childbirth. A donor's artery was opened. The spurting blood was caught in a cup and sent through tubing inserted into the patient's vein. The amount transfused was estimated by timing. Usually the blood clotted in the tubing before it could get into the patient, probably a fortunate occurrence because of the high rate of blood group incompatibility. Blundell had some success with this type of transfusion, but animal blood was used into the 20th century.

It was realized that, to insure successful transfusions, drawn blood must not be exposed too long or it would clot. A variety of ingenious items were used to increase the flow, which included stopcocks, valves, syringes and quills.

At the turn of the century, vein-to-vein transfusion was practised by highly skilled surgeons, but the amount of blood being transfused could still only be estimated.

Some relief from this difficulty followed the discovery in 1911 that paraffin slowed the clotting process. A Y-shaped tube coated with wax was connected to a vein of the patient and one of the donors. A syringe was attached to this apparatus and by alternating the opening and closing of several valves, blood was transfused. Thus, the amount of blood being received by the patient could actually be measured for the first time.

In 1901, an Austrian Scientist, Karl Landsteiner, found that reactions between substances present on the surface of red cells (antigens) and other substances in plasma (antibodies) sometimes caused the red blood cells to clump together, causing adverse reactions in recipients. He was able to show that there were at least three major types of human blood that vary according to the kinds of sugar-containing substances, known as antigens, attached to the plasma membrane of the red blood cells. Based on this he divided human blood into three groups: A, B, and C (later O).

Two of his co-workers, the clinicians Alfred von Decastello (Decastello-Rechtwehr, born 1872) and Adriano Sturli (1873–1964), examined additional persons and in 1902 found the fourth blood group, later named AB.

He investigated the reactions between the red cells and serum of the blood in 22 of his co-workers and found that the serum from some subjects clumped the red cells of certain other subjects. This observation eventually led to blood group classification in the ABO system known as O, A, B or AB types.

For his experiments, Landsteiner drew blood from himself and collected blood samples from his colleagues, doctors Jakob Erdheim (1874–1937), Pletschnig, Oskar Stoerk (1870–1926) and Adriano Sturli.

Landsteiner officially retired in 1939, at the age of seventy-one, but went on working. In 1940, he and his co-workers Alexander Wiener (1907–1976) and Philip Levine (1900–1987) made an important discovery. In a paper published that year, they described a new factor in the human blood, the Rh factor, (Rh for Rhesus monkey, (Figure 17.9) in which the factor was first discovered). Levine was the first to see the connection between this factor and jaundice occurring in newborn children.

In 1914, it was found that sodium citrate would keep the blood from clotting when added to donate blood. This permitted donated blood to be stored in a bottle for later transfusion.

Figure 17.9 Rhesus monkey

In 1912, Roger Lee defined the terms "universal donor" and "universal recipient". He demonstrated that group O blood could be transfused in patients having any one of the four blood groups, while group AB patients could receive blood having any one of the four blood groups.

As mentioned earlier in 1914, several anti-coagulants were being introduced. In 1916, Francis Rous and J.R. Turner introduced a citrate-glucose solution, which was added to the collected blood. This allowed blood to be stored in containers and refrigerated for several days before being transfused. In the years to follow, establishments where blood was collected and stored were being introduced. Later, these came to be known as "blood banks", the first being introduced in a Leningrad Hospital in 1932. However, the term "blood bank" originated in 1937, when Bernard Fantus established the first blood bank at the Cook County Hospital in Chicago. In the following years blood banks spread throughout the United States.

Several other important discoveries helped bring blood transfusion to its present-day standards. A second human blood group system, called Rh, was discovered in 1943. Plastic donor bags, instead of bottles for blood storage, were developed by 1949, alleviating the high cost of re-sterilization of bottle and tubing previously used. Recently, with the addition of adenine to CPD solution, the expiration date of whole blood and red cells was extended to 35 days. Adenine supplies the red cells with a sufficient source of energy to permit longer storage.

Today, one unit of whole blood can be separated into components which can be used to treat several patients with different blood needs. This practice of transfusing patients with components provides maximum efficiency for blood resources management.

GRAFT REJECTION

There are 3 basic types of recognition which allows the host to know that the transplanted tissue is foreign.

1. recognition by antibody
2. recognition of foreign MHC by T cells (direct recognition)

3. recognition of minor histocompatibility loci by T cells (indirect recognition)

Thus the 3 recognitions may lead to very different time scales of destruction of the transplanted cells/tissue and trigger distinct effector mechanisms.

There are three important types of graft rejection according to the same. They are:

1. Hyperacute rejection
2. Acute rejection
3. Chronic rejection

Hyperacute Rejection

This type of rejection occurs very rapidly, resulting in necrosis of the transplanted tissue within minutes or a few hours of contact. It always results from the reactivity of the donor cells with pre-existing antibodies.

Hyperacute rejection occurs usually within the first 24 hours after transplantation. This response occurs so quickly that the tissue never becomes vascularized. It is characterized by thrombotic occlusions and haemorrhage of the graft vasculature that begins minutes to hours after the graft is placed. Hyperacute rejection is caused by pre-existing host antibodies that bind to antigens present in the graft endothelium. Antigen recognition activates the complement system. There is also an influx of neutrophils. Endothelial cells and platelets are also induced to shed lipid particles from their membrane that promote coagulation. The resulting inflammation prevents vascularization of the graft. The graft then suffers irreversible damage from ischaemia.

There are several explanations for the pre-existing antibodies that initiate hyperacute rejection.

The most common situation in which this occurs is in transplants incompatible with ABO blood group. Recipients of blood transfusions sometimes develop antibodies to MHC antigens from the transfused blood. If some of these antigens match those in a graft, then hyperacute rejection may result. Multiple pregnancies may also expose the woman to the paternal antigens of the foetus, resulting in the creation of antibodies. Finally, prior recipients of transplants may have already formed antibodies to other MHC antigens, so they may be present at the time of a second transplant. Most of the time, hyperacute rejection can be avoided by screening for anti-graft antibodies.

Acute Graft Rejection

Acute rejection usually begins after the first week of transplantation, and most likely occurs to some degree in all transplants (except between identical twins). It is caused by mismatched HLA antigens that are present on all cells. HLA antigens are polymorphic. Therefore the chance of a perfect match is extremely rare. The reason for acute rejection to occur only after a week after transplantation is that the T cells involved in rejection must differentiate and the antibodies in response to the allograft must be produced before rejection is initiated.

Allogeneic MHC is recognized by either $CD8^+$ T cells (class I) or $CD4^+$ T cells (class II); up to 10% of T cells can recognize a given allogeneic MHC because it resembles self HC foreign peptide (Figure 17.10).

Host T cells are activated in the draining lymph nodes to graft antigens by two mechanisms.

Grafts contain passenger leucocytes and APC bearing both MHC and co-stimulatory molecules. Passenger leucocytes travel to the draining lymph nodes and activate recipient T cells. This is called as direct alloreactivity. Direct activation of recipient T cells is responsible for acute graft rejection that occurs in the first weeks following transplantation. The effector cells are primarily CTLs. Symptoms of acute rejection include fever, a skin rash, impaired organ function (such as decreased urine output from a transplanted kidney) and a mononuclear (T cell) infiltrate into the graft, visible on biopsy.

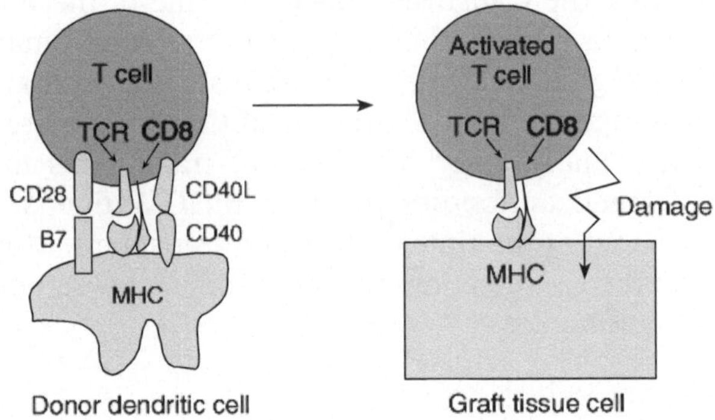

Figure 17.10 Mechanism of acute graft rejection

Indirect alloreactivity comes from uptake of graft antigens by recipient APC and presentation on self MHC. Peptides from both MHC and minor H antigens are presented by recipient APC. Effectors are usually T_H1 cells that activate macrophages to cause tissue injury and scarring that can cause chronic rejection or organ failure (Figure 17.11).

First Rejection

Transplantation of a second graft, which shares a significant number of antigenic determinants with the first one, results in a rapid (2–5 days) rejection. It is due to the presence of T lymphocytes sensitized during the first graft rejection. Accelerated rejection is mediated by immediate production of lymphokines, activation of monocytes and macrophages and induction of cytotoxic lymphocytes. The rejection mechanism that occurs during the first graft transplant is called first set rejection.

These T cells cause the graft cells to lyse or produce cytokines that recruit other inflammatory cells, eventually causing necrosis of allograft tissue. Endothelial cells in vascularized grafts such as kidneys are some of the earliest victims of acute rejection. Damage to the endothelial lining is an early predictor of irreversible acute graft failure. The risk of acute rejection is highest in the first 3 months after transplantation and is lowered

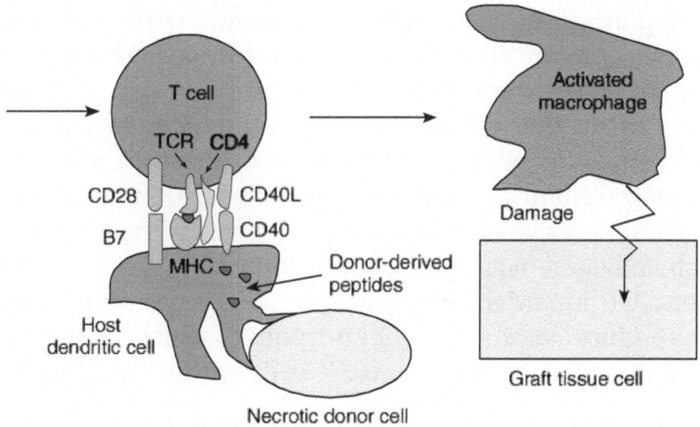

Figure 17.11 Mechanism involved in the graft rejection in which minor H antigen is involved

by immunosuppressive agents in maintenance therapy. The onset of acute rejection is combated by episodic treatment.

> **Minor Histocompatibility Antigens**
>
> There are many histocompatibility (H) antigens encoded outside the MHC, the so-called minor H antigens.
>
> The two important antigens identified so far are
>
> - H-Y, an antigen encoded on the Y chromosome and thus present in male, but not female tissue
> - HA-2, an antigen derived from the contractile protein myosin

Chronic Rejection

Chronic rejection is a third type of rejection and occurs months to years following transplantation. It is characterized by graft arterial occlusions (vasculopathy), which results from the proliferation of smooth muscle cells and production of collagen by fibroblasts. This process, termed accelerated or graft arteriosclerosis, results in fibrosis which can cause ischaemia and cell death. These fibrous lesions occur without evidence of an overt cause (such as vascular injury or infection), although it is hypothesized that chronic rejection is really the result of continued prolonged multiple acute rejections. This hypothesis is based on the knowledge that the resulting fibrosis is similar to the fibrosis that accompanies natural healing of a wound. Its pathogenesis probably involves both humoral and cellular immune mechanisms. Chronic rejection may be mediated by a low-grade, persistent delayed type hypersensitivity response in which activated macrophages secrete mesenchymal cell growth factors. Of potential importance are persistent viral infections which induce cellular immune responses which may synergize with donor-specific alloreactive T cells within the allograft.

GRAFT-VERSUS-HOST DISEASE (GVHD)

Graft-versus-host disease (GVHD) is a complication that is observed after allogeneic stem cell or bone marrow transplant. GVHD occurs when immunocompeting cells from the donor recognize the recipient's body as being foreign. These immunocompeting cells then attack tissues in the patient's body just as if they were attacking an infection.

Acute and Chronic Graft-versus-Host Disease

Acute GVHD Acute GVHD appears within the first 100 days after transplant. Acute GVHD can range from mild symptoms to life threat. Symptoms depend on the parts of the body affected and how severe it is. It may appear as a rash on the skin. It often starts on the palms of the hands and soles of the feet, and can spread later to other parts of the body. If GVHD is severe, the skin may blister and peel. There may be cramping, nausea or diarrhoea if it affects the stomach or intestines. There will be yellowing of the skin and eyes (jaundice), if it affects the liver.

Transplant patients are usually given drugs to try and prevent GVHD for the first six months after transplant.

Chronic GVHD Chronic GVHD can begin any time during or after the third month post-transplant. Transplant patients who get acute GVHD are more likely to also get chronic GVHD, but it can also appear in patients who did not get acute GVHD. Chronic GVHD can range from mild symptoms to life threat. Some transplant survivors have problems with chronic GVHD for many years.

Exposing the skin for a long period of time can trigger GVHD. Thus avoiding the exposure of sunlight is the best way to prevent the GVHD.

Signs of chronic GVHD The most common symptoms of chronic GVHD are:

- Rash or changes in skin colour or texture
- Dry or irritated eyes
- Pain, dryness or sensitivity in the mouth

Other less common symptoms of chronic GVHD include:

- Thinning hair
- Brittleness or changes in the texture of fingernails
- Dry or irritated vagina (women)
- Nausea, vomiting, diarrhoea, loss of hunger or an unexplained drop in weight

In more advanced GVHD, the skin could feel tight or hard in more advanced GVHD. The joints could feel stiff or become hard to straighten (fingers, wrists, elbows, ankles or knees).

ACUTE VASCULAR REJECTION

Acute vascular rejection is found during transplantation of xenografts. Even though initial hyperacute rejection is avoided in discordant species through transgenic and knock-out techniques in gene recombination, acute vascular rejection still occurs 4–8 days after transplantation. This form of rejection is characterized by endothelial activation and coagulation, causing cell damage, thrombosis and eventual graft rejection. Although it occurs more slowly than in hyperacute rejection, the end result of transplant failure is the same.

METHODS FOR TESTING FOR HLA ANTIGENS

Lymphocytotoxicity Test

In the lymphocytotoxicity test, lymphocytes are added to recipient's sera which may or may not have antibodies directed to HLA antigens. This is also called as tissue typing.

The purpose of the tissue typing laboratory is to assess donor–recipient compatibility for HLA and ABO to analyse patient serum for lymphocytotoxic antibodies which may be specific for the potential transplant donor. Most relevant is the crossmatch assay whereby patient sera are tested for their reactivity with donor lymphocytes. This is usually done by lymphocytotoxicity testing whereby donor lymphocytes are first incubated with patient serum, then with rabbit complement. Lysis of lymphocytes is assessed by the uptake of an extravital dye like trypan blue or eosin red. A positive crossmatch is a contraindication for organ transplantation because of the risk for antibody-mediated (hyperacute) rejection.

The cells used for the test are lymphocytes because of their excellent expression of HLA and ease of isolation compared to most other tissue.

The most important use of this test is to detect specific donor-reactive antibodies present in a potential recipient prior to transplantation.

Historically, this test has long been used to type HLA class I and class II antigens, using antisera of known specificity. However, the problems of cross-reactivity and non-availability of certain antibodies has led to the introduction of DNA-based methods. Currently, many laboratories have changed to molecular genetic methods for HLA class typing.

Mixed Lymphocyte Culture (MLC)

When lymphocytes from two individuals are cultured together, each cell population is able to recognize the "foreign" HLA class II antigens of the other. As a response to these differences, the lymphocytes transform into blast cells, with associated DNA synthesis. Radiolabelled thymidine added to the culture will be used in this DNA synthesis. Therefore, radioactive uptake is a measure of DNA synthesis and the difference between the HLA class II types of the two people.

This technique can be refined by treating the lymphocytes from one of the individuals to prevent cell division, for example by irradiation. It is thus possible to measure the response of T lymphocytes from one individual to a range of foreign lymphocytes. It has thus proved possible by using the mixed lymphocyte culture (MLC) test to use T lymphocytes to define what were previously called HLA-D antigens. The HLA-D antigen defined in this way is actually a combination of HLA-DR, DQ and DP.

An important use of the MLC is in its use as a "cellular crossmatch" prior to transplantation, especially bone marrow. By testing the prospective donor and recipient, an *in vitro* transplant model is established which is an extremely significant indicator of possible rejection or graft-versus-host reaction.

Molecular Genetic Techniques

RFLP Restriction fragment length polymorphism (RFLP) methods rely on the ability of certain enzymes to recognize exact DNA nucleotide sequences and to cut the DNA at each of these points. Thus the frequency of a particular sequence will determine the lengths of DNA produced by cutting with a particular enzyme.

The DNA for one HLA class II antigen, e.g. DR15, will have these particular enzyme cutting sites (or restriction sites) at different positions to another antigen, e.g. DR17. So the lengths of DNA seen when DR15 is cut by a particular enzyme, are characteristic of DR15 and different to the sizes of the fragments seen when DR17 is cut by the same enzyme.

Polymerase chain reaction The polymerase chain reaction (PCR) is a recently developed and revolutionary new system for investigating the DNA nucleotide sequence of a particular region of interest in any individual. Very small amounts of DNA can be used as a starting point such that it is theoretically possible to tissue-type using a single hair root. Sequencing DNA has been transformed from a long and laborious exercise to a technique that is essentially automatable in the not too distant future.

The first step in this technique is to obtain DNA from the nuclei of an individual. The double-stranded DNA is then denatured by heat into single-stranded DNA. Oligonucleotide primer sequences are then chosen to flank a region of interest. The oligonucleotide primer is a short segment of complementary DNA which will associate with the single-stranded DNA to act as a starting point for reconstruction of double-stranded DNA at that site.

If the oligonucleotide is chosen to be close to a region of special interest like a hypervariable region of HLA-DRB then the part of the DNA and only that part, will become double-stranded DNA when DNA polymerase and deoxyribonucleotide triphosphates are added. From one copy of DNA, it is thus possible to make two copies. Those two copies can then, in turn, be denatured, reassociate with primers and produce four copies. This cycle can then be repeated until there are sufficient copies of the selected portion of DNA to isolate on a gel and then sequence or type.

There are a number of PCR-based methods in use. Some of them are:

Sequence-specific priming (SSP) In this test, the oligonucleotide primers used to start the PCR have sequences complimentary to known sequences which are characteristic to certain HLA specificities.

The primers which are specific to HLA-DR15, for example, will not be able to instigate the PCR for HLA-DR17. Typing is done by using a set of different PCRs, each with primers specific for different HLA antigens.

Sequence-specific oligonucleotide (SSO) typing By this method, the DNA for a whole region (e.g. HLA-DR gene region) is amplified in the PCR. The amplified DNA is then tested by adding labelled (e.g. radioactive) oligonucleotide probes, which are complementary for DNA sequences, characteristic for certain HLA antigens. These probes will then "type" for the presence of specific DNA sequences of HLA genes.

POINTS TO REMEMBER

- Transplantation is the introduction or replacement of biological material such as organs, tissue, cells and fluids into a living being.
- The tissue compatibility is mainly determined by the major histocompatibility complex (MHC).
- The ABO blood groups were the first to be discovered by Karl Landsteiner in 1900 and is most important in assuring safe blood transfusions.
- There are 3 basic types of recognition which allow the host to know that the transplanted tissue is foreign—recognition by antibody, recognition of foreign MHC by T cells (direct recognition) and recognition of minor histocompatibility loci by T cells (indirect recognition).
- There are three important types of graft rejection according to the same. They are hyperacute rejection, acute rejection and chronic rejection.

REVIEW QUESTIONS

1. Write short notes on:
   i. Syngeneic transplants
   ii. Allogeneic transplants
   iii. Xenogeneic transplants
   iv. H antigen

v. "A" antigen
vi. "B" antigen
vii. Rh antigens
viii. Rhogam
ix. Karl Landsteiner
x. Universal donor and universal recipient
xi. Hyperacute rejection
xii. Acute rejection
xiii. Chronic rejection
xiv. Second set rejection
xv. Minor H antigens
xvi. Graft-vs-host disease (GVHD)
xvii. HLA typing

2. Write a detailed note on the immunology of allograft rejection.

18

TUMOUR IMMUNOLOGY

INTRODUCTION

Cancer is a state of uncontrolled cell growth. In healthy cells, cell growth is regulated by proto-oncogenes and tumour suppressor proteins. Mutations caused by a variety of factors can stimulate the development of proto-oncogenes into oncogenes and can affect the availability of tumour suppressor proteins. Mutations in proto-oncogenes can result in the constant stimulation of the cell growth pathway or the constant activation of growth receptors, even in the absence of stimulating signals. Mutations can also affect tumour suppressor proteins which normally control the stimulatory signals necessary for cell growth.

The theory of immune surveillance says that the immune system continuously recognizes and eliminates tumour cells. However, when a tumour escapes immune surveillance and grows too large for the immune system to kill, it will result in cancer. Immune surveillance is most likely to be successful against virus-induced tumours which express foreign peptides. Tumours vary greatly in their immunogenicity and even tumours with antigens which can be recognized by the host immune system can evade immune elimination.

Immune Surveillance

Immune surveillance refers to the idea that as T cells circulate through the tissues, they recognize and destroy virally infected cells and cancer cells. More recent evidence indicates that immune surveillance is more important against virally infected cells and may be relatively ineffective against malignant cells.

There are many types of cancer. Severity of symptoms depends on the site and character of the malignancy and whether there is metastasis, a process through which cancer cells are transported through the blood or lymphatic system to other parts of the body.

TYPES OF CANCER

Cancers are classified by the type of cell that resembles the tumour and, therefore, the tissue presumed to be the origin of the tumour. The following general categories are usually accepted.

Carcinoma They are malignant tumours derived from epithelial cells. This group represents the most common cancers, including the common forms of breast, prostate, lung and colon cancer.

Lymphoma and leukemia They are malignant tumours derived from blood and bone marrow cells.

Sarcoma They are malignant tumours derived from connective tissue or mesenchymal cells.

Mesothelioma These tumours are derived from the mesothelial cells lining the peritoneum and the pleura.

Glioma These tumours are derived from brain cells.

Germ-cell tumours These tumours are derived from germ cells, normally found in the testicle and ovary.

Choriocarcinoma They are malignant tumours derived from the placenta.

ONCOGENES

They were discovered in cancer-causing viruses. Most oncogenes were actually present in the host cell, where they function in regulated cell growth. The host cell gene was called a **proto-oncogene**. When transduced by the virus and expressed under the control of a viral promoter, the gene product contributes to the unregulated growth of the tumour cell. Since proteins encoded by proto-oncogenes are expressed by normal cells, their over-expression on tumour cells (Table 18.1) would qualify them as tumour-associated antigens.

Table 18.1 Oncogenes

Genes for growth factors and their receptors	
PDGF	Codes for platelet-derived growth factor which is involved in glioma
erb-B	Codes for epidermal growth factor receptor which is involved in glioblastoma and breast cancer
erb-B2	Codes for a growth factor receptor which is involved in breast, salivary gland and ovarian cancers
Genes for cytoplasmic relay in stimulatory signalling pathways	
Ki-ras	Involved in lung, ovarian, colon and pancreatic cancers
N-ras	Involved in leukemias
Genes for transcription factors that activate growth-promoting genes	
c-myc	Involved in leukemias and breast, stomach and lung cancers
N-myc	Involved in neuroblastoma and glioblastoma
L-myc	Involved in lung cancer

(Contd.)

Table 18.1 (Continued)

Genes for other kinds of molecules	
Bcl-2	Codes for a protein that normally blocks cell suicide
Bcl-1	Codes for cyclin D1, a component of the cell cycle clock
MDM2	Codes for an antagonist to the p53 tumour suppressor protein
Tumour Suppressor Genes	
Cytoplasmic Protein Genes	
APC	Involved in colon and stomach cancers
DPC4	Codes for relay molecule that inhibits cell division
NF-1	Codes for protein inhibitor of stimulatory (Ras) protein
NF-2	Involved in meningioma, ependyoma, and schwannoma
Nuclear Protein Genes	
MTS1	Codes for p16 protein, a component of the cell cycle clock
RB	Codes for pRB protein, a component of the cell cycle clock
p53	Codes for p53 protein which can induce apoptosis

TUMOUR ANTIGENS

Two types of antigens have been identified on tumour cells.

1. Tumour-specific transplantation antigens (TSTAs) which are unique to cancer cells.
2. Tumour-associated transplantation antigens (TATAs) which are found on both cancer and normal cells.

However, TATAs may be expressed at increased levels on cancer cells compared to normal cells and thus have been able to play a role in inducing immune responses. Peptides encoded by oncogenes can also serve as tumour markers and are being assessed for their ability to induce an immune response.

Tumour-associated transplantation antigens (TATAs) are tumour antigens that may be present in very low levels in normal cells or in foetal cells. TATAs that are normally present during foetal development but are otherwise expressed only in cancerous cells are called oncofoetal tumour antigens. Alpha-foeto proteins (AFPs) and carcinoembryonic antigens (CEAs) are two examples of oncofoetal tumour antigens that have been associated with various cancers.

TATAs may be recognized as foreign but they are less immunogenic than TSTAs because they are found in some self cells.

TUMOUR-ASSOCIATED ANTIGENS

There are many tumour-associated antigens that are exploited as the tumour markers for the diagnosis of certain kinds of tumour. Tumour markers are molecules occurring in blood or tissue that are associated with cancer and whose measurement or identification is useful in patient diagnosis or clinical management.

Tumour markers can be used for one of the following four purposes.

- Screening a healthy population or a high risk population for the presence of cancer
- Making a diagnosis of cancer or of a specific type of cancer
- Determining the prognosis in a patient
- Monitoring the course in a patient in remission or while receiving surgery, radiation or chemotherapy

Some of the tumour markers are discussed in the following sections.

Carcinoembryonic Antigen

The carcinoembryonic antigen (CEA) was one of the first oncofoetal antigens to be described and exploited clinically.

Carcinoembryonic antigen (CEA) consists of a number of related cell-surface glycoproteins (45 to 55 per cent carbohydrate; MW: 150 to 300 kDa).

Although CEA was first identified in colon cancer, an abnormal CEA level in blood is specific neither for colon cancer nor for malignancy in general. Elevated CEA levels are found in a variety of cancers other than colonic cancer, including pancreatic, gastric, lung, and breast cancers. It is also detected in benign conditions including cirrhosis, inflammatory bowel disease, chronic lung disease, and pancreatitis. The CEA was found to be elevated in up to 19 per cent of smokers and in 3 per cent of a healthy control population. Thus the test for CEA cannot substitute for a pathological diagnosis.

As a screening test, the CEA is also inadequate. Since cancer prevalence in a healthy population is low, an elevated CEA has an unacceptably low positive predictive value, with excess false positives. Also, since elevated CEA occurs in the advanced stage of incurable cancer but is low in early, curable disease, the likelihood of a positive result affecting a patient's survival is diminished.

Alpha-foetoprotein

Alpha-foetoprotein is a normal foetal serum protein synthesized by the liver, yolk sac and gastrointestinal tract that shares sequence homology with albumin. It is a major component of foetal plasma, reaching a peak concentration of 3 mg/ml at 12 weeks of gestation. Following birth, it clears rapidly from the circulation, having a half-life of 3.5 days and its concentration in adult serum is less than 20 ng/ml.

The most widely used biochemical blood test for liver cancer—hepatocellular carcinoma (HCC)—is alpha-foetoprotein. At birth, infants have relatively high levels of AFP, which fall to normal adult levels by the first year of life. Also, pregnant women carrying babies with neural tube defects may have high levels of AFP. (A neural tube defect is an abnormal foetal brain or spinal cord that is caused by folic acid deficiency during pregnancy.)

In adults, high blood levels (over 500 ng/ml) of AFP is seen in only three situations:

- HCC
- Germ-cell tumours (cancer of the testes and ovaries)
- Metastatic cancer in the liver (originating in other organs)

Several assays (tests) for measuring AFP are available. Generally, normal levels of AFP are below 10 ng/ml. Moderate levels of AFP (even almost up to 500 ng/ml) can be seen in patients with chronic hepatitis. Moreover, many patients with various types of acute and chronic liver diseases without documentable HCC can have mild or even moderate elevations of AFP.

CA-125

Cancer antigen-125 (CA-125) is found on the surface of many ovarian cancer cells. It also can be found in other cancers and in small amounts in normal tissue.

CA-125 is an antigen present on 80 per cent of non-mucinous ovarian carcinomas. It is defined by a monoclonal antibody (OC-125) that was generated by immunizing laboratory mice with a cell line established from human ovarian carcinoma. It circulates in the serum of patients with ovarian carcinoma and was therefore investigated for possible use as a marker.

CA-125 is used as a tumour marker, which means some types of cancer are present. Most often CA-125 test is used to check how well treatment for ovarian cancer is working or to see if ovarian cancer has returned.

CA-125 is often elevated in patients with ovarian cancer, its level following the patient's clinical course. With surgical re-section or chemotherapy, the level correlates with patient response. Thus, it is superior to other markers such as CEA.

The CA-125 is elevated in other cancers including endometrial, pancreatic, lung, breast, and colon cancer, and in menstruation, pregnancy, endometriosis and other gynaecological and non-gynaecological conditions.

Because of the low prevalence of ovarian cancer, the test is not useful in screening.

CA19-9

CA19-9 is an oncofoetal antigen, expressed by several different cancers but especially carcinomas of the gastrointestinal tract. Along with CEA, it may be a useful marker to determine prognosis and tumour recurrence.

It is found to be elevated in 21 to 42 per cent of cases of gastric cancer, 20 to 40 per cent of colon cancer and 71 to 93 per cent of pancreatic cancer and has been proposed to differentiate benign from malignant pancreatic disease but this capability remains to be established.

Prostate-specific Antigen (PSA)

Prostate-specific antigen (PSA, also known as kallikrein III, seminin, semenogelase, γ-seminoprotein and P-30 antigen) is a protein manufactured almost exclusively by the prostate. PSA is produced in the ejaculate where it liquifies the semen and allows sperms to "swim" freely. It is also believed to be instrumental in dissolving the cervical mucous cap, allowing the entry of sperm.

Higher-than-normal levels of PSA are associated with both localized and metastatic prostate cancer (PCa). PSA is normally present in the blood at very low levels. Increased levels of PSA may suggest the presence of prostate cancer.

PSA seems to have the capability of achieving at least one of the characteristics of ideal tumour marker-tissue specificity. It is found in normal prostatic epithelium and secretions but not in other tissues. PSA is highly sensitive for the presence of prostatic cancer. The elevation correlates with the stage and tumour volume. It is predictive of recurrence and response to treatment. Finally, the antigen has prognostic value in that patients with very high values prior to surgery are likely to relapse.

Unfortunately, PSA is detectable in normal men and often is elevated in benign prostatic hypertrophy, which may limit its value as a screening tool for prostate cancer. A recent study has shown that PSA combined with rectal examination is a better method of detecting prostate cancer than rectal examination alone.

Human Chorionic Gonadotropin

The human chorionic gonadotropin (hCG) is a glycoprotein composed of 244 amino acids with a molecular mass of 36.7 kDa. It is heterodimeric, with an α (alpha) subunit identical to that of luteinizing hormone (LH), follicle-stimulating hormone (FSH) and thyroid-stimulating hormone (TSH), and β (beta) subunit that is unique to hCG.

The human chorionic gonadotropin (hCG) test is done to measure the amount of the hormone hCG in blood or urine to determine whether a woman is pregnant. hCG is produced by the placenta during pregnancy.

hCG may also be produced abnormally by certain tumours, especially those that develop from an egg or sperm (germ-cell tumours). Therefore, hCG levels are usually tested in a woman who may have cancer of the ovary or abnormal tissue growing in her uterus (molar pregnancy) instead of a normal foetus. In a man, hCG levels may be measured to help and determine whether he has cancer of the testicles.

The level of hCG is occasionally elevated in other cancers including those of breast, lung and gastrointestinal tract, but in these diseases it has found little clinical application.

p53

p53, also known as tumour protein 53 (TP53), is a transcription factor that regulates the cell cycle and hence functions as a tumour suppressor. It is very important for cells in multicellular organisms to suppress cancer. TP53 has been described as "the guardian of the genome", referring to its role in conserving stability by preventing genome mutation.

TP53 is 393 amino acids long and has three domains.

- A domain that activates transcription factors.
- A domain that recognizes specific DNA sequences (core domain). Mutations which deactivate p53 in cancer usually occur here. Contains zinc molecules.
- A domain that is responsible for the tetramerization of the protein. Tetramerization greatly increases the activity of p53 *in vivo*.

The p53 tumour antigen is a protein found in increased amounts in a wide variety of transformed cells. It is also detectable in many proliferating non-transformed cells, but it is undetectable or present at low levels in resting cells. It is frequently mutated or inactivated in many types of cancer.

TAG-72 (Tumour-associated Glycoprotein-72)

It is a high molecular weight glycoprotein (220–400 kDa) complex expressed by a wide variety of human adenocarcinomas. This antigen is expressed by a majority of invasive ductal breast carcinomas and most colon, pancreatic, gastric, oesophageal, lung (non-small cell), ovarian and endometrial cancers and adenocarcinomas. It is not expressed by leukemias, lymphomas, sarcomas, mesotheliomas, melanomas or benign tumours. TAG-72 is also expressed on normal secretory endometrium, but not on other normal tissues.

IMMUNITY AGAINST CANCER

As the cancerous cells are abnormal cells they can elicit immune reaction in the body. NK cells and tumour-specific CTLs can be induced to kill tumour cells *in vitro*, however not exactly *in vivo*.

The following may be the mechanism by which cancer cells can evade the body's immune system.

- Tumour cells may present only self peptides or down-regulate class I MHC expression.
- Tumour cells often lack co-stimulatory molecules like B7 or adhesion molecules that are necessary for them to interact with $CD8^+$ T cells.
- Tumours also shed their tumour antigens or change their structure spontaneously (antigenic variation) to avoid elimination by the immune system.
- Antibodies to tumour surface antigens may promote tumour survival (enhancing antibodies) if they bind without being cytotoxic, hiding the tumour antigens from T cells and inducing the tumour to down-regulate tumour antigen expression.
- Some tumours actively suppress the immune response by producing TGF-β, a suppressive cytokine that inhibits cellular immunity.
- Some tumours, including myeloma and HTLV-1 T-cell leukemia, also produce cytokines that stimulate their own proliferation.

Tumour cells that do have the B7 protein can bind to either CD28 or CTLA-4, both of which are T-cell receptors. The B7 protein on the tumour cell must bind to CD28 in order to activate the host immune system. When $CD8^+$ T cells receive both of these signals, they cause a CTL response to occur. Cytotoxic T lymphocytes release perforin which forms pores in the tumour cell membrane. These pores affect ion concentration within the cancerous cell and provide a port of entry for tumour necrosis factor.

Natural killer (NK) cells and macrophage cells can also be involved in tumour destruction. NK cells are capable of lysing a wide variety of tumour cells because they are not MHC restricted. When the Fc receptor on an NK cell binds to a tumour cell, the NK cell facilitates antibody-dependent cell-mediated cytotoxicity

(ADCC). When macrophages are activated by IFN-γ and macrophage activation factor (MAF), they secrete lytic enzymes into cancerous cells. These lytic enzymes can inhibit tumour growth.

IMMUNOTHERAPY FOR CANCER

Immunotherapy for cancer is essentially the stimulation of the immune system via a variety of reagents such as

- Vaccines
- Infusion of T cells or
- Immunostimulants

They act through any one of the following mechanisms

- by stimulating the anti-tumour response, either by increasing the number of effector cells or by producing one or more soluble mediators such as lymphokines,
- by decreasing suppressor mechanisms,
- by altering tumour cells to increase their immunogenicity and make them more susceptible to immunologic defence or
- by improving tolerance to cytotoxic drugs or radiotherapy, such as stimulating bone marrow function with granulocyte colony-stimulating factor (G-CSF).

Immunotherapy for cancer generally involves conferring either passive or active immunity.

In passive immunity, antibodies and cytotoxic T cells are administered rather than activating the immune system directly. These approaches have met with some success, in spite of some disadvantages. Any element infused this way has a half-life, and thus therapy has to be repeated many times. Here the immunity is not complete. For example, when cytotoxic T cells are introduced, there will not be any expansion *in vivo* with helper T cells.

Active immunity may be ideal for immunotherapy. In cancer, active immunity aims to achieve an endogenous immune response, where the immune system is primed to recognize the tumour as foreign. This approach has not been successful in patients with widespread disease as their immune systems are unable to mount a sufficient response. In the past several years, efforts have focused on using active immune therapies in patients with minimal disease. However, we have seen that the immune system can be quite functional despite advanced stage cancer when the patient has been treated to maximal response. That is, the cancer patient can be vaccinated.

ANTIBODIES AS IMMUNOTHERAPY AGENTS

Antibodies have been the focus of research in the past 20 years or so, as researchers discovered that antibodies recognize and respond to antigens produced by cancer cells. Thus antibodies can elicit the humoral immunity to eliminate the cancerous cells.

A number of monoclonal antibodies show promise against cancer, especially cancers of white blood cells.

Monoclonal antibodies achieve their therapeutic effect through various mechanisms. The important among them are the following.

- They can have direct effects in producing apoptosis or programmed cell death.
- They can have indirect effects.
- They can block growth factor receptors, effectively arresting proliferation of tumour cells.

Indirect effects include recruiting cells that have cytotoxicity, such as monocytes and macrophages. This type of antibody-mediated cell killing is called antibody-dependent cell-mediated cytotoxicity (ADCC). Monoclonal antibodies also bind complement, leading to direct cell toxicity, known as complement-dependent cytotoxicity (CDC).

An example is Rituximab (IDEC-C2B8), a chimeric antibody that targets the CD20 antigen. This antigen is expressed on a significant number of B-cell malignancies. Rituximab is an IgG monoclonal antibody. The Fc fragment of the monoclonal antibody binds the Fc receptors found on monocytes, macrophages and natural killer cells. These cells in turn engulf the bound tumour cell and destroy it. Natural killer cells secrete cytokines that lead to cell death and they also recruit B cells.

Conjugated Monoclonal Antibodies

A major problem with chemotherapy is the damage the drugs cause to all tissues where rapid cell division is going on. Conjugated monoclonal antibodies are those which are joined to drugs, toxins, or radioactive atoms. They are used as delivery vehicles to take those substances directly to the cancer cells. mABs act as a homing device, circulating in the body until it finds a cancer cell with a matching antigen. It delivers the toxic substance to where it is needed most, minimizing damage to normal cells in other parts of the body.

Conjugated mABs are also sometimes referred to as "tagged," "labelled or "loaded" antibodies.

- mABs with chemotherapy drugs attached are generally referred to as **chemolabelled**.
- mABs with radioactive particles attached are referred to as radiolabelled, and this type of therapy is known as **radioimmunotherapy (RIT)**.
- mABs attached to toxins are called **immunotoxins**.

Immunotoxins

Immunotoxins are made by attaching toxins (poisonous substances from plants or bacteria) to monoclonal antibodies. Various immunotoxins have been made by attaching monoclonal antibodies to bacterial toxins such as diphtherial toxin (DT) or pseudomonal exotoxin (PE40) or to plant toxins such as ricin A or saporin.

Improving Monoclonal Antibody (mAB) Therapy

The effectiveness of mAB therapies was limited by the fact that the antibodies were produced by mouse hybridoma cells. In some cases, the patient's immune system recognizes the mouse antibodies as "foreign" after a while and starts destroying them as soon as they enter the body.

For this reason, scientists combine the part of the mouse antibody gene responsible for recognizing a specific tumour antigen with other parts from a human antibody gene. The product of this mouse–human antibody gene is called a "chimeric" or "humanized" monoclonal antibody. It looks more like a normal human antibody, so there is a better chance it will not be destroyed by the patient's own immune system. This means that antibody therapy may still be effective if used more than once.

CANCER THERAPY WITH T CELLS

Another important strategy in immunotherapy is the introduction of T cells. T cells are either cytotoxic ($CD8^+$) or helpers ($CD4^+$). Unlike antibodies which react to intact proteins only, the $CD8^+$ T cells react to peptide antigens expressed on the surface of a cell. Peptide antigens are proteins that have been digested by the cell and presented as protein fragments or peptides and displayed in the MHC. The peptide and the MHC together attract T cells. $CD8^+$ T cells are specific for class I MHC molecules, while $CD4^+$ T cells are specific for class II MHC molecules.

After attaching to the MHC–peptide complex expressed on a cell, the $CD8^+$ T cells destroy the cell by perforating its membrane with enzymes or by triggering an apoptotic or self-destructive pathway. The $CD8^+$ T cells will then move to another cell expressing the same MHC–peptide complex and destroys it as well. In this manner cytotoxic T cells can kill many invasive cells. Ideally, $CD8^+$ T cells could engender a very specific and robust response against tumour cells.

Like the $CD8^+$ T cells, $CD4^+$ T cells also recognize MHC–peptide complexes in the context of class II MHC. $CD4^+$ T cells augment the immune response by secreting cytokines that stimulate either a cytotoxic T cell response (T_H1 helper T cells) or an antibody response (T_H2 helper T cells). These cytokines can initiate B cells to produce antibodies or enhance $CD8^+$ T cell production. The function of the $CD4^+$ T cell depends upon the type of antigen it recognizes and the type of immune response required.

IMMUNOSTIMULANTS

There is considerable evidence that cancer patients do have T cells that are capable of attacking their tumour cells. In fact, it may be that the appearance of cancer is a failure of immune surveillance: the ability of one's own immune system to destroy cancer cells as soon as they appear.

Immunostimulants are non-specific agents that can stimulate the immune system. There have been some successes with

- Injecting adjuvant-like agents directly into the tumour. The only one that succeeds often enough to remain in use is the bacterial preparation, BCG. Introduced into the bladder, it can help eradicate early-stage bladder tumours.
- Oral therapy with levamisole, a drug widely used for deworming, seems to have helped some patients with kidney cancer.
- Interleukin-2 (IL-2), a potent growth factor for T cells.
- Alpha-interferon (IFN-α).

DENDRITIC-CELL VACCINES

Dendritic cells are the most potent antigen-presenting cells. They engulf antigen, process it into peptides and "present" these to T cells.

To make a dendritic cell vaccine, dendritic cells are harvested from the patient. They are exposed *in vitro* to antigens associated with the type of tumour in the patient. The antigens are found in normal as well as cancerous cells of that tissue (e.g. tyrosinase in melanocytes, prostatic acid phosphatase [PAP] in prostate cells). They may be fused with a stimulatory molecule such as granulocyte-macrophage colony-stimulating factor (GM-CSF). These "pulsed" dendritic cells are injected back into the patient. They may elicit a strong cell-mediated immune response, e.g. by cytotoxic T lymphocytes (CTL).

Dendritic cell vaccines have shown some promise against melanoma, prostate cancer, lymphoma.

POINTS TO REMEMBER

- Cancer is a state of uncontrolled cell growth.
- The theory of immune surveillance says that the immune system continually recognizes and eliminates tumour cells.
- Oncogenes are present in the host cell, where they function in regulated cell growth.
- Tumour-specific transplantation antigens (TSTAs) are unique to cancer cells.
- Tumour-associated transplantation antigens (TATAs) are found on both cancer and normal cells.
- Immunotherapy for cancer generally involves conferring either passive or active immunity.

REVIEW QUESTIONS

1. Write short notes on:
   i. Immune surveillance
   ii. Oncogenes
   iii. Tumour-specific transplantation antigens (TSTAs)

iv. Tumour-associated transplantation antigens (TATAs)
v. Carcinoembryonic antigen
vi. Alpha-foetoprotein
vii. CA-125
viii. CA19-9
ix. Prostate-specific antigen
x. Human chorionic gonadotropin

2. Write a detailed note on immunotherapy in cancer.

19

MOLECULAR IMMUNOLOGY

INTRODUCTION

Our immune system is a complex and fascinating system that relies on numerous protein interactions and signalling capabilities in order to develop the required immune response. Many of these pathways require more than one signal before the cascade can begin to ensure that the appropriate response to a given circumstance can be provided.

The manipulation of immune system for the treatment of various diseases is the fantasy for many clinical immunologists. It can be done only by the proper understanding of the molecules involved in the immune response.

CO-STIMULATION

The stimulation of a T-cell antigen receptor is necessary in the immune response, but not sufficient, to induce complete T-cell activation. It does not lead to cell proliferation or cytokine secretion. Complete T-cell activation requires a second signal. This second signal is called as co-stimulation. These so-called co-stimulatory signals depend on the interaction of non-polymorphic proteins and serve to initiate, maintain, and regulate the activation cascade. Studies of co-stimulatory pathways have provided knowledge of immune diseases and opened up new possibilities for prophylaxis and therapy.

Signalling through the T-cell antigen receptor in the absence of a co-stimulatory signal may or may not affect T cells. Sometimes, the T cell simply ignores peptide–MHC complexes presented to it in the absence of co-stimulators. Many times, however, recognition of peptide–MHC complexes in the absence of co-stimulators can induce apoptotic death of the T cell; it can also render the T cell anergic, a condition in which the T cell is refractory (i.e., unable to respond to antigens).

B7 Molecules

One of the best characterized co-stimulatory pathways involves the B7 molecules.

The first signal for T-cell activation is provided by the binding of a naive T cell through its receptor to its specific peptide–MHC complex on an APC. Before T-cell activation can be complete, however, a second, co-stimulatory signal must be provided. This is the role of the B7 molecules. The B7 molecules are found on cells that activate T cells.

When a T cell binds to the peptide–MHC complex, the B7 molecules bind to their ligand, CD28, which is found on the surface of resting and activated T cells. This second signal allows the T cell to be fully activated.

There are two B7 molecules. They are B7-1 (also called CD80) and B7-2 (CD86), both of which are members of the immunoglobulin superfamily.

Both B7-1 and B7-2 are homodimers and each has an extracellular V-like domain and C-like domain. Each also has a transmembrane anchor and a short cytoplasmic tail. The two molecules have very similar structures, although B7-2 is slightly larger with 304 amino acids rather than the 254 amino acids of B7-1. The regions of the B7 molecules used for binding to their ligands also differs. Their extracellular domains have only 27 per cent amino acids in common.

The B7 molecules are primarily found on cells that activate T cells, namely activated dendritic cells, activated B cells and activated macrophages. The B7 molecules are also found constitutively on memory B cells. These cells will start to express the B7 molecules after they engulf the antigen. Once the B7 molecules have begun expression on the cell surfaces, several cytokines contribute to their continued expression. These include BM-CSF in dendritic cells, IFN-γ in macrophages and IL-7 in B cells.

The B7 molecules (when bound to their CD28 ligand of the T cell) provide the co-stimulatory signal necessary for T-cell activation. When a naive T cell binds to its specific peptide–MHC complex, the CD28 on the T cell's surface binds to the B7 molecules on the antigen-presenting cell. These two signals together lead to the production of several cytokines. The importance of B7 co-stimulation is evident in incomplete T-cell stimulation when anti-B7 molecules are present.

An extremely important effect of B7-CD28 signalling is the synthesis of interleukin-2 (IL-2). IL-2 is the cytokine primarily responsible for the proliferation and differentiation of activated T cells. Activated T cells express a high affinity for IL-2 receptors, to which binds the cytokine IL-2. This further activates the T-cell proliferation.

Thus, B7-1 or B7-2 can co-stimulate the proliferation of T cells, the production of interleukin-2 and the expression of interleukin-2 receptors on the cell surface.

To further promote T-cell proliferation, the B7–CD28 signalling pathway also induces production of anti-apoptotic proteins.

The B7–CD28 pathway can co-stimulate both type 1 and type 2 effector CD4 T cells. The B7–CD28 pathway appears to be more important in the generation of type 2 helper T cells than in the generation of type 1 helper T cells.

CTLA-4 (CYTOTOXIC T LYMPHOCYTE-ASSOCIATED ANTIGEN-4)

After the effector function has been carried out, T-cell activation should be down-regulated. Many observations indicate that pathogenic processes occur when negative signals are not functional. Inhibitory receptors identified on T cells include.

- CTLA-4 (Cytotoxic T-lymphocyte-associated antigen-4)
- PD-1 (Programmed death 1) and
- KIRs (Killer inhibitory receptors)

Many reports described negative regulation of cell cycle progression and cytokine production by CTLA-4 engagement, showing that CD28 and CTLA-4 exert opposite effects on T-cell activation (Figure 19.1).

The cytotoxic T lymphocyte-associated antigen-4 gene (CTLA-4) encodes the T-cell-surface molecule. This receptor protein is a specific T lymphocyte surface antigen that is detected on cells only after antigen presentation.

When induced CTLA-4 is transported to an activated T cell's surface, it positively inhibits co-stimulatory signals to activated T cells. The binding of CTLA-4 to B7 molecules presented by activated T lymphocytes creates a modification of co-stimulatory signals that are vital in the immune system.

CTLA-4 shares structural homology with CD28, especially in the extracellular domain containing a conserved sequence motif which allows both receptors to bind the same ligands, even if with a different affinity. Following activation, T cells begin to induce the expression of CTLA-4, which is very similar to CD28 in sequence. CTLA-4 has a higher affinity for B7 molecules. Therefore it competes with CD28 molecules and eventually out-competes CD28 to bind to B7 molecules. CTLA-4 binds to most B7 molecules and effectively sends a negative signal to T cells, which requires the elimination of the proliferative phase (Figure 19.2).

Also, subsequent to T-cell activation, the lymphocyte secretes the cytokine IL-2, which drives proliferation and differentiation of the cell. Activated T cells express a high affinity for IL-2 receptors, which bind to the cytokine IL-2. However, it was discovered that when CTLA-4 is induced, it decreases IL-2 production. The extra cellular domain of CTLA-4 can be fused with the antibody Ig. CTLA-4 Ig is a protein that is induced to block the production of IL-2 and CD28–B7 co-stimulation.

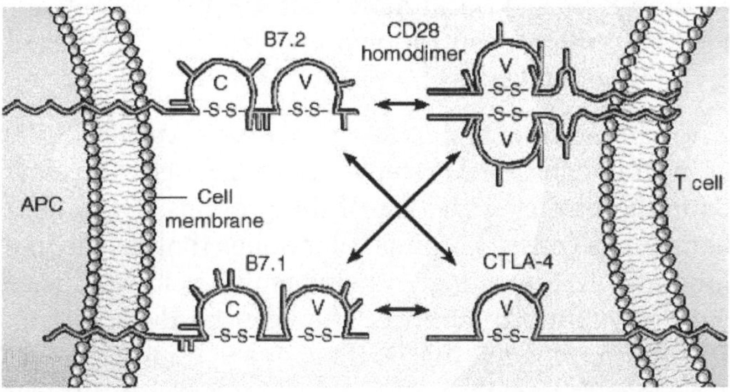

Figure 19.1 Interaction of B7 molecule with CD28 and CTLA-4

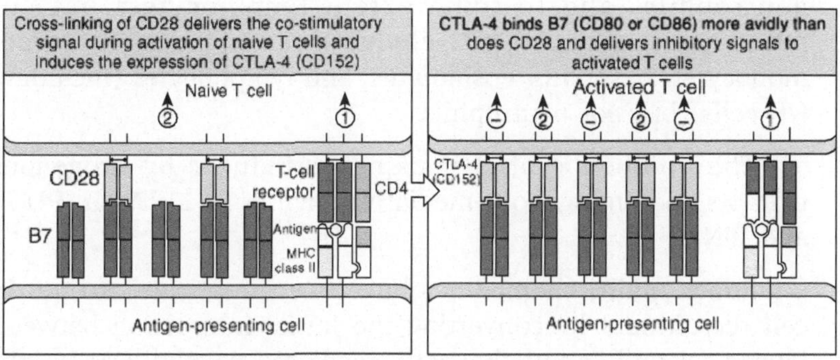

Figure 19.2 The consequence of binding of B7 molecule with CD28 and CTLA-4

Thus, CD28 and CTLA-4 play opposite roles in the tuning of T-cell activation and immune responses. Thus, CD28 and CTLA-4 are directly involved in both immune and autoimmune responses and may be involved in the pathogenesis of multiple T-cell-mediated autoimmune disorders.

CHEMOKINES

Chemokines (shortening of *chemo*attractant cyto*kines*) represent a superfamily of about 30 chemotactic cytokines acting as vital initiators and promulgators of inflammatory reactions. They range from 8 to 11 kDa in molecular weight, are active over a 1 to 100 ng/ml concentration range and are produced by a wide variety of cell types.

The chemokine molecules share structural similarities, including four conserved cysteine residues which form disulphide bonds in the tertiary structure of the proteins. The superfamily of chemokines consists of four subfamilies that display between two and four NH_2-terminal cysteine amino acid residues. Based on these structural differences they can be differentiated into the CXC (or α), the CC (or β), the C (or γ) and the CX3C (or δ) family. Most chemokines known in humans fall into the CXC and CC groups. Most CXC chemokines are chemoattractants for neutrophils (and to some extent lymphocytes) but not monocytes, whereas CC chemokines appear to attract monocytes, basophils, eosinophils, and lymphocytes (including NK cells) but not neutrophils.

The production of chemokines is induced by exogenous irritants and endogenous mediators such as IL-1, TNF-α, PDGF and IFN-γ.

Functionally, chemokines have been implicated to aid in cell recruitment by converting the initial interaction between leucocytes and endothelial cells, mediated through CAMs, into a stable binding and by directing cell migration along a chemokine gradient.

During an inflammatory response to an infectious agent, many different leucocyte populations are recruited to the site. Each population must cross the endothelial barrier and this process is dependent on the presentation of numerous inflammatory chemokines. Different leucocyte populations respond to different chemokines due to differential expression of chemokine receptors. This allows distinct waves of leucocytes to be recruited to specific tissues during an immune response.

The chemokines are not only involved in leucocyte migration. The chemokines have been reported to have other roles.

- They are involved in haematopoietic precursor cell cycling regulation and differentiation.
- They are involved in processes like leucocyte trafficking and inflammatory processes.
- The chemokines are important in a number of disease states. It is now clear that certain CC chemokines, namely RANTES, MCP-1, MCP-3 and MIP-1 exhibit potent pro-migratory and activating potentials for eosinophils, basophils and T cells. These cells are most often associated with respiratory pathologies and allergic disorders. These observations are now being coupled with an emerging body of evidence showing that these mediators can be localized to affected tissues during these pathologies.

IL-8 and MCP-1 are more widely produced than other chemokines (Figure 19.3) and there is a suggestion that they represent the first line of defence.

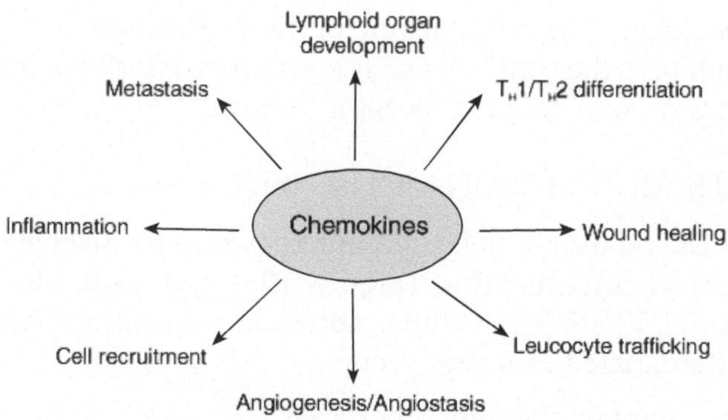

Figure 19.3 Various functions of chemokines

Interleukin-8

It is a polypeptide consisting of 72 amino acids in its mature form. Interleukin-8 (IL-8) belongs to the family of chemokines. It mediates the activation and migration of neutrophils from peripheral blood into tissue and thereby plays a pivotal role in the initiation of inflammation. Thus it is important in inflammatory lung diseases like bronchial asthma or severe infections by respiratory syncytial virus (RSV). IL-8 acts through binding to the IL-8 receptor alpha (IL-8Rα). It is able to induce the full pattern of responses observed in chemotactically stimulated neutrophils, i.e., activation of the motile apparatus and directional migration, expression of surface adhesion molecules, release of lysosomal enzymes and production of reactive oxygen intermediates.

Monocyte Chemoattractant Protein-1 (MCP-1)

It is a chemoattractant for human monocytes. They play a role in the accumulation of monocytes over a period of 24–48 hours after interaction of antigen and sensitized lymphocytes. MCP-1 is nearly as effective as C5a and much more potent than IL-8 in the degranulation of basophils, resulting in histamine release. This may play an important role in the pathogenesis of the late phase of allergic disorders such as atopic food allergies, asthma and chronic urticaria. Histamine release also occurs after stimulation with two other CC chemokines, RANTES and MIP-1α.

Monocyte chemoattractant protein's function is mediated by binding to the CCR2 and CCR4 receptors, which are members of the G protein-coupled receptor family.

CELLULAR ADHESION MOLECULES

Many cell adhesion molecules are known to be involved in the process of inflammation (Figure 19.4 and 19.5). Based on structural differences, cellular adhesion molecules (CAMs) can be differentiated into four groups:

- Immunoglobulin superfamily
- Selectins

- Integrins and
- Cadherins

All of these are involved in lymphocyte recruitment and extravasation.

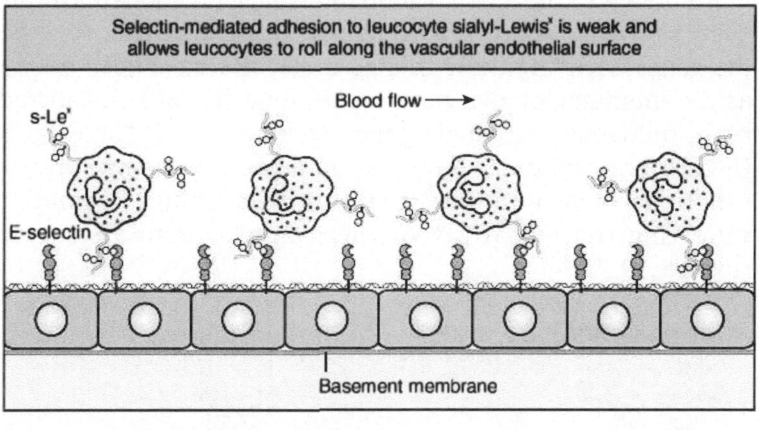

Figure 19.4 Involvement of adhesive molecules in rolling adhesion

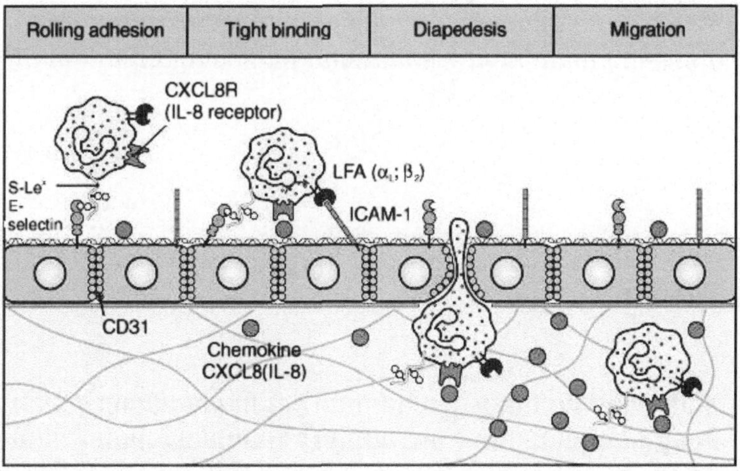

Figure 19.5 Steps involved in movement of cells from the blood capillary to the area of inflammation with the help of adhesion molecules

Immunoglobulin Superfamily

Many of the molecules being involved in the vertebrate immune responses share a common evolutionary precursor—the immunoglobulin homology unit. However, several other molecules with no known immunological functions have also been shown to share this same precursor element. Together, the genes encoding these related molecules have been defined as the immunoglobulin gene superfamily (IgGSF) and include both multigene and single gene representatives. These IgGSF (Figure 19.6) products represent an amazingly diverse array of functions from immune receptors to cartilage formation, reflecting the versatility of the shared common structure (Figure 19.7).

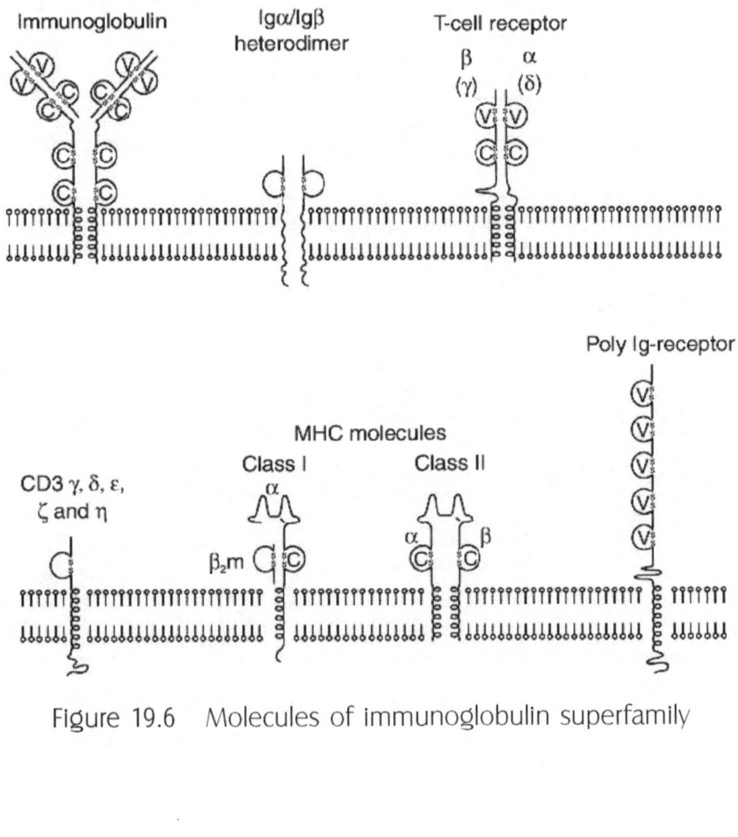

Figure 19.6 Molecules of immunoglobulin superfamily

These molecules can be classified into six categories (Table 19.1).

1. *Non-antigen-presenting, β2-microglobulin-associated molecules* The β2-microglobulin (the light chain of the MHC class I molecule) is a single C homology unit. It is probably divergently related to the MHC class II α chain and may be considered functionally an orphan MHC gene. It is encoded by a single non-polymorphic gene.
2. *T-cell-associated molecules* Besides the TCR, T cells express other accessory molecules that are presumably involved in signal transduction, cell adhesion, and even the facilitation of antigen/MHC targeting. The CD4 and CD8 molecules are accessory molecules of T cells that appear to play an important role in facilitating T-cell interaction with target cells.
3. *Molecules expressed on both T cells and nervous system cells* The Thy-1 molecule, possessing a single V-like homology unit, found in abundance on thymocytes and neurons as well as fibroblasts and a variety of other cells is an example. It may possibly be involved in signal transduction.
4. *Nervous-system-associated molecules* The N-CAM gene encodes five H-type N terminal homology units, a long connecting sequence, a transmembrane region and a very large cytoplasmic domain is an example. It is generally involved in cell-to-cell interaction or adhesion in neuronal morphogenesis. Interestingly, from the number of genes expressed in both the brain and the immune system, the possibility of shared cell-surface recognition functions, as well as the involvement of related molecules in some of the intriguing phenomena linking mental states and immune response, could be hypothesized.
5. *Ig-binding molecules* The poly(Ig) receptor (p-IgR), whose function is to shuttle polymeric IgM and IgA antibodies from the blood side to the serosal side of

mucous membranes is an example for this category. Its external portion is released during the process and is known as the secretory component, SC. The SC molecule has five V homology units. Comparison of the individual units indicates that they are each more closely related to each other than to other IgGSF members. Hence, they are likely to be the product of a series of internal duplication events that resulted in the expansion of a single unit sequence.

6. *Growth factor/kinase receptors* PDGFR and CSF-1R, both possessing H-type homology units are examples. It seems that the IgGSF receptor-kinase coupling took place through an exon shuffling event and then they diverged to generate a family of growth factor receptors.

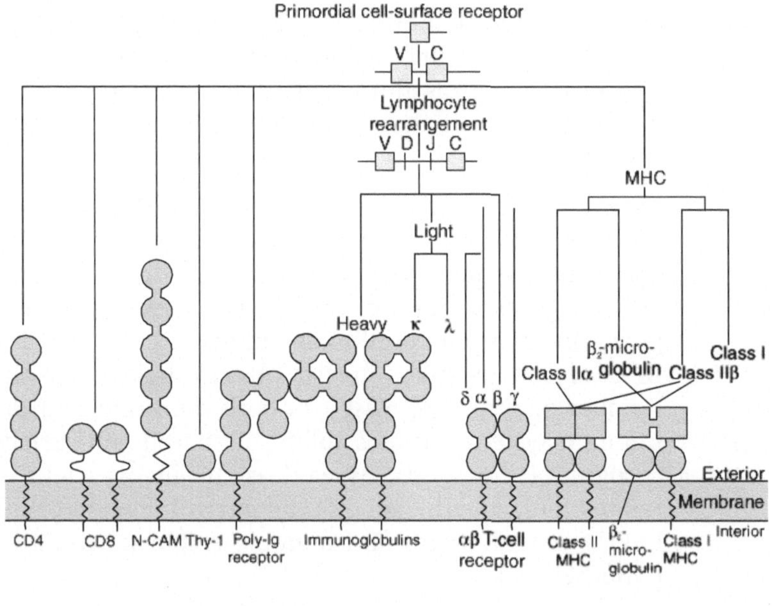

Figure 19.7 An illustration of common origin of immunoglobulin superfamily

Table 19.1 Types of immunoglobulin superfamily molecules

NEURAL-SPECIFIC IgCAMS		
Molecule	**Ligands**	**Distribution**
Adhesion molecule on glia (AMOG)		Glial Neural migration
L1CAM	Axonin	Neural
Myelin-associated glycoprotein (MAG)	MAG	Myelin
Myelin-oligodendrocyte glycoprotein (MOG)		Myelin; Oligodendrocytes
NCAM-1 (CD56)	NCAM-1 via polysialic acid Modulated by sialyl transferase X Polysialyl transferase	Neural cells
NrCAM	Ig superfamily	Neural
OBCAM	Opioids (μ); Acidic lipids	Brain
P_0 protein	P_0	Myelin
PMP-22 protein	PMP-22	Myelin
SynCAM	Ig superfamily	Neural; Synapse
ALCAM (CD166)	CD6, CD166, NgCAM, 35 kD protein	Neural, leucocytes
Basigin (CD147)		Leucocytes, RBCs, platelets, endothelial cells

(Contd.)

Table 19.1 (Continued)

SYSTEMIC IgCAMS		
Molecule	**Ligands**	**Distribution**
BL-CAM (CD22)	Sialylated glycoproteins LCA (CD45)	B lymphocytes
CD44	Hyaluronin, Ankyrin, Fibronectin, MIP1β Osteopontin	Lymphocytes, epithelial, WM perivascular astrocytes, glial tumours (malignant), metastases (CD44v splice variant)
ICAM-1 (CD54)	αLβ2, LFA-1	Leucocytes, endothelial cells, dendritic cells, fibroblasts, epithelium, synovial cells
ICAM-2 (CD102)	αLβ2 (LFA-1)	Endothelial cells, lymphocytes; monocytes
ICAM-3 (CD50)	αLβ2	Leucocytes
Lymphocyte function antigen-2 (LFA-2) (CD2)	LFA-3	Lymphocytes, thymocytes
LFA-3 (CD58)	LFA-2	Leucocytes, stromal endothelial cells, astrocytoma

(Contd.)

Table 19.1 (Continued)

SYSTEMIC IgCAMS		
Molecule	**Ligands**	**Distribution**
Major histocompatibility complex (MHC) molecules		
MAdCAM-1	α4β7, L-selectin	Mucosal endothelial cells
PECAM (CD31)	CD31, αvβ3	Leucocytes, synovial cells, endothelial cells
T-cell receptor (C-region)		
VCAM-1	α4β1, α4β7	Satellite cells, monocytes, synovial cells, activated endothelial cells

Selectins

Selectins are a family of transmembrane molecules, expressed on the membrane surface of leucocytes and activated endothelial cells. They belong to the C-type (calcium-dependent) lectin family and a similarity in the domain organization (and homology in amino acid sequence) allows them to be recognized as a family of cell–cell adhesion molecules. They have a single C-type lectin domain (L) at their extracellular amino termini, followed by an epidermal growth factor (EGF)-like domain (E), two to nine consensus repeats (CR), transmembrane domain and a short cytoplasmic tail (Figure 19.8).

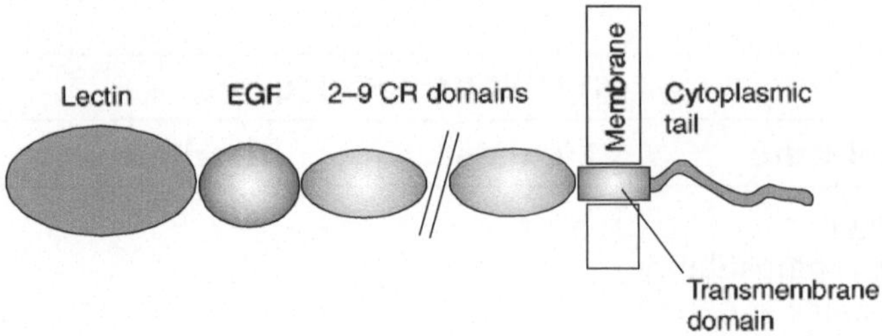

Figure 19.8 Basic structure of selectin molecule

Selectins contain an N-terminal extracellular domain with structural homology to calcium-dependent lectins, followed by a domain homologous to epidermal growth factor and two to nine consensus repeats (CR) similar to sequences found in complement regulatory proteins. Each of these adhesion receptors is inserted via a hydrophobic transmembrane domain and possesses a short cytoplasmic tail. The initial attachment of leucocytes, during inflammation, from the bloodstream is afforded by the selectin family and causes a slow downstream movement of leucocytes along the endothelium via transient, reversible, adhesive interactions called leucocyte roll. Each of the three selectins (Figure 19.9) can mediate leucocyte rolling given the appropriate conditions.

Three types of selectins (Table 19.2) have been discovered so far. They are

1. L-selectins which are generally expressed on almost all leucocytes,
2. E-selectins which are inducible on vascular endothelium upon stimulation with cytokines due to transcriptional activation and
3. P-selectins which were originally found on activated platelets; however, their expression is also induced on activated vascular endothelium.

Table 19.2 Types of selectin molecules

Molecule	Ligands (Receptor)	Distribution
L-selectin (CD62L)	Sulphated: GlyCAM-1; CD34; MAdCAM-1	Leucocytes (Homing receptor)
E-selectin (CD62e)	Tetrasaccharides-Sialyl-Lewis^x; Sialyl-Lewis, a cutaneous lymphocyte-associated antigen E-selectin ligand-1 (ESL-1)	Endothelial cells
P-selectin (CD62P)	Tetrasaccharides- Sialyl-Lewis^x P-selectin glycoprotein ligand-1	Endothelial cells Platelets

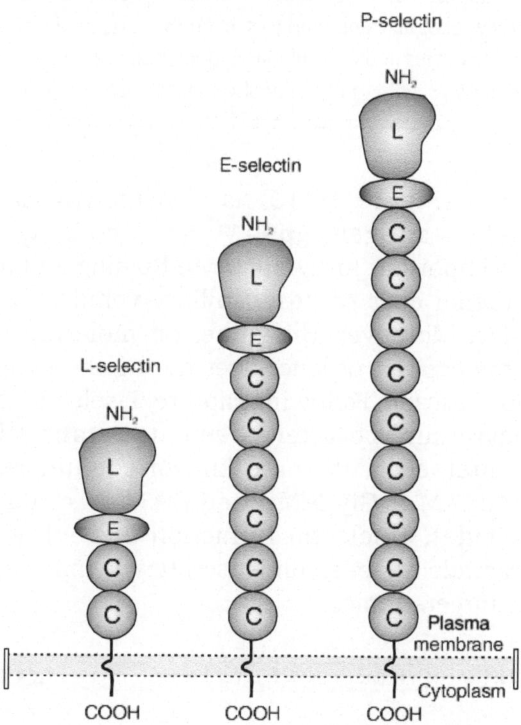

Figure 19.9 Structure of various selectin molecules

L-selectins (Leucocyte selectins)

It is the smallest of the vascular selectins (74–100-kDa molecule). It is expressed at the tips of microfolds on granulocytes, monocytes and a vast array of circulating lymphocytes.

Sialyl-Lewis^x [NeuNAc-alpha-2, 3-Gal-beta-1, 4-(Fuc-alpha-1, 3) GlcNAc]

Sialyl-Lewis^x (sLex), the ligand for selectins, is the fucosylated carbohydrate. Sialyl Lewis^x binds to a trio of proteins called selectins (E, L and P). Recognition in this case typically occurs when body tissues are damaged or injured in some way. When damage occurs, selectins appear on the surfaces of cells lining nearby blood vessels. There they recognize and bind to sLe^x on the surface of white blood cells or leucocytes in the circulating bloodstream, thereby recruiting these defensive cells to fight infection at the site of injury.

L-selectin (Figure 19.10) is also known as LECAM-1, LAM-1, Mel-14 antigen, gp90^{mel}, and Leu8/TQ-1 antigen. L-selectin is important for lymphocyte homing and adhesion to high endothelial cells of post-capillary venules of peripheral lymph nodes. Moreover, this adhesion molecule contributes greatly to the capture of leucocytes during the early phases of the adhesion cascade. Following capture, L-selectin is shed from the leucocyte surface after chemoattractant stimulation. L-selectin interacts with three known counter receptors or ligands, MAdCAM-1, GlyCAM-1 and CD34. In conjunction with other molecules, L-selectin's function and influence in the adhesion cascade has been under scrutiny in many experiments using gene-targeted mice.

Figure 19.10 Structure of L-selectin

P-selectins (Platelet selectins)

It is the largest of the known selectins (140-kDa molecule). It contains nine consensus repeats (CR) and extends approximately 40 nm from the endothelial surface. Other names for P-selectin include CD62P, granule membrane protein 140 (GMP-140), and platelet activation-dependent granule to external membrane protein (PADGEM). P-selectin is expressed in α-granules of activated platelets and granules of endothelial cells.

Within minutes of stimulation of the endothelial cells by inflammatory mediators such as histamine, thrombin or phorbol esters, P-selectin is expressed on the surface. The expression is short-lived, reaching its peak only after ten minutes. Additional synthesis of P-selectin is brought about within two hours by cytokines such as interleukin-1 (IL-1) or tumour necrosis factor- α (TNF-α).

> The sLex-bearing glycoprotein that binds to P-selectin is PSGL-1, a 220-kDa protein with sugar chains projecting in a bottle-brush fashion along its length. But P-selectin does not attach just anywhere along this length. For binding to occur, P-selectin requires not only the presence of sLe^x but also a particular stretch of peptide carrying a sulphated tyrosine residue. L-selectin is somewhat less choosy, while no one yet knows for sure whether E-selectin binds directly to the sLe^x structure.

The primary ligand for P-selectin (Figure 19.11) is PSGL-1 (P-selectin glycoprotein ligand-1) which is found on all leucocytes. Other ligands for P-selectin include CD24 and uncharacterized ligands.

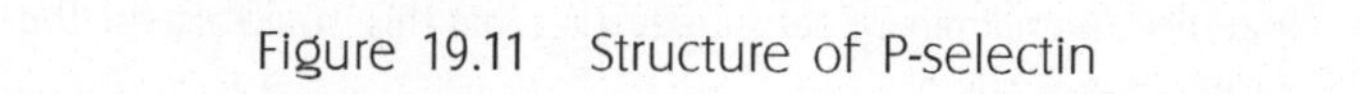

Figure 19.11 Structure of P-selectin

The transient interactions between P-selectin and PSGL-1 allow leucocytes to roll along the venular endothelium.

Accordingly, P-selectin is largely responsible for the rolling phase of the leucocyte adhesion cascade. P-selectin can also mediate capture when L-selectin is not present.

E-selectin (Endothelial selectins)

It is expressed on inflamed endothelial cells in response to treatment with inflammatory cytokines. It functions in mediating leucocyte rolling along with the P-selectin.

In addition to mediating leucocyte rolling, E-selectin (Figure 19.12) participates in the conversion of rolling to firm adhesion.

Figure 19.12 Structure of E-selectin

E-selectin is expressed in skin microvessels under baseline conditions and there is some evidence that E-selectin is of importance in skin inflammation, because it supports the recruitment of skin-specific T lymphocytes.

High endothelial venules (HEV)

High endothelial venules (HEV) lined by the high endothelium are the sites where leucocytes enter into the lymph nodes from the blood. Lymphocyte homing into lymph nodes is organ-selective, i.e., different molecules are involved in the lymphocyte homing to peripheral nodes compared with mucosa-associated lymphoid tissue. The traffic into peripheral nodes is regulated by the expression of L-selectin on leucocytes and its ligand on HEVs. The ligand for L-selectin is suggested to be a 50-, 90- or 105-kDa glycoprotein, which is sulphated, fucosylated, and sialylated. The two other members of the selectin family (E- and P-selectin) recognize sialyl-Lewis^x and -Lewis^a (sLe^x and sLe^a, respectively) carbohydrate motifs and there is preliminary data suggesting that this would also be the case for L-selectin.

Integrins

Integrins are cell-surface receptors. They mediate adhesion to the extracellular matrix (ECM) and cell–cell interactions. Many cells express several integrins (Figure 19.13) that recognize a range of cell-surface-associated and ECM-associated ligands. Individual integrins often bind more than one ligand. They play multiple roles in differentiation and cell communication.

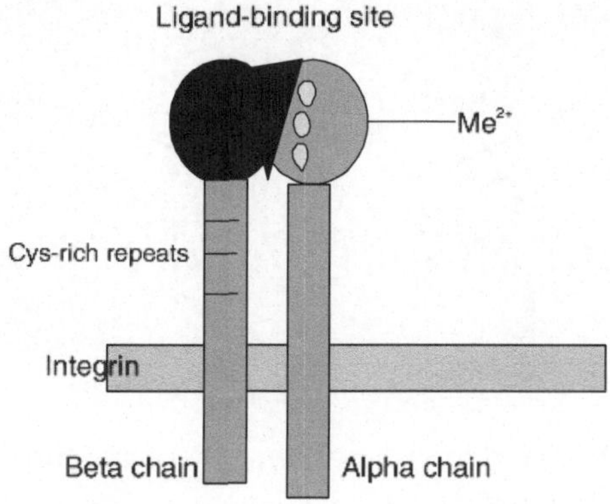

Figure 19.13 Structure of integrin

Functional integrins consist of two transmembrane glycoprotein subunits that are non-covalently bound. These subunits are called alpha and beta. The alpha subunits all have some homology to each other, as do the beta subunits. The receptors always contain one alpha chain and one beta chain and are thus called heterodimeric. Both the subunits contribute to the binding of ligand. Till now, 16 alpha and eight beta subunits have been identified. From these subunits, some 22 integrins are formed in nature, which implies that not all possible combinations exist. The beta-4 subunit for instance can only form a heterodimer with the alpha-6 subunit. On the other hand, the beta-1 subunit can form heterodimers with ten different alpha subunits (Figure 19.14). As not all the beta-1 alpha heterodimers have the same ligand specificities, it is believed that the alpha chain is at least partly involved in the ligand specificity. The diversity of the integrins is increased by the alternative splicing of some integrin messenger RNAs.

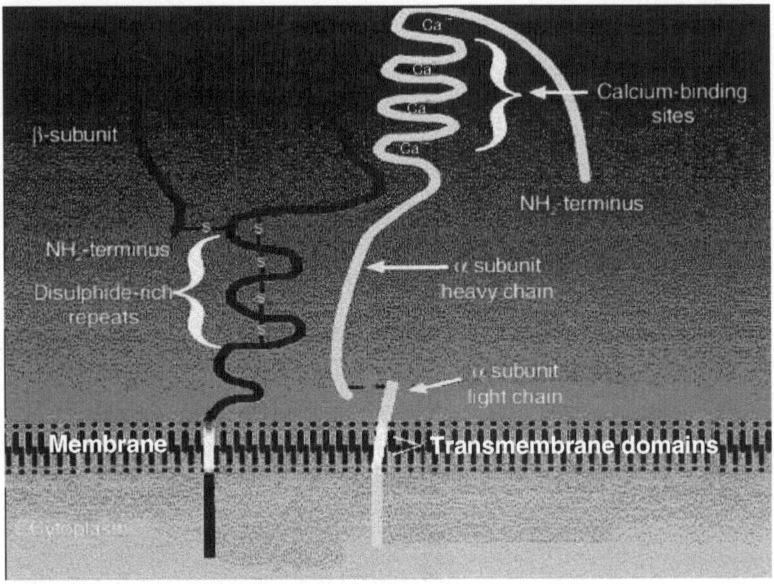

Figure 19.14 Two-dimensional structure of integrin showing various domains

The structure of the alpha subunits is very similar. All contain 7 homologous repeats of 30–40 amino acids in their extracellular domain, spaced by stretches of 20–30 amino acids. The three or four repeats that are mostly extracellular, contain sequences with cation-binding properties. These sequences are thought to be involved in the binding of ligands, because the interaction of integrins with their ligand is cation-dependent. All the alpha subunits share the 5 amino acid motif GFFKR, which is located directly under the transmembrane region. The precise function of this motif is not known yet, but a protein called calreticulin has been isolated on a column loaded with this fragment. Also an association between calreticulin and integrins (alpha-3 and alpha-5) has been demonstrated. The protein may play a role in the assembly of integrin dimers. The alpha subunits are subdivided into two groups based on some structural differences. The first group is formed by alpha-1, alpha-2, alpha-L, alpha-M and alpha-X. The members of the first group all contain a so-called inserted domain (I-domain). This domain of about 180 amino acids is situated between the second and the third repeat. The function of this domain is not known, but it has been speculated that it is involved in ligand binding because the domain has a resemblance with domains found in von Willebrand factor which is important for its binding to collagen for example. The second group is formed by alpha-3, alpha-5, alpha-6, alpha-7, alpha-8, alpha-IIb, alpha-V and alpha-IEL. Members of this group all share a post-translational cleavage of their precursor into a heavy and a light chain. The light chain is composed of the cytoplasmic domain, the transmembrane region and a part of the extracellular domain (about 25 kDa), while the heavy chain contains the rest on the extracellular domain (about 120 kDa).

Some of the subunits are expressed exclusively on one type of cell, for example:

- Beta 2 on leucocytes—lymphocytes can interact via integrins with proteoglycans and glycoseaminoglycans which are the major components of ECM. Such interactions are important for lymphocyte migration, recognition, activation and differentiation.
- Integrins—alpha3 beta1 and alpha6 beta4 are abundant receptors on keratinocytes for laminin-5 which is the major component of the basement membrane in the skin.

Integrins participate in cell–cell adhesion and are of great importance in binding and interactions of cells with components of the extracellular matrix such as fibronectin and laminin. Importantly, integrins facilitate communication between the cytoskeleton and extracellular matrix, allowing each to influence the orientation and structure of the other. It is clear that interactions of integrins with the extracellular matrix can have profound effects on cell function, and events such as clustering of integrins activates a number of intracellular signalling pathways. Another feature of integrins is that they exist in "active" or "inactive" states. For example, a group of integrins responsible for binding of white blood cells to endothelium are normally inactive, allowing the blood cells to circulate freely, but become activated in response to inflammatory mediators, resulting in the white cells being "pulled" from blood into inflamed tissues. It follows that deficits in expression of certain integrins can result in diseases characterized by abnormal inflammatory responses (Table 19.3).

Many integrins recognize the sequence RGD (Arg-Gly-Asp).

Table 19.3 Types of integrins and their properties

Molecule	Ligands	Distribution
$\alpha1\beta1$	Laminin, collagen tenascin	NK, B and activated T cells, fibroblasts, glial perineurium, Schwann cells, endothelium
$\alpha2\beta1$	Laminin, collagen	NK, B and activated T cells, platelets, endothelial, fibroblasts, epithelium, astrocytes, Schwann cells, ependymal cells
$\alpha3\beta1$	Laminin, collagen, fibronectin	Activated T cells; thymocytes; endothelium; fibroblasts; epithelium; astrocytes
$\alpha4\beta1$	$\alpha4\beta1$, $\alpha4\beta7$, fibronectin, VCAM-1, MAdCAM-1, TSP-1	NK, B and T cells, eosinophils, endothelium, muscle, fibroblasts, neural-crest derived *Function* T cell transendothelial migration
$\alpha5\beta1$	Fibronectin, murine L1	Activated B and T cells, memory T-cells, thymocytes, fibroblasts, epithelium, platelets, endothelium, astrocytes

(Contd.)

Table 19.3 (Continued)

Molecule	Ligands	Distribution
α6β1	Laminin	Leucocytes, thymocytes, epithelium, T cells (memory and activated) glial, fibroblasts, endothelium
α7β1	Laminin	Skeletal and cardiac muscle, melanoma
α8β1 VLA-8	Fibronectin, vitronectin tenascin, common form	Epithelium, neurons, oligodendroglia
α9β1		Epithelium (airway), muscle
αvβ1	Vitronectin, fibronectin, collagen von Willebrand factor, fibrinogen	Oligodendroglia
αLβ2 (LFA-1α; CD11a)	ICAM-1, ICAM-2, ICAM-3	Leucocytes, thymocytes, macrophages, T cells, microglia
αMβ2 (CD11b)	ICAM-1, Factor X, iC3b, fibrinogen	Myeloid, B cells (activated), NK cells, macrophages, microglia, B leukemic cells
αXβ2 (CD11c)	iC3b, fibrinogen	Myeloid, dendritic cells, B cells (activated), macrophages, microglia, B leukemic cells

(Contd.)

Table 19.3 (Continued)

Molecule	Ligands	Distribution
αIIbβ3	Fibronectin, vitronectin, von Willebrand's factor, thrombospondin	Platelets
αvβ3 (CD51/CD 61)	Fibronectin, Osteopontin, von Willebrand's factor, PE-CAM-1 vitronectin, fibrinogen, human L1 thrombospondin, collagen	B cells (activated), T cells (activated and γδ), endothelium, monocytes, tumours, glia, Schwann cells, endothelium
α6β4	Laminin	Schwann cells, perineurium endothelium, epithelium, fibroblasts not in immune cells
αvβ5	Vitronectin, fibronectin, fibrinogen	Fibroblasts, monocytes, macrophages, epithelium, oliogodendroglia, tumours
αvβ6	Fibronectin	
αvβ8	Fibronectin	Oligodendroglia, Schwann cells, brain synapses
α4β7	Fibronectin, VCAM-1, MAdCAM-1	NK, B and T cells Not in neural cells
13αIELβ7 (CD103)	E-cadherin	Intraepithelial T lymphocytes (IEL) (Intestinal)
α11		Uterus, heart, skeletal muscle, smooth muscle containing tissues

Cadherins

Cadherins (Figure 19.15) are integral transmembrane glycoproteins which mediate Ca^{2+}-dependent cell–cell adhesion in most tissues. The three most common cadherins are neural (N)-cadherin, placental (P)-cadherin, and epithelial (E)-cadherin. All three belong to the classical cadherin subfamily. There are also desmosomal cadherins and proto-cadherins.

Figure 19.15 Structure of cadherins

The cadherins are synthesized as precursor polypeptides which require a series of post-translational modifications (glycosylation, phosphorylation and proteolytic cleavage) to form a protein which is between 723 and 748 amino acids long. The extracellular domain contains 3–5 internal repeats of approximately 110 amino acids.

The N-terminal has 113 amino acids which contain a conserved HAV sequence that has been shown to be important in ligand binding and specificity. The extracellular domain is anchored to the cell membrane by a transmembrane domain of approximately 24 amino acids. The short cytoplasmic domain is the most highly conserved region of homology between cadherins and is particularly important for cadherin function.

Desmosomes

Desmosomes are localized patches that hold two cells tightly together. They are common in epithelia (e.g. the skin). Desmosomes are attached to intermediate filaments of keratin in the cytoplasm.

Cadherins (Figure 19.16) generally mediate monotypic cell–cell adhesion, although heterotypic binding between different cadherin molecules is possible. They act as both receptor and ligand. They are responsible for the selective cell–cell adhesion or cell sorting which is necessary to allocate different cell types to their proper positions during development. They also play a fundamental role in maintaining the integrity of multicellular structures. During embryonic morphogenesis, the expression of multiple members of the cadherin family is spatio-temporally regulated, and correlates with a variety of morphogenetic events that involve cell aggregation or disaggregation.

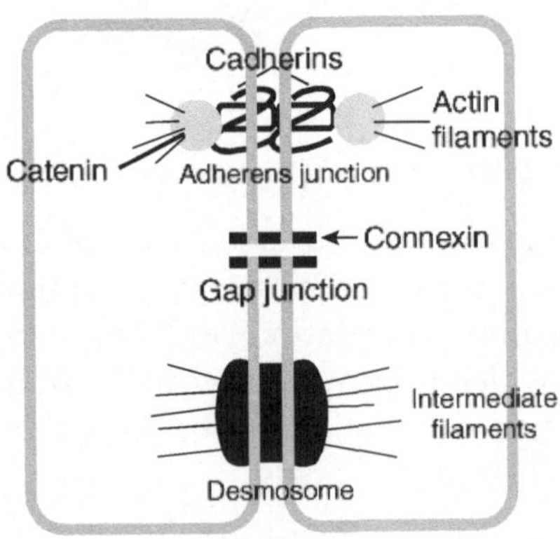

Figure 19.16 Cell junction showing various molecules associated with it, including cadherins

It is crucial for the mutual association of vertebrate cells. For example in epithelial cells, E-cadherins stabilize the junction by interacting with E-cadherin molecules from same adjacent cells. This interaction is called as homophilic adhesion. E-cadherin adhesion system is impaired in cancer cells (Figure 19.17). Thus cancer cells can spread throughout the body.

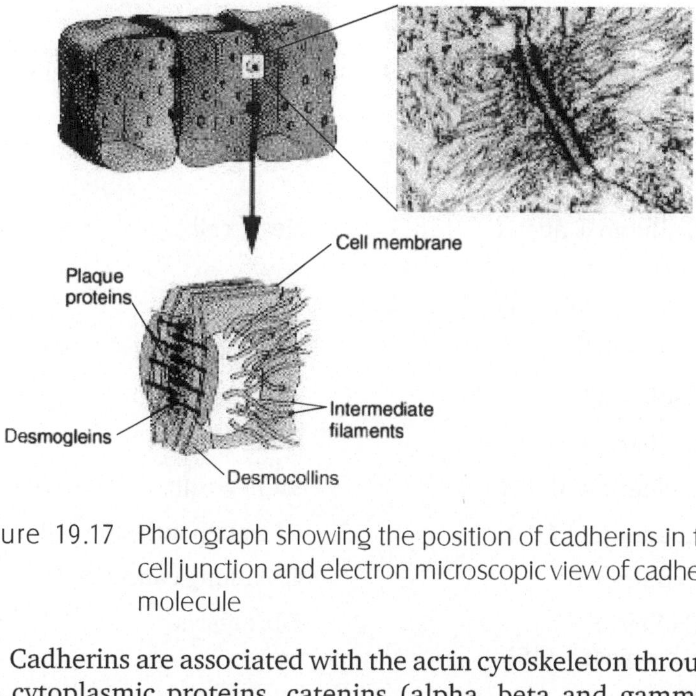

Figure 19.17 Photograph showing the position of cadherins in the cell junction and electron microscopic view of cadherin molecule

Cadherins are associated with the actin cytoskeleton through the cytoplasmic proteins, catenins (alpha, beta and gamma). The gamma-catenin is also called as *plakoglobin*. The beta-catenin or gamma-catenin is associated directly with the cadherin, and alpha-catenin is bound to beta or gamma catenin and to the actin cytoskeleton.

The anchorage of cadherins (Table 19.4) to the cytoskeleton appears to be regulated by tyrosine phosphorylation.

Table 19.4 Types of cadherin molecule and its properties

Molecule	Ligands	Distribution
Cadherin E (1)	**H**	Epithelial cell
Cadherin N (2)	**O**	Neural cell
Cadherin BR (12)	**M**	Brain cell
Cadherin P (3)	**O**	Placental cell
Cadherin R (4)	**P**	Retinal cell
Cadherin M (15)	**H**	Muscle, activated satellite cells
Cadherin VE (5) (CD144)	**I**	Endothelial cell, brain cell
Cadherin T and H (13)	**L**	Heart cell
Cadherin OB (11)	**I**	Osteoblast
Cadherin K (6)	**C**	Brain; kidney cell
Cadherin 7		
Cadherin 8		Brain cell
Cadherin KSP (16)		Kidney cell
Cadherin LI (17)		GI tract pancreatic cell
Cadherin 18		CNS, lung cancer cell
Cadherin, Fibroblast 1 (19)		Fibroblasts
Cadherin, Fibroblast 2 (20)		Fibroblasts
Cadherin, Fibroblast 3 (21)		Fibroblasts
Cadherin 23		Ear
Desmocollin 1		Skin

(Contd.)

Table 19.4 (Continued)

Molecule	Ligands	Distribution
Desmocollin 2		Epithelium, mucosa myocardium lymph nodes
Desmoglein 1		Epidermis, tongue
Desmoglein 2		All
Desmoglein 3		Epidermis, tongue, antibody target in pemphigus
Protocadherin 1, 2, 3, 7, 8, 9		Brain cells

T-CELL RECEPTOR (TCR) SIGNALLING

One of the first steps in the generation of the immune response is the recognition by T lymphocytes of peptide fragments (antigens) derived from foreign pathogens that are presented on the surface of antigen-presenting cells (APC). This event is mediated by the T-cell receptor (TCR), which transduces these extracellular signals by initiating a wide array of intracellular signalling pathways (Figure 19.18).

One of the first biochemical events following TCR activation is the activation of Src family tyrosine kinases.

Src was the first proto-oncogenic protein described. Since then, many other tyrosine kinases have been identified and characterized. Src remains the "grandfather" non-receptor tyrosine kinase and serves as a prototype for understanding the function and regulation of other tyrosine kinases.

The Src family is composed of nine members in vertebrates: Src, Yes, Fgr, Yrk, Fyn, Lyn, Hck, Lck and Blk.

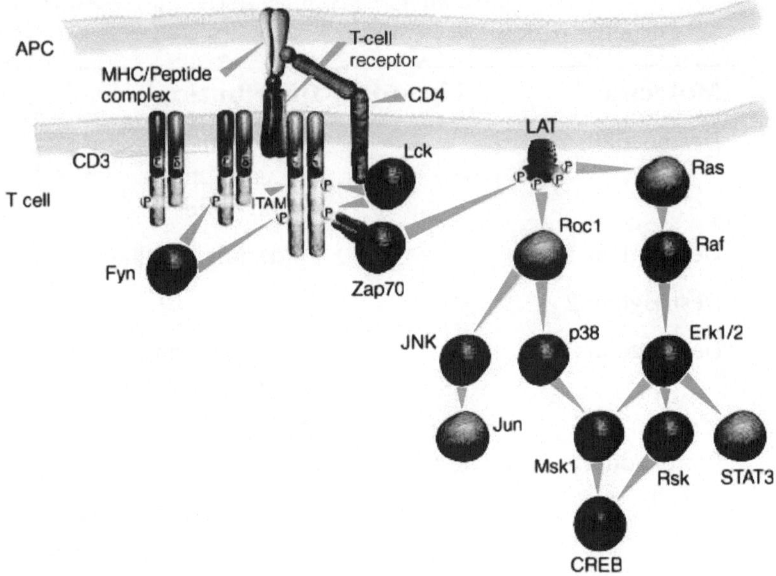

Figure 19.18 TCR signalling pathway

One of the key steps that initiate receptor activation is the tyrosine phosphorylation of TCR subunits by the Src family protein kinases Lck and Fyn.

T-cell receptor phosphorylated by Lck and Fyn recruits the ZAP-70 protein kinase to the receptor complex, which becomes activated and stimulates downstream pathways including an increase in intracellular free calcium, PLCG1, PKC, NFkB and Ras-MAPK activation. An increased calcium level activates the calcium-binding molecule calmodulin, which binds and activates other signalling molecules.

These pathways activate transcription factors such as AP-1, NFAT, and Rel which ultimately lead to the expression of genes that control cellular proliferation, differentiation, anergy or apoptosis.

The CD45 protein tyrosine phosphatase activates Lck and Fyn by dephosphorylating these proteins and is required for TCR activation. Association of the CD4 co-receptor with Lck may also be involved in Lck activation.

MAP Kinase/Erk Pathway

Mitogen-activated protein kinases (MAP kinases) are a group of protein serine/threonine kinases that are activated in response to a variety of extracellular stimuli and mediate signal transduction from the cell surface to the nucleus. In combination with several other signalling pathways, they can differentially alter phosphorylation status of the transcription factors. A controlled regulation of these cascades is involved in cell proliferation and differentiation, whereas an unregulated activation of these MAP kinases can result in oncogenesis.

Phospholipase C (PLC) Signalling Pathway

PLC comes in multiple forms and plays a key role in the signal transduction process for many receptors. Its main function is to hydrolyse PIP_2 (phosphatidylinositol 4, 5-bisphosphate) into DAG (diacylglycerol) and IP_3 (inositol trisphosphate). DAG is necessary for further activation of PKC while IP_3 leads to the release of intracellular calcium ions.

NF-κB

The eukaryotic nuclear factor κB (NF-κB) plays an important role in inflammation, autoimmune response, cell proliferation, and apoptosis by regulating the expression of genes involved in these processes. Five members of the NF-κB family have been identified: NF-κB1 (p50/p105), NF-κB2 (p52/p100), RelA (p65), RelB, and c-Rel. They share a highly conserved Rel homology domain (RHD), which is responsible for DNA binding, dimerization, and interaction with IκB. The p50/RelA (p65) heterodimer is the major Rel/NF-κB complex in most cells. RelB can act both as a transcriptional activator and as a repressor of NF-κB-dependent gene expression.

Protein Kinase C (PKC)

PKC was originally identified as a serine/threonine kinase that was maximally active in the presence of diacylglycerols (DAG) and calcium ion. It is now known that there are at least ten proteins of the PKC family. Each of these enzymes exhibits specific patterns of tissue expression and activation by lipid and calcium. PKCs are involved in the signal transduction pathways initiated by certain hormones, growth factors and neurotransmitters. The phosphorylation of various proteins, by PKC, can lead to either increased or decreased activity. Of particular importance is the phosphorylation of the EGF receptor (epidermal growth factor receptor) by PKC which down-regulates the tyrosine kinase activity of the receptor. This effectively limits the length of the cellular responses initiated through the EGF receptor.

POINTS TO REMEMBER

- Complete T-cell activation requires a second signal. This second signal is called as co-stimulation.
- One of the best characterized co-stimulatory pathways involves the B7 molecules.
- There are two B7 molecules. They are B7-1 (also called CD80) and B7-2 (CD86), both of which are members of the immunoglobulin superfamily.
- The binding of CTLA-4 to B7 molecules presented by activated T lymphocytes creates a modification of co-stimulatory signals that are vital in the immune system.
- Chemokines have been implicated to aid in cell recruitment by converting the initial interaction between leucocytes and endothelial cells.
- Selectins are a family of transmembrane molecules, expressed on the membrane surface of leucocytes and activated endothelial cells.
- Integrins mediate adhesion to the extracellular matrix (ECM) and cell–cell interactions.
- Cadherins are integral transmembrane glycoproteins which mediate Ca^{2+}-dependent cell–cell adhesion in most tissues.

REVIEW QUESTIONS

1. Write short notes on:
   i. Co-stimulation
   ii. B7 molecules
   iii. CTLA-4 (Cytotoxic T-lymphocyte-associated antigen-4)
   iv. Chemokines
   v. Immunoglobulin superfamily
   vi. Selectins
   vii. Integrins
   viii. Cadherins
2. Write a detailed note on the role of adhesion molecules in immune response.
3. Write in detail on TCR signalling pathway.

20

IMMUNOLOGICAL TOLERANCE

INTRODUCTION

Immunological tolerance is the failure to mount an immune response to an antigen. It is a state of unresponsiveness to a specific antigen or group of antigens to which a person is normally responsive. Immunological tolerance is achieved under conditions that suppress the immune reaction and is not just the absence of an immune response. Thus immunological tolerance is not simply a failure to recognize an antigen but it is an active response to a particular epitope and is just as **specific** as an immune response.

Both B cells and T cells can be made tolerant, but it is more important to make the T cells tolerant than the B cells because B cells cannot make antibodies to most antigens without the help of T cells.

The immunological tolerance can be

- **Natural** or "**self**" **tolerance** which is the failure of the immune system to attack the body's own proteins and other antigens. If the immune system responds to "self", an **autoimmune disease** may result.
- **Induced tolerance** which is tolerance to external antigens that has been created by deliberately manipulating the immune system. It is done

1. to protect allergic reactions due to various allergens,
2. to enable transplanted organs (e.g. kidney, heart, liver) to survive in their new host and
3. to reveal the mechanisms of autoimmunity in the hope of designing treatments for diseases like systemic lupus erythematosus (SLE) and multiple sclerosis (MS).

Immunological tolerance may be of two kinds. They are:

- **Central tolerance** This occurs during lymphocyte development.
- **Peripheral tolerance** This occurs after lymphocytes leave the primary organs.

T-CELL TOLERANCE

Central Tolerance

T cells develop in the thymus. As they mature, recombination of gene segments creates the two chains that make up the T-cell receptor (TCR) for antigen. Although the receptors on a single T cell are all alike, there is a virtually unlimited repertoire of receptor specificities created in the population of T cells within the thymus. In the thymus, the T cells undergo a phenomenon called as negative selection. The T cells will encounter the body's own protein along with the self MHC.

In the thymus, the epitopes recognized by the T-cell receptors consist of a small molecule, usually a peptide of 6–8 amino acids derived from body proteins, i.e., "self" proteins such as proteins within the cytosol or serum proteins, i.e., proteins circulating in the blood and lymph.

T cells whose receptors bind these epitopes along with self MHC molecules with high affinity will be destroyed by apoptosis. Thus the T cells having the receptors specific for the self antigens will be eliminated leading to tolerance. The T cells that survive this **negative selection** leave the thymus and migrate throughout the immune system (lymph nodes, spleen, etc.).

Peripheral Tolerance

The T cells that leave the thymus may have receptors (TCRs) that can respond to self antigens.

However, it is clear that there are mechanisms for maintaining T-cell tolerance throughout the body. What is not so clear is how many and how important each is.

The following are some possibilities by which the T cell may develop peripheral tolerance.

Lack of Co-stimulation

As it has been already discussed, the binding of a T cell to an antigen-presenting cell (APC) is by the T-cell receptor which is not enough to activate the T cell. It needs co-stimulation elicited by the co-stimulatory molecules (Figure 20.1).

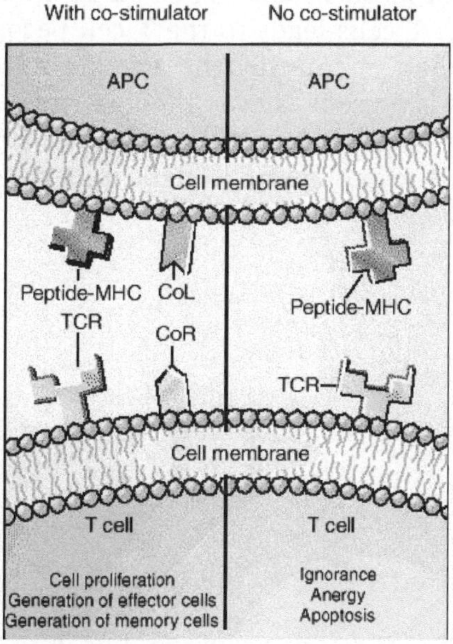

Figure 20.1 Co-stimulation leading to the production of effector T cells and lack of co-stimulation leading to anergy or apoptosis of T cells

Among the most important of these co-stimulators are molecules on the APC designated as B7 and their ligands on the T cell designated as CD28. The binding of CD28 to B7 provides the second signal needed to activate the T cell.

Although T cells may encounter self antigens in body tissues, they will not respond unless they receive a second signal. In fact, binding of their TCR ("signal one") without "signal two" causes them to self-destruct by apoptosis. Most of the time, the cells presenting the body's own antigens fail to provide second signal and self-tolerance results.

Further the expression of the co-stimulatory molecules is restricted so that most tissue cells lack either B7.1/B7.2 or CD40 or both. Such cells also normally lack class II MHC molecules (Figure 20.2). Thus tissue cells normally will present a spectrum of peptides from their endogenously synthesized proteins on self MHC class I in the absence of co-stimulation. Interaction of such cells with T cells leads to the T cell becoming refractive. Thus when they encounter the same antigen even when co-stimulation is present, they will not respond to the stimulus. This refractory state is termed anergy (Figure 20.3). The anergic cells are unable to proliferate in the absence of IL-2 and fail to secrete cytokines.

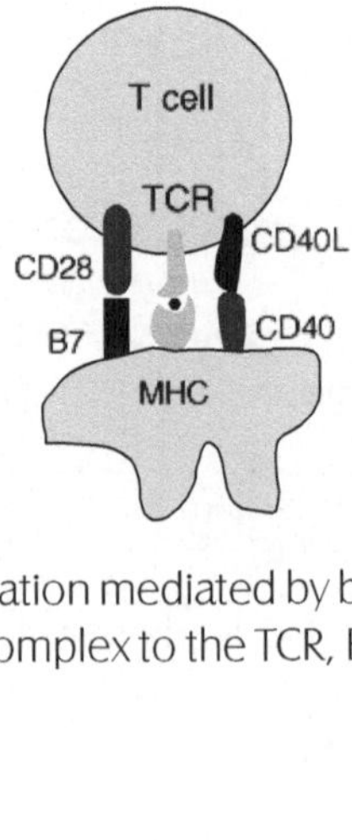

Figure 20.2 T-cell activation mediated by binding of MHC molecule–antigen complex to the TCR, B7 to CD28 and CD40 to its ligand

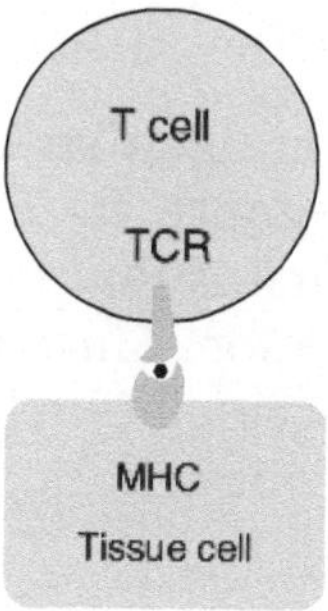

Figure 20.3 Attachment of MHC–antigen complex with TCR without co-stimulation

NFAT (Nuclear factor of activated T cells)

The ubiquitous transcription factor NFAT (nuclear factor of activated T cells) is associated with other proteins to bind DNA and induce genes responsible for cell–cell interactions. NFAT is expressed in a variety of lymphocytes like T cells, B cells, natural killer (NK) cells, monocytes and non-immune-related cells like muscle, cardiac, and neuronal. NFAT transcriptional activity is modulated by cytoplasmic Ca^{2+} concentration through various Ca^{2+}-associated signalling pathways. Increase in cytoplasmic Ca^{2+} concentration induce NFAT dephosphorylation and NFAT translocation to the nucleus where it binds to *cis* regulatory elements of target genes as a monomer.

NFAT activation regulates a variety of immune processes. They are apoptosis, anergy, T-cell development, and ageing of the immune system. In particular, NFAT isoforms are responsible for regulating IL-2, IL-3, IL-4, IL-5, granulocyte macrophage colony-stimulation factor (GM-CSF), IFN-γ, tumour necrosis factor (TNF) alpha and the cell-surface receptors CD40L, CTLA-4 and FasL expression.

The process of T-cell anergy, which renders cells unresponsive to antigens, is important for the prevention of autoimmunity. Central to this process is signalling through Ca^{2+} and calcineurin to the transcription factor NFAT, resulting in the up-regulation of a set of anergy-associated genes.

Ignorance

Even if T cells of autoantigen specificity is present in the circulation they may ignore it. This may arise due to 2 different reasons. The first is that the antigen may simply be present in low concentration. Since all T lymphocytes have a threshold for receptor occupancy which is required to trigger a response, very low concentration of antigen will not be sensed.

The second possibility is more interesting. Some antigens are sequestered from the immune system in locations which are not freely exposed to surveillance. These are termed immunologically privileged sites and the antigens are called sequestrated antigens. Examples of such sites are the eye, CNS and testis.

Receipt of Death Signals

Some cells of the body express the Fas ligand, FasL. Activated T cells always express Fas. When they encounter these cells, binding of Fas to FasL triggers their death by apoptosis. Some examples are given below.

- Cells within the eye always express FasL and are thus ready to kill off any rogue T cells that might gain entry.
- Macrophages infected with HIV express FasL and thus kill any anti-HIV T cells that try to kill them. This may account for the disastrous decline in $CD4^+$ T cells late in the development of AIDS.

FasL is a 40-kDa type II membrane protein belonging to the tumour necrosis factor (TNF) family. It is expressed on activated lymphocytes, NK cells, platelets, certain immune-privileged cells and some tumour cells. The expression of FasL on either a neighbouring cell, or on the Fas-bearing cell, induces trimerization of Fas, which then initiates a signal-transduction cascade, leading to apoptosis of the Fas-bearing cell.

Role of tolerance by IL-10

IL-10 induces the down-regulation of MHC class II and decreases the expression of intercellular adhesion molecule 1 (ICAM-1), CD80 and CD86 on APCs, each of which might be expected to prevent T-cell priming and promote "tolerance".

IL-10 also inhibits IL-1α and IL-1β, IL-6, IL-8 and TNF-α. In addition, chemokines such as macrophage inflammatory protein 1α (MIP-1α) and growth factors such as granulocyte colony-stimulating factor (G-CSF) and granulocyte-macrophage colony-stimulating factor (GM-CSF) from activated monocytes and macrophages are also inhibited and IL-10 counter-regulates the production of IL-12.

B-CELL TOLERANCE

The problem of B-cell tolerance is not so acute because B cells cannot respond to most antigens unless they receive help from T helper cells. However B cells too become tolerant to self components and, like T cells, this occurs both in the bone marrow (central tolerance) and elsewhere in the body (peripheral tolerance) (Figure 20.4).

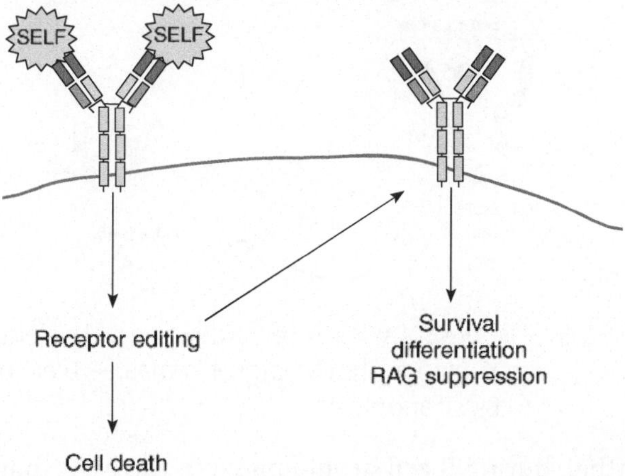

Figure 20.4 Mechanism of central tolerance by B cells

Central Tolerance

B cells are formed and mature in the bone marrow. In humans, over half of the developing B cells produce a BCR able to bind self components.

Any cell that produces a receptor for an antigen (BCR) that would bind self components too tightly undergoes a process of receptor editing.

Once a B cell produces an antigen receptor, it is normally prevented from further rearrangement of the heavy and light chain sequences (allelic exclusion). In the process of receptor editing, however, a B cell re-expresses the RAG proteins and can then produce alternate light chain sequences. Replaced light chains are paired with the existing heavy chain and the modified BCR is once again subjected to antigen selection. If receptor editing results in a BCR unresponsive to self antigen, the B cell continues along the development pathway. If receptor editing results in a different BCR that is still autoreactive, rearrangement of the light chain locus will continue. Autoreactive B cells which cannot re-express their RAG proteins will be deleted by apoptosis (Figure 20.5).

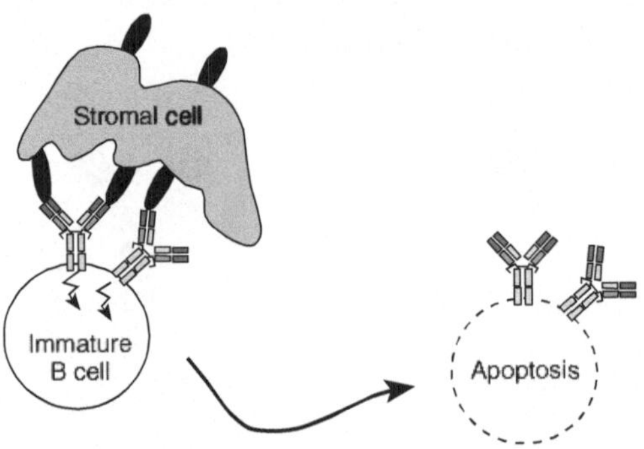

Figure 20.5 The body's self cell behaving as a polyvalent antigen, cross-binding the receptor of immature B cell and leading to B-cell apoptosis

Further, during B cell development in the bone marrow, the complete antigen receptor (IgM) is first expressed on "immature"

B cells. If those cells encounter their target antigen in a form which can cross-link their IgM, then such cells are programmed to die and will be deleted from the repertoire. It was shown that injection of a polyclonal anti-IgM from birth prevented the development of B cells, resulting in a "B-cell-less mouse". The requirement for cross-linking means that the antigen has to be polyvalent, the most obvious example of this being cell-surface molecules.

Despite these mechanisms, some of the B cells that migrate out of the bone marrow continue to express self-reactive BCRs and may still be able to produce anti-self antibodies. So a mechanism is needed to make them tolerant out in the tissues (peripheral tolerance).

Peripheral Tolerance

B cells with a potential for attacking self can be kept in check by the absence of the T helper cells they need. T-cell tolerance is probably the most important mechanism for maintaining B-cell tolerance.

Suppressor Cells

Both low and high doses of antigen may induce suppressor T cells which can specifically suppress immune responses of both B and T cells, both directly or by production of cytokines, most importantly, TGF-beta and IL-10.

T-suppressor cell, T8 cell

The existence of these cells is a relatively recent discovery and hence their functioning is still somewhat debated. The basic concept of suppressor T cells is a cell-type that specifically suppresses the action of other cells in the immune system, notably B cells and T cells, thereby preventing the establishment of an immune response. How this is done is not known with certainty, but it seems that certain specific antigens can stimulate the activation of the suppressor T cells. Discrete epitopes have been found, that display suppressor activity on killer T cells, T helper cells and B cells. This suppressor effect is thought to be mediated by some inhibitory factor, secreted by suppressor T cells. It is not any of the known lymphokines. A fact that renders the study of this cell-type difficult is the lack of a specific surface marker. Most suppressor T cells are CD8-positive as are cytotoxic T cells.

Anti-idiotype Antibody

Anti-idiotype antibodies produced experimentally have been demonstrated to inhibit immune response to specific antigens. Anti-idiotype antibodies are produced during the process of tolerance and such antibodies have been demonstrated in tolerant animals. These antibodies prevent the receptor from combining with antigen.

Anergy of B Cells

There is a mechanism for B-cell tolerance to soluble autoantigens if they are present at sufficiently high concentration. It is apparent that the critical parameter is receptor (surface Ig) occupancy. When more than 5 per cent of the sIgM molecules are normally occupied by monomeric soluble antigen, the B cell becomes anergic. This anergic state can be recognized in the case of B cells by the down-regulation of surface IgM.

If an anergized, self-reactive B cell subsequently encounters a T cell specific for a self peptide from the relevant autoantigen, Fas ligand on the T-cell surface binds to Fas on the B cell, inducing apoptosis in the B cell. If an anergized B cell takes up a foreign antigen as a result of immunological cross-reactivity or formation of complexes between foreign and self molecules, this mechanism aborts activation of the self-reactive B cell by T cells that recognize the foreign moiety.

Note that the level of surface IgD remains unaffected and the precise explanation for why such B cells are refractory to stimulation even when T-cell help is available is not known.

POINTS TO REMEMBER

- Immunological tolerance is the failure to mount an immune response to an antigen.
- Immunological tolerance can be: natural or "self" tolerance and induced tolerance.
- Immunological tolerance may be of two kinds. They are: central tolerance and peripheral tolerance.
- Central tolerance occurs during lymphocyte development.
- Peripheral tolerance occurs after the lymphocytes leave the primary organs.

REVIEW QUESTIONS

1. Write short notes on:
   - i. Natural or "self" tolerance
   - ii. Induced tolerance
   - iii. T-cell tolerance
   - iv. B-cell tolerance

21

MONOCLONAL ANTIBODIES

INTRODUCTION

It is known that the B cells are responsible for antibody production. Each B cell in an organism synthesizes only one kind of antibody. In an organism, there is an entire population of different types of B cells and their respective antibodies that were produced in response to the various antigens that the organism had been exposed to.

For many diagnostic and therapeutic purposes, the immunologists are in need of antibodies of same specificities. This can be achieved when a particular B cell is cloned into multiple copies of plasma cells so that they will produce only one kind of antibodies. Thus the antibodies obtained from a single clone of B cell are called as monoclonal antibodies. In contrast, antibodies obtained from the blood of an immunized animal are called polyclonal antibodies as the antibodies are the outcome of multiple clones of various B cells.

The production of monoclonal antibodies were pioneered by Georges Köhler and Caesar Milstein in 1975.

Georges Köhler (1946–1995)

Georges Köhler was born in Munich in 1946. He joined the University of Freiburg in 1965 to study Biology. In 1971, he received a Diploma in Biology for his work on repair-deficient strains of *Escherichia coli* and in 1974 he obtained a Ph. D. for immunological studies on the enzyme β-galactosidase under Fritz Melchers.

In 1974, he joined Caesar Milstein's group at the Laboratory of Molecular Biology, and in 1975, they published in *Nature* the results of their work leading to the production of monoclonal antibodies as "Continuous cultures of fused cells secreting antibody of predefined specificity."

In 1976, he became a member of the Basel Institute for Immunology, working on lymphocyte hybrids.

Georges Köhler died of inhaling smoke and flames in a lab accident in 1995.

Caesar Milstein (1927–)

Born in Bahía Blanca, Argentina, Caesar Milstein graduated from Buenos Aires University with an undergraduate degree in chemistry in 1945. From there he went to England and joined Frederick Sanger's lab at Cambridge, where he received his Ph.D. in 1960. In 1961, he returned to Argentina to become Head of the Division of Molecular Biology at the National Institute of Microbiology. When dozens of faculty members were dismissed following the military coup, Milstein resigned in protest and returned to Cambridge. He joined the staff of the Medical Research Council at the Laboratory of Molecular Biology in 1963.

Milstein conducted groundbreaking work into the synthesis of antibodies, proteins that are produced by the cells of the immune system in response to attacks by foreign bodies called antigens. His work was instrumental in the development of monoclonal antibody technology. By fusing antibody-producing B lymphocyte cells with tumour cells that are "immortal", his lab was able to produce a "hybridoma", which could continuously synthesize antibodies. All of the antibodies produced by this type of hybridoma were identical, the same as those produced by the B cell before it was fused.

Because the antibodies that are produced by this process all come from a single clone of hybridoma cells, they are called monoclonal antibodies. This technique of monoclonal antibody production, developed in 1975 with Georges Köhler, has been used extensively in the commercial development of new drugs and diagnostic tests. For his efforts, Milstein was awarded the Nobel Prize for Physiology or Medicine with Georges Köhler and Niels Jerne in 1984.

PRODUCTION OF MONOCLONAL ANTIBODIES

A sample of B cells is extracted from the spleen of the mouse and added to a culture of myeloma cells (tumour cells) (Figure 21.1). The intended result is the formation of hybridomas, cells formed by the fusion of a B cell and a myeloma cell. The fusion is done by using polyethylene glycol, a virus or by electroporation. Thus this technique is called as hybridoma technology.

The antigen for immunization should be prepared. At least 500 micrograms of antigen should be available for ordinary intraperitoneal immunization.

Balb/c ByJ female mice should be immunized by the antigen. Immunization is done with the antigen in Freund's or other appropriate adjuvant. Serum samples should be collected 2–3 times during the course of the immunization regimen for determination of antibody titre. An ELISA screening assay can be developed to determine serum titres and for later use in screening hybridomas.

In most cases, the spleens from 2 mice will be fused, with the additional mice held in reserve and used for a second fusion one or two weeks later. Spleen cells are fused with Sp2/0 myeloma cells using polyethylene glycol (PEG) or Sendai virus as the fusion agent.

The myeloma cells will lack a specific enzyme (hypoxanthine-guanine phosphoribosyl transferase, HGPRT) and therefore cannot grow under certain conditions.

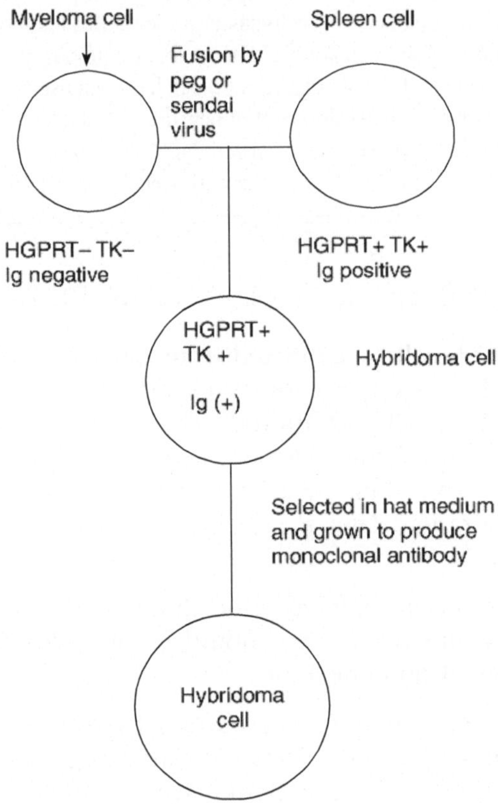

Figure 21.1 Strategy of fusion of myeloma cell and spleen cell and selection of hybridoma cell

Hypoxanthine-guanine phosphoribosyl transferase (HGPRT) is an enzyme in purine metabolism. It converts hypoxanthine to inosine monophosphate, adenosine monophosphate to adenine and xanthine monophosphate to xanthine.

The enzyme primarily functions to salvage purines from degraded DNA to renewed purine synthesis. In this role, it acts as a catalyst in the reaction between guanine and phosphoribosyl pyrophosphate (PRPP) to form GMP.

HAT Medium (Hypoxanthine–Aminopterin–Thymidine medium) is a medium used for preparation of monoclonal antibodies. The myeloma cells are HGPRT– and the B cells are HGPRT+. HGPRT and thymidine kinase are important enzymes for the salvage pathway and therefore HGPRT– myeloma cells cannot synthesize nucleotides via salvage pathway. Fused cells are incubated in the HAT medium. Aminopterin in the medium blocks the de novo pathway. Hence, unfused myeloma cells die, as they cannot produce nucleotides by de novo or salvage pathway. Unfused B cells die as they have a short lifespan. Hybridoma cells (produced by successful fusions) are able to grow indefinitely because the spleen cell partner supplies HGPRT and the myeloma partner is immortal.

In this way, only the B cell-myeloma hybrids survive. These cells produce antibodies (a property of B cells) and are immortal (a property of myeloma cells). The incubated medium is then diluted into multiwell plates to such an extent that each well contains only 1 cell. Then the supernatant in each well can be checked for desired antibody. Since the antibodies in a well are produced by the same B cell, they will be directed towards the same epitope and are known as monoclonal antibodies (Figure 21.2).

Hybridoma cultures can be maintained indefinitely as follows:

- *in vitro*, i.e., in culture vessels. Expand the antibody-secreting cultures by transferring them to progressively larger culture vessels (bioreactors). Once large cultures are established, cyropreserve several vials from each hybridoma cell line. The yield ranges from 10–60 μg/ml.

- *in vivo*, i.e., growing in mice as ascites. Here the antibody concentration in the serum and other body fluids can reach 1–10 mg/ml.

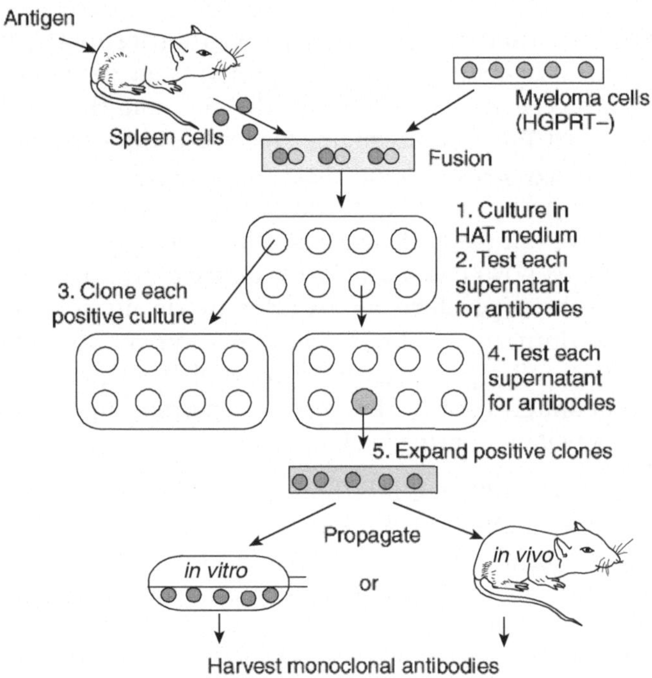

Figure 21.2 Overall methodology of monoclonal antibody production

CHIMERIC OR HUMANIZED MONOCLONAL ANTIBODIES

With an isolated monoclonal antibody, one can easily clone the immunoglobulin genes that encode it and genetically tailor the monoclonal antibody for a particular use. Because many monoclonal antibodies are generated in rodent species, the immune system can recognize the rodent antibody as foreign and mount a potent response against it. As a result of this issue, there is considerable interest to "humanize these antibodies" by replacing the rodent immunoglobulin structures with human counterparts, perhaps allowing the reagent to be ignored by a human immune system. The anti-TNF antibody, infliximab for example, has very little mouse protein, although antibodies are still generated against it.

APPLICATIONS OF MONOCLONAL ANTIBODIES

Generally, monoclonal antibodies are used as invaluable reagents in diagnostics. They have played a major role in deciphering the functions of various biomolecules in cryptic biosynthetic pathways. These have also become the reagents of choice for identification and characterization of tumour-specific antigens and have become a valuable tool in the classification of cancer. Some of the important applications are given below:

Identification of Cell-surface Markers

Monoclonal antibodies are used in defining cell-surface structures such as CD molecules and HLA molecules on the surface of various cells involved in immune response. Monoclonal antibodies that define CD molecules and flow cytometry have been used to identify the normal components of the immune system and to determine if these components are under-represented (in the case of immunodeficiency diseases) or over-produced (in some cancers). It is well known that T helper cells are important for normal immune function. Monoclonal antibodies reactive with the CD4 molecule expressed on T helper cells were used to demonstrate that a decrease in CD4 cells is a feature of AIDS and the levels can be used to stage the disease. Thus, monoclonal antibodies have proven in understanding, diagnosis and management of disease-related immune system.

Detection Assays

Monoclonal antibodies have also been extensively used in the design of sensitive detection immunoassay techniques such as ELISA. Monoclonal antibodies increase the specificity of many diagnostic tests.

Purification Techniques

Another area in which monoclonal antibodies are extensively used is the isolation and purification of molecules. A given monoclonal antibody can be coupled to an insoluble surface and used to affinity-purify the molecule of interest.

Gene Identification

The field of molecular genomics has benefited from the availability of monoclonal antibody technology. A known monoclonal antibody that recognizes a molecule of interest can be used to identify its gene. Alternatively, if one has a newly identified gene with an unknown function, monoclonal antibodies can be generated against the predicted protein encoded by the gene. These reagents can then be used for expression and function studies. These avenues also open up new windows of investigative research. For example, one can use monoclonal antibodies to determine if the gene is abnormally expressed in certain disease states or has a different structure in different individuals.

Disease Therapy

Monoclonal antibodies against bioactive cytokines have been used in therapy for many immunology-based disease processes such as the control of transplantation rejection and the modulation of autoimmune diseases. Further, monoclonal antibodies are utilized for cancer therapy.

POINTS TO REMEMBER

- The antibodies obtained from a single clone of B cell are called as monoclonal antibodies.
- The production of monoclonal antibodies was pioneered by Georges Köhler and Caesar Milstein in 1975.
- B cells are extracted from the spleen of the mouse and added to a culture of myeloma cells (cancer cells). The intended result is the formation of hybridomas, cells formed by the fusion of a B cell and a myeloma cell. This technology is called as hybridoma technology.
- Monoclonal antibodies are used as invaluable reagents in diagnostics.

REVIEW QUESTIONS

1. Write short notes on:
   - i. Polyclonal antibodies
   - ii. Hypoxanthine-guanine phosphoribosyl transferase (HGPRT)
   - iii. HAT Medium (Hypoxanthine–Aminopterin–Thymidine medium)
   - iv. Chimeric or humanized monoclonal antibodies
   - v. Applications of monoclonal antibodies
2. Write a detailed account of hybridoma technology.

22

ANTIGEN–ANTIBODY REACTIONS

INTRODUCTION

It is known that a homologous antigen and antibody react in a specific manner. Interactions between antigen and antibody involve non-covalent binding of an antigenic determinant site of an antigen (epitope) to the variable region (complementarity determining region, CDR) of both the heavy and light immunoglobulin chains. The attachment almost resembles the lock-and-key mechanism of enzyme–substrate attachment.

The bonds that hold the antigen in the antibody-combining site are non-covalent in nature. These include hydrogen bonds, electrostatic bonds, van der Waals forces and hydrophobic bonds. Multiple bonding between the antigen and the antibody ensures that the antigen will be bound tightly to the antibody. However, antigen–antibody reaction is a reversible reaction as it involves only non-covalent bonds. It follows the basic thermodynamic principles of any reversible bimolecular interaction.

$$K_A = \frac{[Ab - Ag]}{[Ab][Ag]}$$

where,

K_A is the affinity constant,

Ab and Ag are the molar concentrations of unoccupied binding sites on the antibody or antigen respectively and

Ab–Ag is the molar concentration of the antibody–antigen complex.

CROSS REACTIVITY

Cross reactivity (Figure 22.1) refers to the ability of an individual antibody-combining site to react with more than one antigenic determinant site of the antigen. Cross reactivity can occur when two or more antigens share similar structural features.

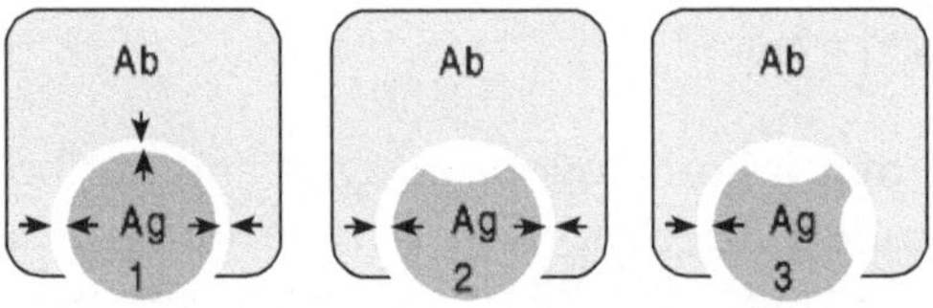

Figure 22.1 Specificity and cross reactivity of different antigens with same antigen-binding site of antibody. Ag1 is highly specific, Ag2 is less specific and Ag3 is a cross reacting antigen

Consider three different antigens—Ag1, Ag2 and Ag3. Antibody produced in response to Ag1 is very specific and would therefore have a large affinity when combining with Ag1. However, Ag2 is similar in shape to Ag1 and is capable of interacting with anti-Ag1 antibody via two of three sites. The interaction between Ab and Ag2 is not as strong as the interaction between Ab and Ag1 but is still strong enough to allow binding. Hence, Ag1 and Ag2 are said to cross react. Ag3, in contrast, cannot interact very well with anti-Ag1 antibody and significant binding would not occur. Ag3, therefore, would not cross react with Ag1.

Cross reactivity also forms the basis for several diagnostic tests. For example, infection with *Treponema pallidum* (syphilis) causes the production of antibodies that cross react with a substance found in cardiac muscle, cardiolipin. Since it is much easier to obtain pure cardiolipin than pure treponemal antigens, this cross reaction is used to test for syphilis (Wassermann test). Similarly, antibodies produced against certain *Rickettsia* cross react with antigens from *Proteus*. Since the latter are much easier to obtain, they can be used to test for the former.

AFFINITY OF ANTIBODY

Antibody affinity is the strength of the reaction between a single antigenic determinant site of antigen and a single combining site on the antibody. It is the sum of the attractive and repulsive forces operating between the antigenic determinant and the combining site of the antibody.

The higher the affinity of the antibody for the antigen, the more stable will be the interaction. Thus, the ease with which one can detect the interaction is enhanced.

High affinity antibodies tend to bind more quickly and to dissociate more slowly than low affinity antibodies. High affinity antibodies also tend to bind a larger proportion of the available antigen.

AVIDITY OF ANTIBODY

The term avidity is often used to indicate the overall ability of antibodies to interact with antigen. Avidity is a measure of the overall strength of binding of an antigen with many antigenic determinants and multivalent antibodies. Functionally, the "strength" of an antibody in a particular application depends on the avidity of the antigen–antibody complex.

Affinity refers to the strength of binding between a single antigenic determinant and an individual antibody-combining site whereas avidity refers to the overall strength of binding between multivalent antigens and antibodies. Avidity is influenced by both the valence of the antibody and the valence of the antigen. Avidity is more than the sum of the individual affinities. Avidity is affected not only by the affinity of the antibody but also by the valency of binding and the three-dimensional features of the complex. Thus IgG has higher affinity than IgM, but IgM has higher avidity.

SPECIFICITY OF ANTIBODY

Specificity refers to the ability of an individual antibody-combining site to react with only one antigenic determinant or the ability of a population of antibody molecules to react with only one antigen.

ZONE PHENOMENON

Let us take a system of soluble antigen and antibody interaction. When the antigen solution is added to the antibody solution there will not be any visible reaction even though they are homologous. Here, the antibody is in excess and in this condition it will not form any visible interaction. However when the antigen solution is constantly added, it will reach the concentration equal with that of the antibody. Now there will be a visible reaction in the form of precipitation. Subsequently when the antigen is added in excess, again the visible reaction will disappear. Thus it is clear that for an antigen–antibody reaction, an optimal amount of the antigen and antibody is required. The phenomenon that occurs at different antigen and antibody concentration is called as zone phenomenon. Three distinct zones can be observed. They are:

1. Zone of antibody excess or pro-zone phenomenon
2. Zone of equivalence
3. Zone of antigen excess or post-zone phenomenon

The zone phenomenon can be depicted in the form of a graph by plotting antigen concentration on X-axis and precipitation concentration on Y-axis (Figure 22.2).

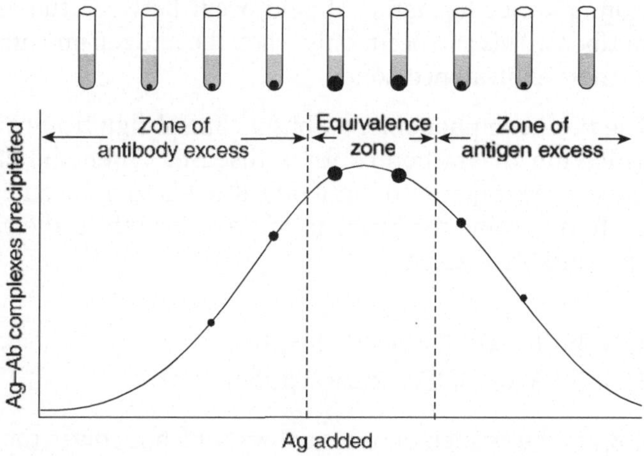

Figure 22.2 Graph showing the zone phenomenon

The zone phenomenon is explained by the principle of lattice formation (Figure 22.3). According to this principle, for a visible

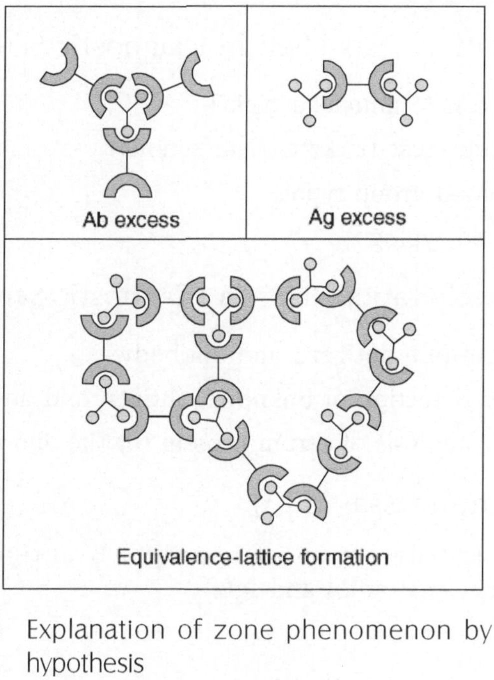

Figure 22.3 Explanation of zone phenomenon by the lattice hypothesis

reaction, a lattice formation should occur between the antigen and antibody. This can occur only when the antigen and antibody are in their equivalence zone.

The pro-zone phenomenon has a clinical significance in the identification of antibodies for a disease. When the sample contains excess amount of antibody, it leads to a false negative result. To overcome the problem of pro-zone effect, the sample may be suitably diluted.

DIAGNOSTIC APPLICATIONS OF ANTIGEN–ANTIBODY REACTIONS

Serology is the branch of immunology which involves the study of antigen–antibody reactions. It involves the measurement of antibodies and antigens in serum.

Serological tests most widely used are agglutination, precipitation and complement-fixation reactions.

Agglutination Tests Used in Diagnostic Serology

- Widal test (*Salmonella typhi)*
- Weil–Felix test (rickettsial infections)
- ABO blood group typing
- Rh blood typing

Use of Precipitation Tests in Diagnostic Serology

- To quantitate antigens and antibody
- For the detection of unknown antigen and antibody
- For the analysis of serum protein for the abnormality

Radioimmunoassay (RIA)

Used for the detection of hepatitis B antigen, insulin, testosterone, oestradiol and IgE.

Complement Fixation Tests (CFT)

- Used for bacterial, viral, protozoan, and fungal infections
- Wasserman test for syphilis

Other Serological Tests

- *Fluorescent-antibody techniques* It is used to detect the tissue antigens.
- *ELISA (Enzyme-linked immunosorbent assay)* It is used to detect a range of biochemical molecules, antigens and antibodies.
- *Intracutaneous diagnostic tests* These include
  - Schick test for diphtheria (*Corynebacterium* sp.)
  - Dick test for scarlet fever (*Streptococcus* sp.)
  - Skin tests involved in hypersensitivity (allergic reactions)

POINTS TO REMEMBER

- Cross reactivity refers to the ability of an individual antibody-combining site to react with more than one antigenic determinant site of the antigen.
- Antibody affinity is the strength of the reaction between a single antigenic determinant site of antigen and a single combining site on the antibody.
- The term avidity is often used to indicate the overall ability of antibodies to interact with antigen.
- Specificity refers to the ability of an individual antibody-combining site to react with only one antigenic determinant.

REVIEW QUESTIONS

1. Write short notes on:
    - i. Cross reactivity
    - ii. Antibody affinity
    - iii. Avidity
    - iv. Zone phenomenon
    - v. Pro-zone phenomenon
    - vi. Serology and its applications

23

PRECIPITATION REACTIONS

INTRODUCTION

When a homologous antigen and antibody meet, they will interact to form a visible reaction. If a soluble antigen is involved, the antigen–antibody complex will lead to precipitation. This precipitation is due to the lattice formed by the antigen–antibody complex and this happens only in the zone of equivalence. The precipitation reaction can be demonstrated either in solution or on a solid support like gel.

MECHANISM OF PRECIPITATION

Antibodies are at least bivalent, that is, they contain at least two combining sites for antigens (polymeric antibodies have 2x their number of basic units). These combining sites are relatively small; they can bind to only about 4–5 sugar residues of a carbohydrate or about 5–6 amino acids of a protein. Since most proteins have a given sequence of 5 amino acids only once on a peptide chain and seldom contain more than two identical chains, a particular antibody will be able to bind to only one or two places on a single protein. However, during an immune response against a protein, antibodies are synthesized that recognize several portions of the antigen.

For example, serum from a mouse immunized with human serum albumin (HSA) will have antibodies to many sections on the protein, so a single HSA molecule would be expected to have several antibodies bound to it. The binding site on the other arm of the antibody would be free to bind to the same site on another HSA molecule. This cross-linking would continue until the complex is no longer soluble and precipitated from solution. If large-enough quantities of antigen and antibody are present, the precipitate would be easily visible. The precipitate will not form if all the antibodies present are specific for only one site on the target molecule as would be the case with monoclonal antibodies.

PRECIPITATION IN SOLUTION

Ring Test

This is the simplest form of precipitin test in which the antigen solution is layered over the antiserum. If the antigen is homologous to the antibody, a line of precipitation is formed in the junction where antigen and antibody meet each other. This will appear in the form of a ring. Hence this test is called as ring test.

Slide Test

When a drop of antigen is added to the corresponding antibody on a slide, the precipitation is seen in the form of floccules. This type of reaction is called as flocculation.

PRECIPITATION IN GELS

Immunodiffusion

When the antigen and antibody are allowed to meet each other in a gel like agar or agarose, they will precipitate in the form of a line. This line is called as precipitation line or precipitin line. The precipitation line thus obtained is stable unlike the precipitation that occurs in solution. Further, this precipitation line can be stained and preserved as a permanent record. Further it has many other applications. Normally the diffusion is performed in 1% agar or agarose. There are four types of immunodiffusion in gel. They are:

1. Single diffusion in one dimension (Oudin technique)
2. Single diffusion in two dimensions (Mancini technique)
3. Double diffusion in one dimension (Oakley–Fulthorpe technique)
4. Double diffusion in two dimensions (Ouchterlony technique)

Single Diffusion in One Dimension (Oudin Technique)

In 1952, Oudin introduced the technique in which the antibody is mixed in molten agar and allowed to set in a test tube. After the agar solidifies, the antigen solution is overlaid on the agar gel. The antigen solution will diffuse into the agar gel and when the concentration reaches the region of equivalent zone, a precipitation ring is formed in the tube. It is called single diffusion test because only antigen is allowed to diffuse in the gel. As the antigen diffuses in only one direction, it is called as one dimension technique (Figure 23.1).

This technique has the following applications.

1. It is used to find unknown antigen.
2. It is used to find the approximate concentration of antigen as the thickness of the precipitation line depends directly on the concentration of the antigen.
3. It is also used to find the number of antigenic components present in the antigen solution against the polyclonal antiserum. Many precipitation lines are formed based on the number of antigenic components present in the given antigenic solution. The number of precipitation lines indicates the number of antigenic components present in the antigen solution.

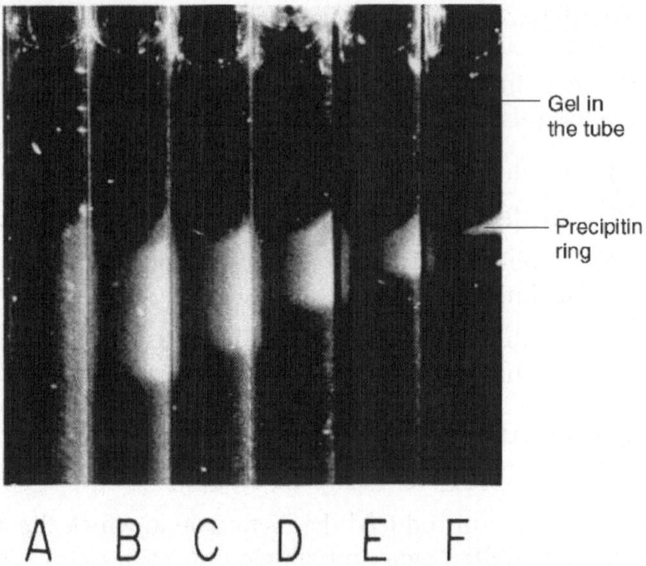

Figure 23.1 Photograph illustrating Oudin technique. Thickness of the precipitin ring is seen decreased as the concentration of antigen is decreased from A to F.

Single Diffusion in Two Dimensions (Mancini Technique)

This technique is also called as radial immunodiffusion (RID) or single radial immunodiffusion (SRID). This technique was introduced in 1965 by Mancini and his co-workers to estimate the concentration of serum proteins. In this technique, only the antigen is allowed to diffuse in the gel that has been incorporated by the antibody. Here the gel is overlaid on a glass slide and thus the antigen can diffuse in two dimensions. Hence the technique is referred to as single diffusion in two dimensions.

This technique is routinely used in hospitals and laboratories to measure the concentration of IgG, IgM and IgA in a patient's serum. Normally the IgG levels will be estimated in the condition of agammaglobulinaemia. The IgM level will be estimated in the case of multiple myeloma condition (Figure 23.2).

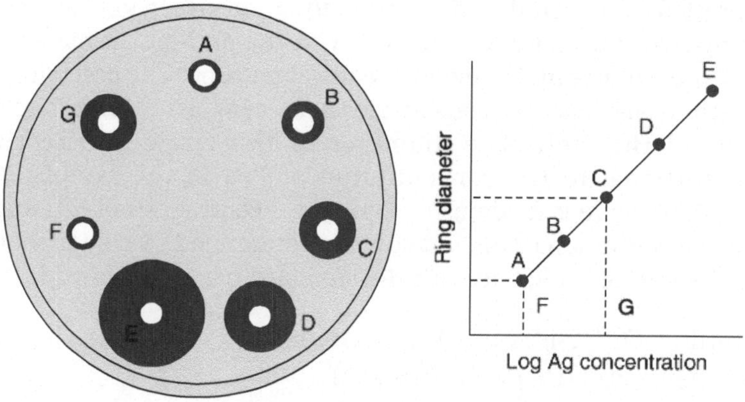

Figure 23.2 SRID showing different diameters of precipitin ring. A to E is the standard of increase in the concentration. G and F are the test antigens.

A small amount of agar is melted and cooled to just above its setting point. Antiserum to human IgG is added to the agar which is mixed and poured onto a small plate. Small holes (1–2 mm diameter) are cut in the gel about 2 cm apart. The well (hole) is filled with serum or a known amount of IgG and

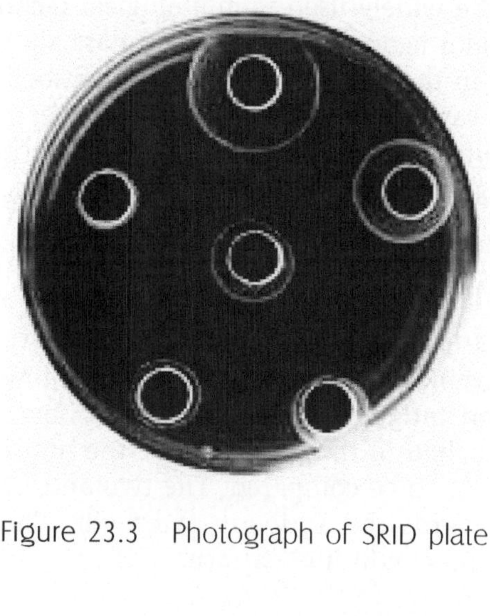

Figure 23.3 Photograph of SRID plate

the plate is covered and allowed to stand overnight. As IgG diffuses out of the well, it reacts with the anti-lgG in the agar and precipitates in the region of equivalent zone. The precipitate is white and can be seen in the clear agar as a white circle around the well. The diameter of the circle is directly proportional to the concentration of IgG in the sample. A standard curve is made using known concentrations of IgG and the concentration of the unknown is determined by comparison of the size of the circle with the standard curve (Figure 23.3).

Double Diffusion in One Dimension (Oakley–Fulthorpe Technique)

This technique is rarely used nowadays. It is similar to Oudin tube method. However, plain agar is overlaid on the antibody-incorporated agar. Thus both the antigen and the antibody are allowed to diffuse into the plain agar (double diffusion). When they meet each other at the equivalent zone, the precipitation line is formed. It has the same application as that of the Oudin technique.

Double Diffusion in Two Dimensions (Ouchterlony Technique)

It is one of the widely used immunological techniques. In this technique, plain agar is overlaid on a glass slide. Small wells are punched in the gel. The antigen and antibody are allowed to diffuse towards each other. When they meet each other at the equivalent zone, they will form a precipitation line on the gel indicating that they are homologous to each other.

The technique has many applications.

1. *To find the identity of two given antigens against the polyclonal antiserum* The aim is to compare two antigens for their antigenic determinant sites, i.e., to know whether they share common antigenic determinant sites or not. For this, three wells are punched in the gel—one for the antiserum and two for the antigens to be compared. The two antigens are allowed to diffuse towards the antibody overnight. Then the slide is analysed for precipitin line pattern.

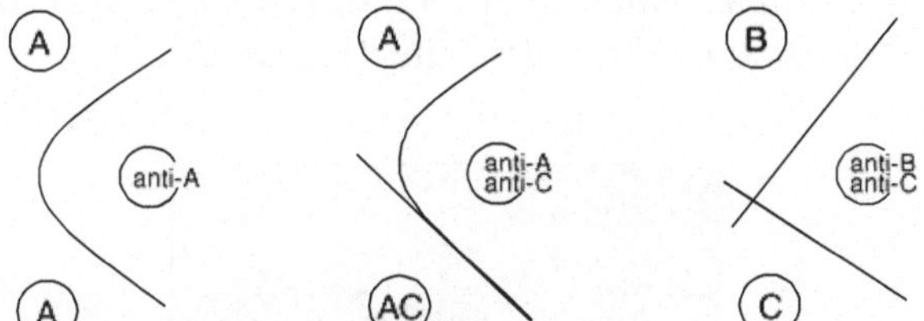

Figure 23.4 Ouchterlony technique showing three types of line patterns

Three different types of precipitin lines can be observed (Figure 23.4). They are:

- *Line of identity* When the two antigens contain identical antigenic determinant sites, a smooth fused precipitin line is observed. This is called as line of identity (Figure 23.5).
- *Line of non-identity* When the two antigens possess entirely different antigenic determinant sites, the lines are formed independently and will cross each other. This precipitin line is called as line of non-identity.
- *Line of partial identity* When the two antigens contain one or few antigenic determinant sites in common then line will meet each other and form a spur-shaped precipitin line. This precipitin line is called as line of partial identity (Figure 23.6).

2. *To find the titre of the antiserum* This technique is very useful to find the titre of the antiserum that is raised in any animal. The titre is the reciprocal of the highest dilution at which the precipitation line is observed. To find titre, the antiserum well is punched in the centre surrounded by the wells to load different dilutions of antigens. After the antigens and antiserum are loaded, the gel is kept overnight for incubation. Then the gel is observed for the precipitin lines. The reciprocal value of

highest dilution at which the precipitin line is observed is taken as the titre value of the antiserum (Figure 23.7).

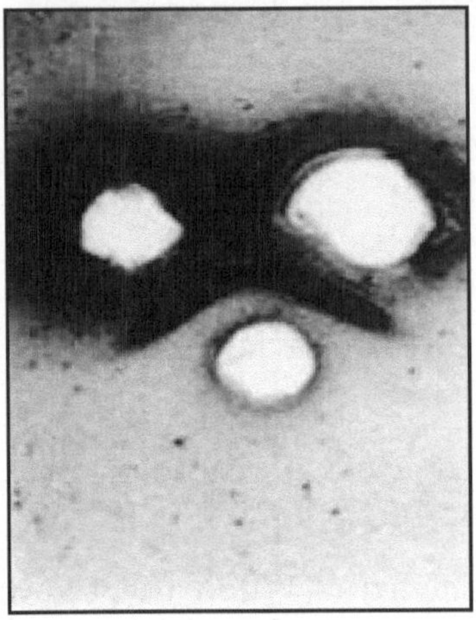

Figure 23.5 Photograph showing line of identity

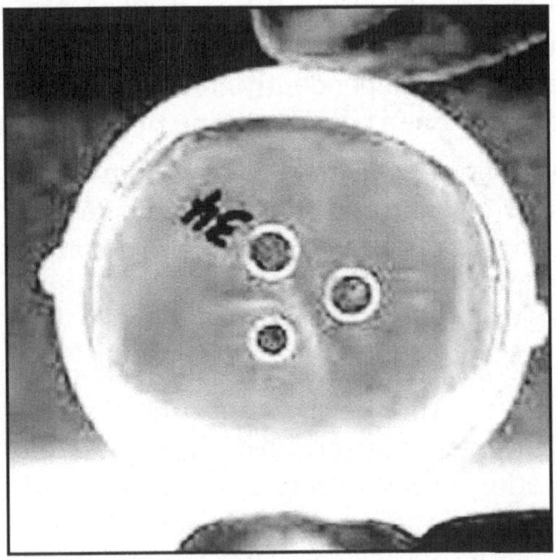

Figure 23.6 Photograph showing the line of partial identity

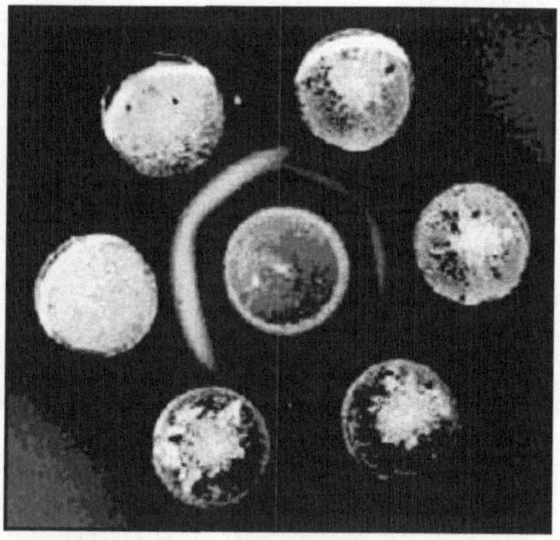

Figure 23.7 Well pattern prepared to find out the titre of the antiserum. The centre well is loaded with the antiserum and the surrounding wells are loaded with the diluted antigen.

Other Interpretations

The relative concentration of antigen and antibody can be understood from the precipitin line pattern.

- If the line is close to the antibody well it can be interpreted that the concentration of antigen is more than the antibody.
- If the line is close to the antigen well it can be interpreted that the concentration of antibody is more than the antigen.
- If the line is exactly at the centre of the antigen and antibody well, it can be understood that the concentration of antigen and antibody are same (Figure 23.8).

The number of antigenic components in a given antigen solution can be determined by the number of precipitin lines observed. Further the thickness of line will give an idea of the concentration of individual antigenic components present in the antigen solution (Figure 23.9).

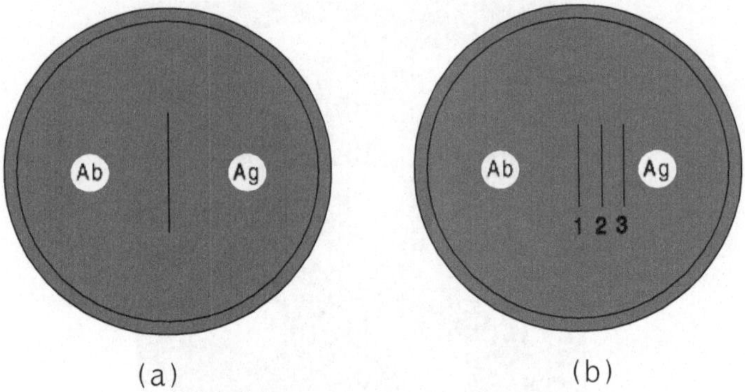

Figure 23.8 Ouchterlony technique. a) Single precipitin line indicating the presence of one antigenic component in the antigen solution. b) Three precipitin lines indicating the presence of three antigenic components in the antigenic solution.

Figure 23.9 Photograph showing Ouchterlony technique with multiple precipitin lines indicating the presence of many antigenic components in the given antigen solution. A good example of such antigen is serum.

IMMUNOELECTROPHORESIS

The immunodiffusion in gel normally takes place overnight to observe the result. In 1953, Grabar and Williams described immunoelectrophoresis, which has facilitated the diagnosis of multiple myeloma. This laid the foundation for immunoelectrophoresis. This facilitates the fast movement of antigen and antibody by the influence of the current.

Counter Immunoelectrophoresis (CIEP)

It is also called as crossover immunoelectrophoresis. In this technique, the antigen is loaded in one well and antibody is loaded in another well. They are allowed to move towards each other by the influence of the current. The pH of the gel is maintained above 8.0 at which almost all the proteins will be negatively charged. Thus the antigens which are negatively charged will move towards the anode. The antibodies, negatively charged but electrophoretically having slow mobility, will move towards the cathode due to electroendosmosis. When the antigen and antibody meet each other, they form a precipitin line (Figure 23.10).

Principle of electroendosmosis

The static support, the stabilizing medium (e.g. the gel) and/or the surface of the separation equipment such as glass plates, tubes or capillaries can carry charged groups, e.g. carboxylic groups in starch and agarose and sulphonic groups in agarose and silicium oxide on glass surfaces. These groups become ionized in basic and neutral buffers and in the electric field they will be attracted by the anode. As they are fixed in the matrix, they cannot migrate. This results in compensation by the counterflow of H_3O^+ ions towards the cathode by electroendosmosis.

In gels, this effect is observed as a water flow towards the cathode, which carries the solubilized substances along. Thus in counter immunoelectrophoresis the antibodies move towards the cathode.

Electroendosmosis is normally seen as a negative effect, yet a few methods take advantage of this effect to achieve separation or detection (MEKC in capillary electrophoresis and counter immunoelectrophoresis).

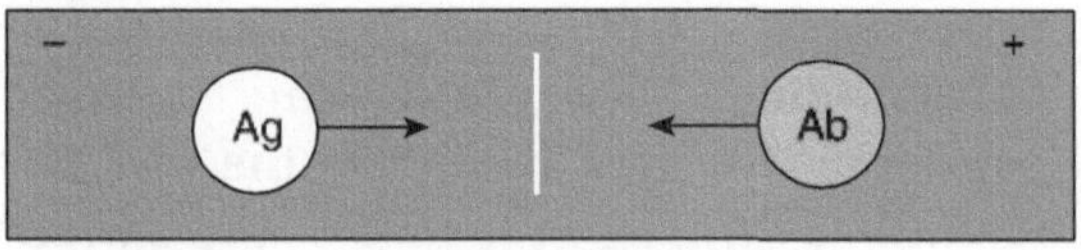

Figure 23.10 CIEP technique showing the movement of antigen and antibody in the electrical field

Serum Immunoelectrophoresis

Serum immunoelectrophoresis, also called gamma globulin electrophoresis or immunoglobulin electrophoresis, is a method of determining the blood levels of three major immunoglobulins: immunoglobulin M (IgM), immunoglobulin G (IgG) and immunoglobulin A (IgA) (Figure 23.11).

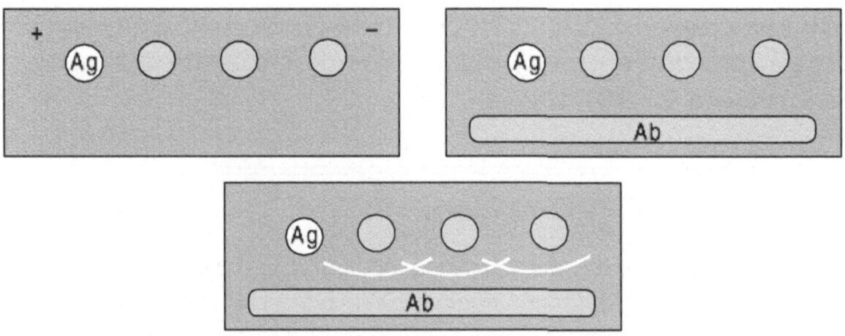

Figure 23.11 Methodology of serum electrophoresis

Immunoelectrophoresis is a powerful analytical technique with high resolving power as it combines separation of antigens by electrophoresis followed by immunodiffusion against an antiserum (Figure 23.12).

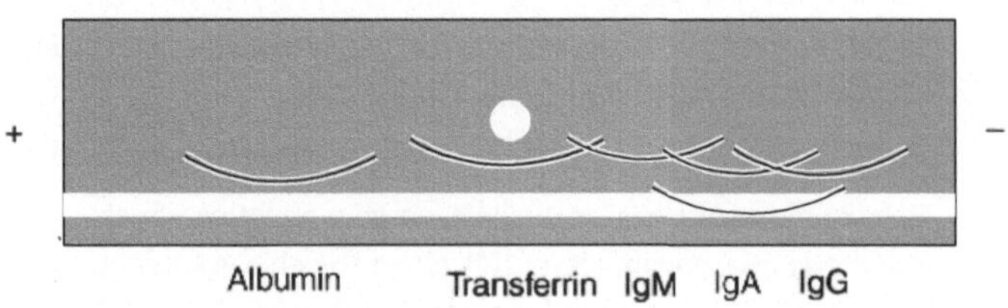

Figure 23.12 Serum electrophoresis showing the position of various precipitin lines for the different serum components

Serum proteins are separated in agar gels under the influence of an electric field into albumin, alpha 1, alpha 2, and beta and gamma globulins. Immunoelectrophoresis is performed by placing serum on a slide containing a gel designed specifically for the test. An electric current is then passed through the gel. The immunoglobulins, which contain an electric charge, migrate through the gel according to the difference in their individual electric charges. Antiserum is placed alongside the slide to identify the specific type of immunoglobulin present. The results are used to identify different disease entities, and to aid in monitoring the course of the disease and the therapeutic response of the patient to conditions such as immune deficiencies, autoimmune disease, chronic infections, chronic viral infections and intrauterine foetal infections (Figure 23.13).

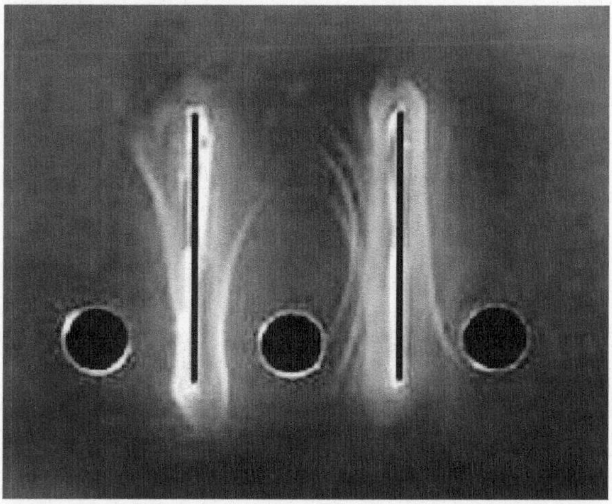

Figure 23.13 Photograph of serum electrophoresis

A modification in serum immunoelectrophoresis is the immunofixation electrophoresis. Specific proteins of interest can be identified by first fixing them onto the gel with antibodies, then washing away all the other proteins prior to staining. This procedure is called immunofixation electrophoresis (IFE). Immunoelectrophoresis was used in the past to identify specific proteins. However, this technique has been largely superceded by IFE, because IFE is easier to perform and interpret (Figure 23.14).

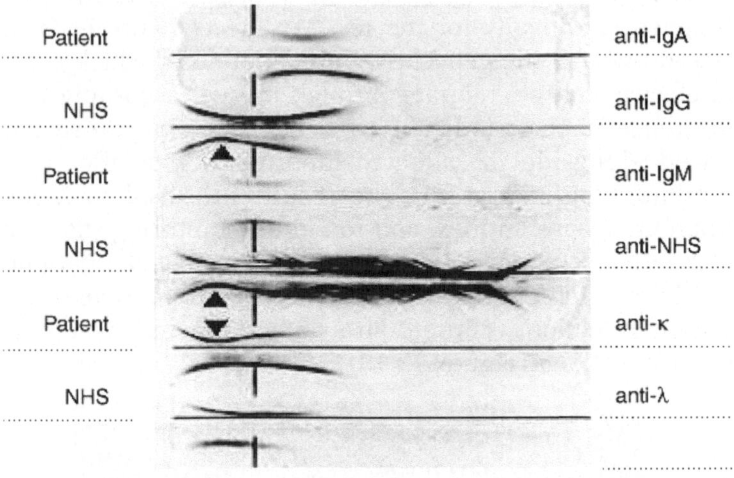

Figure 23.14 Photograph of immunofixation electrophoresis. Different precipitin lines are seen for different antibodies against different protein components of the serum.

Rocket Electrophoresis (Laurell's Technique)

This technique is also referred to as single dimensional immunoelectrophoresis. This is similar to SRID technique. The antiserum is incorporated into the agar which is then poured onto a large plate to cool. Wells are cut along one end and filled with standard solutions of the antigen or with the unknown solution. An electric current is then established across the agar which drives the antigen in the direction of the current. As in SRID, the antigen precipitates with the antibody where its concentration is equal to that in the agar (in equivalent zone). The height of the "rocket" is proportional to the concentration of antigen in the well, and the amount of antigen in the unknown well is determined by referring to the standard curve (Figure 23.15).

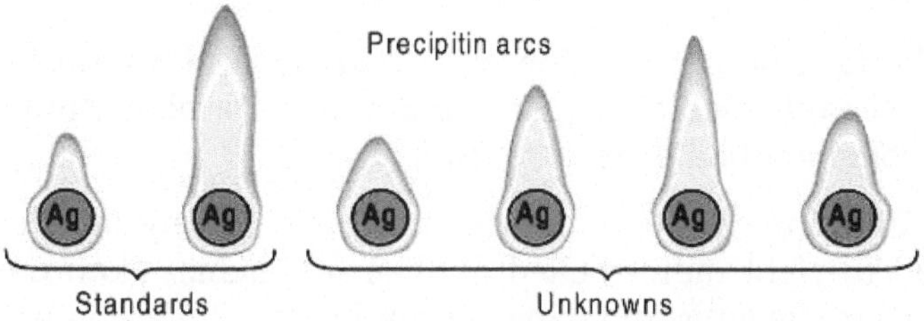

Figure 23.15 Laurell's technique

Two-dimensional Immunoelectrophoresis

This technique is the combination of electrophoretic separation of antigen mixture in first dimension and running a rocket electrophoresis in the second dimension. A square-shaped glass plate in taken. One-fourth of the slide is poured with agarose and a well is punched towards the cathode end. To the well, antigenic mixture is added and it is separated electrophoretically. Then the remaining part of the slide is poured with the agarose mixed with the antiserum. Again the slide is subjected to electrophoresis in the second dimension. The area under the precipitin peak will determine a semi-quantitative estimation of individual antigens (Figure 23.16).

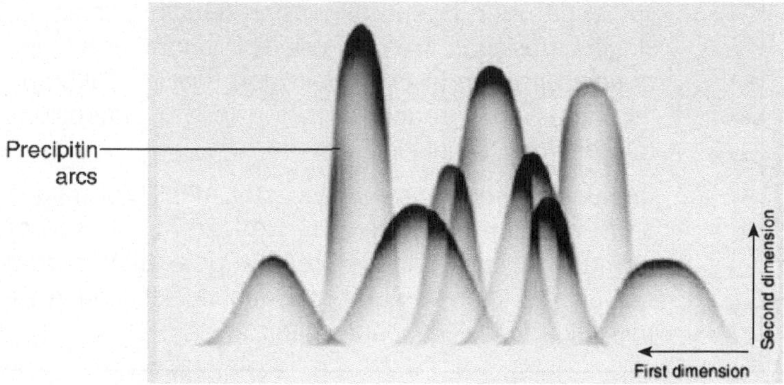

Figure 23.16 Photograph of two-dimensional immunoelectrophoresis

RADIOIMMUNOPRECIPITATION ASSAY (RIPA)

Radioimmunoprecipitation assay (RIPA) is the term used to describe the qualitative assay used as a confirmatory procedure for some antibodies to viral antigens.

Viral-infected cell cultures are radioactively labelled and lysed to yield radiolabelled antigen fragments. Then they are added to the antiserum. Specific antibodies, if present, will bind to these antigen fragments and the resulting antigen–antibody complexes are precipitated using protein A, boiled to free the immune complexes which are then separated by electrophoresis. The pattern of antigenic moieties to which antibodies are present may then be detected using autoradiography (the exposure of sensitive X-ray film by the radioactive emissions of the bound, labelled antigens). Comparing labelled molecular weight standards electrophoresed in the same run allows determination of the molecular weight "bands" of antigen for which antibodies are present.

POINTS TO REMEMBER

- The simplest form of precipitin test is called as ring test.
- There are four types of immunodiffusion in gel. They are single diffusion in one dimension (Oudin technique), single diffusion in two dimensions (Mancini technique), double diffusion in one dimension (Oakley–Fulthorpe technique) and double diffusion in two dimensions (Ouchterlony technique).
- Immunoelectrophoresis enhances the mobility of antigen and antibody. There are different types of immunoelectrophoresis techniques like serum immunoelectrophoresis, counter immunoelectrophoresis, rocket immunoelectrophoresis, etc.

REVIEW QUESTIONS

1. Write short notes on:
   i. Ring test
   ii. Oudin technique
   iii. Single radial immunodiffusion (SRID)
   iv. Ouchterlony technique
   v. Counter immunoelectrophoresis (CIEP)
   vi. Electroendosmosis
   vii. Serum immunoelectrophoresis
   viii. Immunofixation electrophoresis
   ix. Rocket electrophoresis (Laurell's technique)
   x. Radioimmunoprecipitation assay
2. Write in detail about various precipitation techniques.

24

IMMUNOHISTOCHEMISTRY

INTRODUCTION

Immunohistochemistry is the branch of applied immunology which deals with the localization of antigens in tissue sections by the use of labelled antibodies. This involves specific antigen–antibody reactions. The label may be a fluorescent dye, enzyme, radioactive element or colloidal gold.

The first report of an immunohistochemical technique was made in 1942 when Coons and his colleagues detected pneumococcal antigen using a fluorescently tagged antibody. The immunohistochemical techniques for electron microscope (EM) were developed by Singer using ferritin in 1959. This was quickly followed by the first use of an enzyme, horseradish peroxidase in 1966 by Graham and his colleagues. Since then, a variety of enzyme and heavy metal techniques have been developed, the most important of which are colloidal gold for EM by Faulk in 1971, immunoperoxidase assay by Nakane in 1966, peroxidase/antiperoxidase PAP technique by Sternberger in 1970 and the Avidin–Biotin Complex (ABC) technique by Hsu, 1981.

Immunohistochemistry is generally carried out in sectioned tissue, which allows the antibodies free access to the interior of the cells. Immunohistochemistry can also be carried out on cells either in free solution or bound to membranes or on monolayers of cultured cells.

There are numerous immunohistochemical methods that may be used to localize antigens. The selection of a suitable method should be based on parameters such as the type of specimen under investigation and the degree of sensitivity required.

TYPES OF IMMUNOHISTOCHEMICAL METHODS

Direct Method

Direct method is one step staining method and it involves a labelled specific antibody reacting directly with the antigen in tissue sections. This technique utilizes only one antibody and the procedure is short and quick. However, it is insensitive due to little signal amplification and is rarely used since the introduction of indirect method. If the antibody is labelled with fluorescent dyes, the technique is called as direct immunofluorescence technique. Instead of fluorescent dye, horseradish peroxidase can be used as the label. Then the technique is called as direct immunoperoxidase technique. If the antibody is labelled with other enzymes then the technique is called as direct immunoenzyme method. The fluorescent dyes normally used as labels are FITC, rhodamine and Texas red (Figure 24.1).

The enzymes that are normally used as labels are peroxidase, alkaline phosphatase and glucose oxidase.

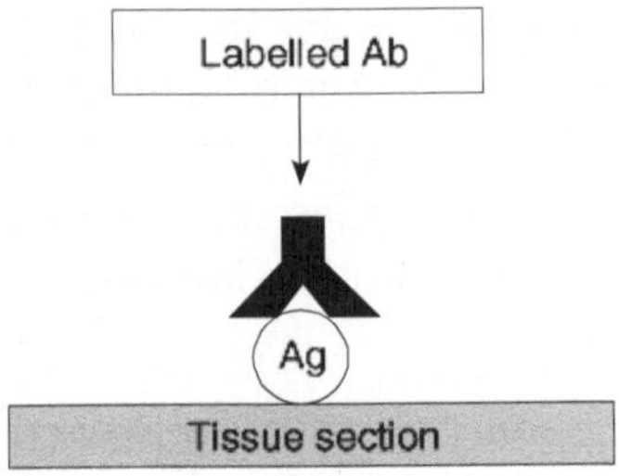

Figure 24.1 Direct method of tissue antigen detection

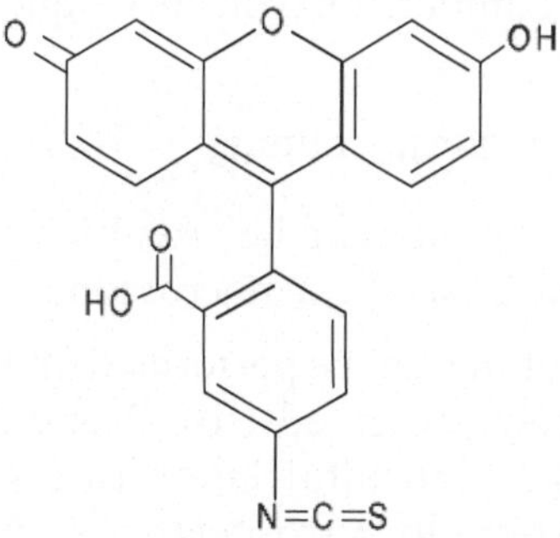

Fluorescein isothiocyanate (FITC)

Fluorescein isothiocyanate (FITC) is currently the most commonly used fluorescent dye in immunoassay techniques. FITC is a small organic molecule, and is typically conjugated to proteins via primary amines (i.e., lysines). Usually, between three and six FITC molecules are conjugated to each antibody; higher conjugations can result in solubility problems as well as internal quenching (and reduced brightness). Thus, an antibody will usually be conjugated in several parallel reactions to different amounts of FITC and the resulting reagents will be compared for brightness (and background stickiness) to choose the optimal conjugation ratio. Fluorescein is typically excited by the 488 nm line of an argon laser and emission is collected at 530 nm. It emits an apple green fluorescence.

Indirect Method

Indirect method involves an unlabelled specific primary antibody (first layer) which reacts with tissue antigen. After a wash, a labelled secondary antibody (second layer) is allowed to react with the primary antibody (the secondary antibody must be against the IgG of the animal species in which the primary antibody has been raised) (Figure 24.2).

This method is more sensitive due to signal amplification through several secondary antibody reactions with different antigenic sites on the primary antibody. In addition, it is also economical since one labelled second layer antibody can be used with different specific first layer antibodies (raised from the same animal species) to different antigens.

When the fluorescent label is used, the technique is called as indirect immunofluorescence technique. When peroxidase is used, the technique is called as indirect immunoperoxidase technique. It is called as indirect immunoenzyme technique when the label is an enzyme.

Advantages of immunoperoxidase staining

1. The antigen–antibody complex can be visualized within minutes by using an ordinary light microscope.
2. Immunoperoxidase staining can be performed on tissues that have been routinely preserved (fixed) and stored (paraffin embedded) by hospital laboratories. This allows the pathologist to do retrospective studies as well as current evaluations.

The labelling (staining) of antigen sites is permanent and the cell or tissue sections can be stored for future evaluations.

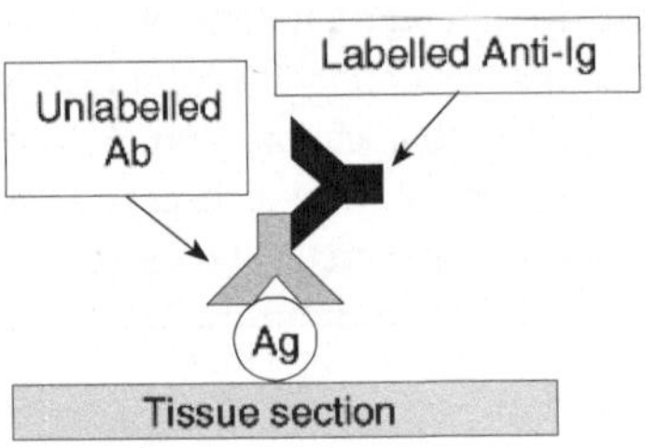

Figure 24.2 Indirect method of detection of tissue antigen

PAP Method (Peroxidase–Antiperoxidase Method)

The second antibody preparations used in indirect immunochemical technique are polyclonal and frequently contain antibodies that react non-specifically with tissue without application of first antibody.

The PAP technique is designed to filter out these contaminating antibodies. In the PAP technique, antibody-specific second antibodies are not labelled. It reacts with first antibody via one of their Fab combining sites. Another Fab end is allowed to react with the peroxidase–antiperoxidase complex (rabbit antibody to peroxidase, coupled with peroxidase to make a very stable peroxidase–antiperoxidase complex).

Contaminating, non-specific antibodies will bind with contaminants with both of its combining sites (Fab). Thus they cannot bind with the antiperoxidase in the PAP complex (Figure 24.3).

The sensitivity is about 100 to 1000 times higher since the peroxidase molecule is not chemically conjugated to anti-IgG but is immunologically bound, and loses none of its enzyme activity. It also allows for much higher dilution of the primary antibody, thus eliminating many of the unwanted antibodies and reducing non-specific background staining.

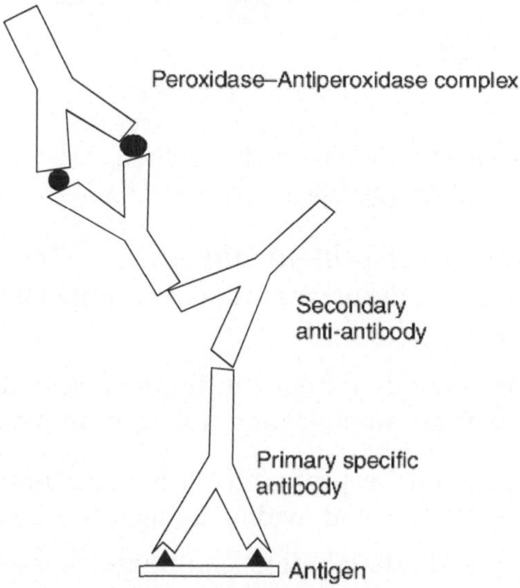

Figure 24.3 Peroxidase–Antiperoxidase method (PAP method)

Avidin–Biotin Complex (ABC) Method

ABC method is a standard immunohistochemical method and one of the widely used techniques for immunohistochemical staining. Avidin, a large glycoprotein, can be labelled with peroxidase or fluorescein and has a very high affinity for biotin (Figure 24.4). Biotin, a low molecular weight vitamin, can be conjugated to a variety of biological molecules such as antibodies.

The technique involves three layers. The first layer is unlabelled primary antibody. The second layer is biotinylated secondary antibody. The third layer is a complex of avidin–biotin peroxidase. The peroxidase is then developed by the DAB or other substrates to produce different colorimetric end products.

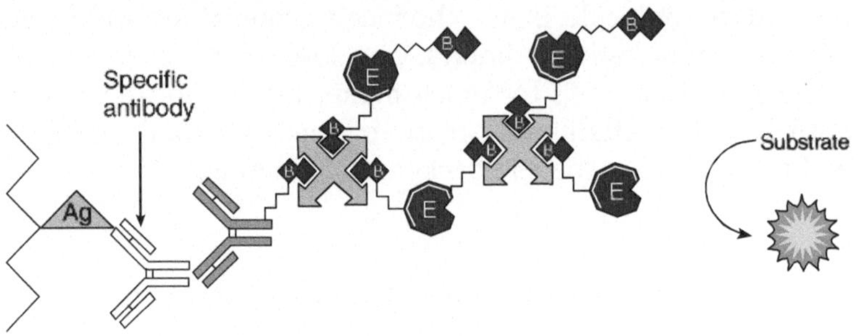

Figure 24.4 Avidin-Biotin complex method, Ag—antigen, B—biotin, A—avidin, E—enzyme

Advantages of the biotin–avidin system The biotin–avidin system has several advantages over direct coupling of the marker to an antibody.

The biotin–avidin system can improve sensitivity because of the potential for amplification due to multiple site binding.

- Avidin can be prepared with high fluorochrome to protein ratios and avidin conjugates are very stable.
- Only a single-labelled conjugate, namely avidin or streptavidin, needs to be kept on hand since it can be used with a variety of biotinylated lectins, antibodies or probes.
- Biotin–avidin system reagents can overcome the problem of background fluorescence sometimes encountered in the use of heavily fluorescein-labelled or rhodamine-labelled antibodies. These conjugates are sometimes “sticky” and adsorb non-specifically to tissues, while fluorochrome-conjugated avidin D does not.

- The extraordinarily high affinity between avidin or streptavidin and biotin assures the user of a rapidly formed and stable complex between the (strept)avidin conjugate and the biotin-labelled protein.
- Simultaneously localizing more than one antigen in the same tissue section can be performed even with two or three primary antibodies from the same species. By using either separate enzyme systems or two different substrates for the same or assorted fluorochrome conjugates, more than one antigen can be localized in the same tissue section.

Avidin–Biotin System

Avidin is an egg-white derived glycoprotein with an extraordinarily high affinity (affinity constant > 1015 M^{-1}) for biotin. Streptavidin is similar in properties to avidin but has a lower affinity for biotin. Many biotin molecules can be coupled to a protein, enabling the biotinylated protein to bind more than one molecule of avidin. If biotinylation is performed under gentle conditions, the biological activity of the protein can be preserved. By covalently linking avidin with different ligands such as fluorochromes, enzymes or EM markers, what we have termed the botin–avidin system can be utilized to study a wide variety of biological structures and processes. The botin–avidin system has proven to be particularly useful in the detection and localization of antigens, glycoconjugates and nucleic acids by employing biotinylated antibodies, lectins, or nucleic acid probes.

Labelled Streptavidin Biotin (LSAB) Method

Streptavidin, derived from *Streptococcus avidini*, is a recent innovation for substitution of avidin. The streptavidin molecule is uncharged relative to animal tissue, unlike avidin which has an isoelectric point of 10 and therefore electrostatic binding to tissue is eliminated. In addition, streptavidin does not contain carbohydrate groups which might bind to tissue lectins, resulting in some background staining (Figure 24.5).

LSAB is technically similar to standard ABC method. The first layer is unlabelled primary antibody. The second layer is biotinylated secondary antibody. The third layer is enzyme–streptavidin conjugates (horseradish peroxidase–streptavidin or alkaline phosphatase–streptavidin) to replace the complex of avidin–biotin peroxidase. The enzyme is then visualized by the application of the substrate to produce different colorimetric end products. The third layer can also be fluorescent dye–streptavidin such as FITC–streptavidin if, fluorescence labelling is preferred.

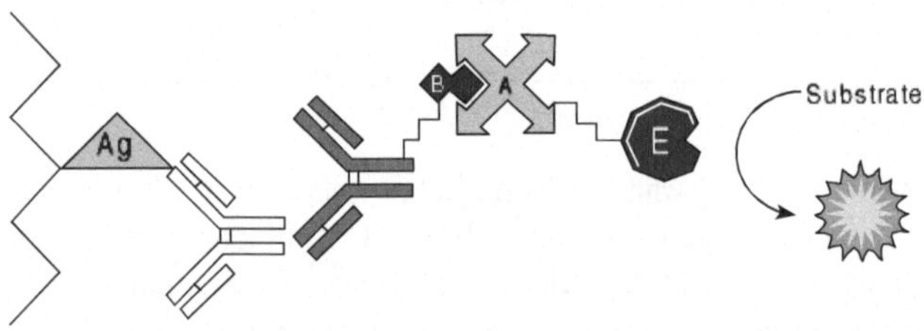

Figure 24.5 Labelled streptavidin-biotin complex method Ag—antigen, B—biotin, A—avidin, E—Enzyme

Multiple Staining Method

It is often required to identify two or more antigens in the given tissue section at a time. This can be achieved by immunofluorescence method using different fluorescent dyes. It can also be done with peroxidase conjugated antibodies developed with different coloured substrates to produce the end products of different colours.

Electron Microscopic (EM) Immunohistochemical Method

Electron microscopic (EM) immunohistochemical techniques can be divided into two groups. They are:

- Pre-embedding—the immunostaining takes place prior to resin embedding.

- Post-embedding—the immunolabelling is undertaken after resin embedding.

The choice of whether to apply pre- or post-embedding method to the detection of an antigen in any particular location will depend to a large extent upon the distribution and liability of the antigen and the characteristics of the primary antibody.

Several recently developed methods rely on labelling with colloidal gold particles. These methods were originally introduced for electron microscopy by Faulk and Taylor in 1971 as the gold particles are easily visible under the electron microscope and they are also useful for light microscopy.

Since gold particles can be made in different sizes from 5 to 30 nm, it is possible to carry out multiple staining at the electron microscopic level, most easily by direct labelling of several first layer antibodies with different sized particles. The indirect techniques can also be used in double or triple labelling by parallel approach if the primary antibodies are from different species and by sequential approach if the primary antibodies are from same species.

ANTIBODY MICROARRAY

The antibody microarray contains over 500 monoclonal antibodies immobilized onto a glass surface, allowing for the comparison of multiple proteins in biological samples. Antibodies are typically spotted on amino-reactive glass slides. As many as 20000 spots can be printed per standard glass slide 25.3 × 75.5 mm, spot diameter of ±300 μm from centre to centre. Several grids can be spotted on the same slide, allowing to perform multiple binding experiments on the same slide. The antibody microarray is a complete analysis system for profiling protein expression in biological samples. This new technology advances the current state of proteomics allowing researchers to compare the relative abundance of hundreds of proteins in a single experiment. This type of multiplex analysis has, until now, been very difficult to perform with proteins using conventional technologies.

Most antibody arrays currently used are based on the ELISA sandwich approach that uses two antibodies to screen for the expression of a limited number of proteins. Also because antigen–antibody interactions are concentration-dependent, antibody microarrays need to normalize the amount of antibody that is used. In response to the limitations with the currently existing technology, we have developed a single antibody-based microarray where the quantity of antibody spotted is used to standardize the antigen concentration. In addition, this new array utilizes an internally controlled system where one colour represents the amount of antibody spotted and the other colour represents the amount of the antigen that is used to quantify the level of protein expression. When compared with median fluorescence intensity alone, normalization for antibody spot intensity decreased variability and lowered the limits of detection.

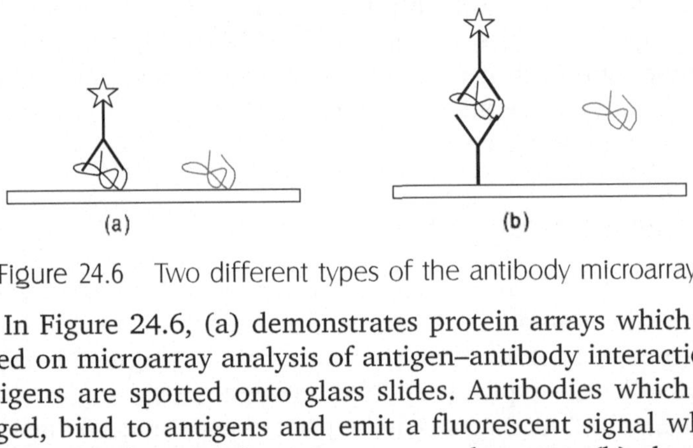

Figure 24.6 Two different types of the antibody microarray

In Figure 24.6, (a) demonstrates protein arrays which are based on microarray analysis of antigen–antibody interactions. Antigens are spotted onto glass slides. Antibodies which are tagged, bind to antigens and emit a fluorescent signal which can then be detected from the spot on the array. (b) shows a protein microarray composed of spots which act as sandwich immunoassays. Antibodies are spotted onto the chips and cell lysate or a solution is applied to the slide. Antibodies are incubated with the solution, with later washing to remove non-specific binding. The sandwich complexes formed emit a detectable signal from the spot.

Antibody microarray is designed as a screening tool for correlating proteins with a physiological or pathological process. Antibodies are carefully selected to cover five major functional categories based on gene ontology: apoptosis, cancer, cell cycle, protein kinases and neurobiology.

POINTS TO REMEMBER

- Immunohistochemistry deals with the localization of antigens in tissue sections by the use of labelled antibodies.
- Direct method is one step staining method and it involves a specific labelled antibody reacting directly with the antigen in tissue sections.
- If the antibody is labelled with fluorescent dyes, the technique is called as immunofluorescence technique.
- Instead of fluorescent dye, horseradish peroxidase can be used as the label. Then the technique is called as immunoperoxidase technique.
- Several recently developed methods rely on labelling with colloidal gold particles.

REVIEW QUESTIONS

1. Write short notes on:
    - i. Immunofluorescence technique
    - ii. Immunoperoxidase technique
    - iii. FITC
    - iv. PAP technique
    - v. Avidin–Biotin complex (ABC) method
    - vi. Biotin–Avidin system
    - vii. Antibody microarray

25

RADIOIMMUNOASSAY

INTRODUCTION

Radioimmunoassay (RIA) is one of the important labelled immunoassay techniques used in the detection of hormones, steroids, drugs and microbial antigens. It was developed by Rosalyn Yalow and Solomon Aaron Berson in the 1950s.

Yalow and Berson developed the first radioisotopic technique to study blood volume and iodine metabolism. They later adopted the method to study how the body uses hormones, particularly insulin, which regulates sugar levels in the blood. The researchers proved that type II (adult onset) diabetes is caused by the inefficient use of insulin. Previously, it was thought that diabetes was caused only by a lack of insulin.

In 1959, Yalow and Berson perfected their measurement technique and named it radioimmunoassay (RIA). RIA is extremely sensitive. It can measure one trillionth of a gram of material per millilitre of blood. Because of the small sample required for measurement, RIA quickly became a standard laboratory tool.

History of the invention of RIA

The creation of the RIA started with investigations concerning the metabolism of ^{131}I-labelled insulin in non-diabetic and diabetic subjects. Berson and Yalow observed that, contrary to their expectation, radioactive insulin disappeared more slowly from the plasma of patients who had previously been treated with insulin than from the plasma of subjects never treated with insulin. Immunologists of the mid-1950s did not believe that insulin was immunogenic, hence the *Journal of Clinical Investigation* rejected their paper. However, Berson and Yalow eventually proved that the retarded rate of insulin disappearance was due to the binding of labelled insulin to anti-insulin antibodies present in the serum of insulin-treated diabetics. Initially they used labelled and unlabelled insulin to examine the characteristics of the antibodies. For the immunoassay, they recognized that antibodies could also be used to examine the hormone and further, that competition between unlabelled insulin in a sample and the ^{131}I- or ^{125}I-labelled insulin for binding to sites on the anti-insulin antibodies could provide the basis of a sensitive and specific assay of the hormone.

RIA involves mixing known quantities of radioactive antigen (frequently labelled with gamma-radioactive isotopes of iodine attached to tyrosine) with antibody to that antigen, then adding unlabelled or "cold" antigen and measuring the amount of labelled antigen displaced.

The technique is both extremely sensitive and specific, but it requires special precautions (because radioactive substances are used) and sophisticated apparatus and is expensive.

PRINCIPLE OF RIA

The technique is based on the competition between the known fixed amount of labelled antigen and unlabelled antigen for the limited number of antigen-binding sites of the fixed amount of antibody.

A calibration curve is made by adding different concentrations of unlabelled antigen. As the concentration of unlabelled antigen is increased, the binding of the labelled antigen to the antibody will decrease. Thus the labelled antigen–antibody complex concentration will be decreased as the concentration of the unlabelled antigen is increased. A graph is plotted having the unlabelled antigen concentration on X-axis and concentration of labelled antigen–antibody complex on Y-axis. A standard graph obtained in this way is called as calibration curve (Figure 25.1). To the same system the unknown antigen (sample) is added and the concentration of labelled antigen–antibody is measured. From the calibration curve the unknown concentration of antigen can be interpreted.

The principle of RIA (Figure 25.2) can be understood by the following equation:

4 Ag* + 4 Ab → 4 Ag*Ab

4 Ag + 4 Ag* + 4 Ab → 2 Ag* Ab + 2 Ag Ab + 2 Ag* + 2 Ag

12 Ag + 4 Ag* + 4 Ab → Ag* Ab + 3 Ag Ab + 3 Ag* + 9 Ag

Ag*—Labelled antigen

Ag—Unlabelled antigen

Ab—Antibody

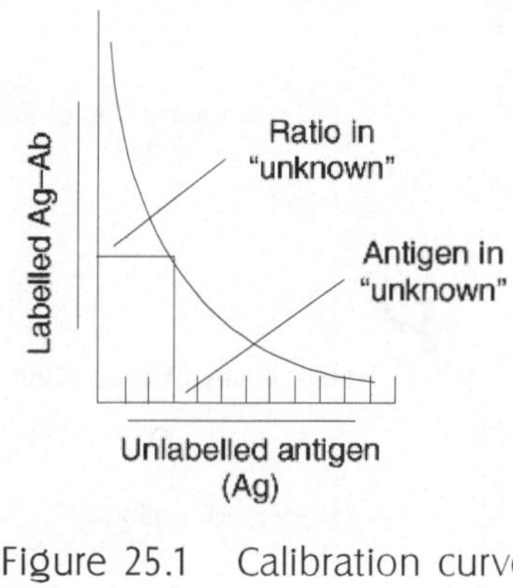

Figure 25.1 Calibration curve

From the above equation it is clear that when the concentration of unlabelled antigen (Ag) is increased, the concentration of labelled antigen–antibody complex (Ag*Ab) decreases.

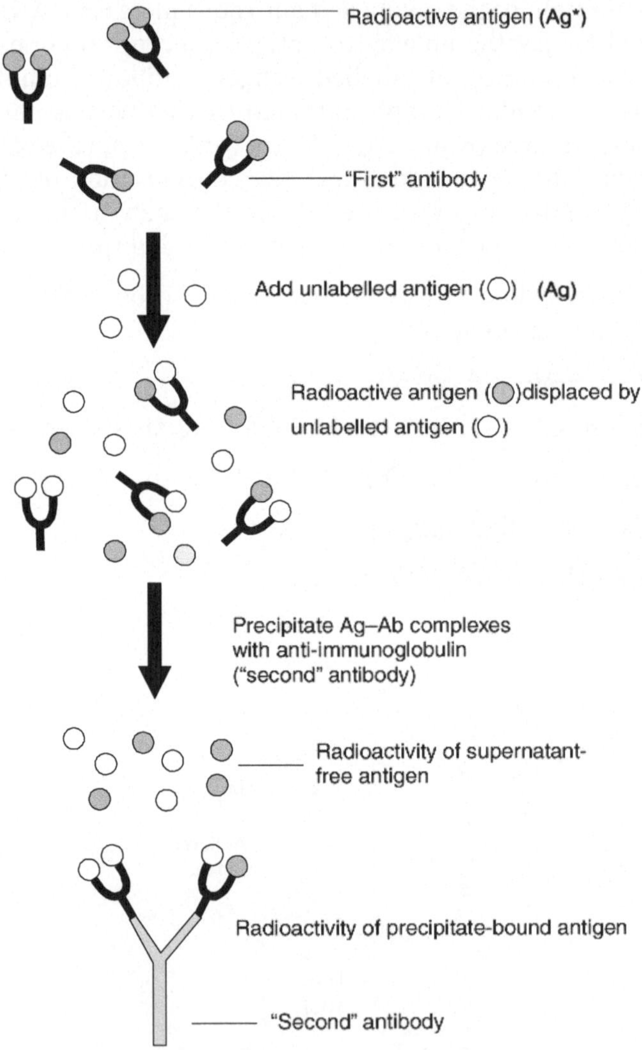

Figure 25.2 The principle of RIA

Yalow, Rosalyn Sussman, (1921–)

Rosalyn Sussman Yalow was born on July 19, 1921 in the Bronx, New York.

By 1945, Yalow had a Ph.D. in nuclear physics. Soon thereafter, she began researching the medical applications of radioactive materials.

In the late 1950s, Yalow and a collaborator invented a technique called "radioimmunoassay" (RIA) which uses radioactive isotopes to detect extremely small quantities of substances in the body. Hormones, enzymes, toxins and other substances that exist in minute amounts could be tested much more easily. RIA spawned great advances in areas of research that had been closed off and the technique has become a standard tool for diagnosing diseases and making drugs.

For her work on RIA, Dr. Yalow was awarded the 1977 Nobel Prize in Medicine.

Solomon Berson (1918–1972)

Solomon Berson was born in New York City in 1918. After being graduated from the City College of New York in 1938 and being rejected by many medical schools, he entered New York University and earned an M.S. degree and a fellowship to teach anatomy in the Dental School. In 1941 he was admitted to New York University School of Medicine, from which he graduated AOA in 1945. He interned at Boston City Hospital from 1945 to 1946, and then joined the Army as a Medical Officer from 1946 to 1948. Subsequently, for 2 years, he served as a resident in medicine at the Bronx VA Hospital, following which he joined Roz Yalow in the Radioisotope Laboratory. In the first year he also moonlighted in the private practice of medicine, which he enjoyed very much. However, due to the demands of the research work in which he had become involved, he discontinued the part-time practice.

RADIOLABELLING OF ANTIGEN

Two radioisotopes are normally used for radiolabelling of antigens in RIA. They are ^{125}I and ^{3}H. The ^{125}I emits gamma rays and hence a gamma counter is used to detect the radioactivity. The ^{3}H emits low-energy beta particles and the radioactivity is detected by using a liquid scintillation counter.

The most popularly used radioisotope is ^{125}I. Radioiodination is done to the tyrosine residue of the antigen. Sometimes the antigen will be tyrosylated for enough attachment of radioisotopes. Sodium iodide is used as the source of ^{125}I. The labelling can be performed by two methods (Figure 25.3). They are:

1. Chloramine-T method
2. Lactoperoxidase method

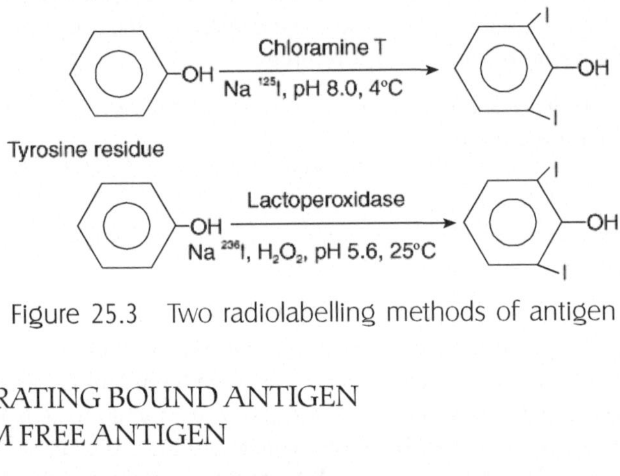

Figure 25.3 Two radiolabelling methods of antigen

SEPARATING BOUND ANTIGEN FROM FREE ANTIGEN

In order to measure the radioactivity of bound labelled antigen–antibody complex, it is imperative to separate the same from the unbound antigen.

There are several ways of doing this. They are

- Double antibody technique where the antigen–antibody complexes are precipitated by adding a "second" antibody directed against the first. For example, if a rabbit IgG is used to bind the antigen, the complex can be precipitated by adding an anti-rabbit IgG antiserum (e.g. raised by immunizing a goat with rabbit IgG).

- The antigen-specific antibodies can be coupled to the inner walls of a test tube. After incubation, the content in the tube is removed, washed and the tube will contain only the bound antigen. The radioactivity can be measured in the tube.
- The antigen-specific antibodies can be coupled to inert particles, like Sephadex. Centrifugation of the reaction mixture separates bound antigen in the form of Sephadex molecules.

APPLICATIONS OF RIA

Radioimmunoassay is widely used because of its great sensitivity. Using antibodies of high affinity, it is possible to detect a few picograms (10^{-12} g) of antigen in the tube.

In medicine, it is especially useful in diagnosing autoimmune diseases such as Hashimoto's thyroiditis and systemic lupus erythematosus.

RIA has many uses, including narcotics (drug) detection, blood bank screening for the hepatitis B virus, early cancer detection, measurement of growth hormone levels, tracking of the leukemia virus, diagnosis and treatment of peptic ulcers, and research with brain chemicals called neurotransmitters.

MAJOR ADVANTAGES OF RIA

The following are some of the important advantages of RIA.

1. It can be used to assay any molecule which is immunogenic.
2. It is highly sensitive.
3. It has high specificity.
4. It can be automated. This enables to process a large number of samples in a short period of time.

MAJOR DISADVANTAGES OF RIA

The following are the main disadvantages of RIA.

1. The equipment and reagents are very costly.
2. The shelf life period of reagents is very less. For example, the half-life period of ^{125}I is only 60 days.
3. It involves radioactive hazards.
4. It requires highly-skilled technical persons to perform the assay.

IMMUNORADIOMETRIC ASSAY (IRMA)

This utilizes the radiolabelled antibody. This technique is more popular in the detection of thyroid stimulating hormone (TSH). TSH contains two polypeptide chains, viz. alpha and beta chains. In IRMA antibody to alpha unit of TSH is bound to a solid support like Sephadex. To this, the serum sample is added (Figure 25.4). The TSH present in the sample will bind to the antibody by its alpha subunit. Then the radiolabelled antibody to the beta subunit is added. It will bind to the beta subunit of the TSH. After the washing procedure the radioactivity is measured to know the level of TSH.

The same principle is applied to detect many other compounds in the field of biochemistry utilizing the two different subunits of a molecule.

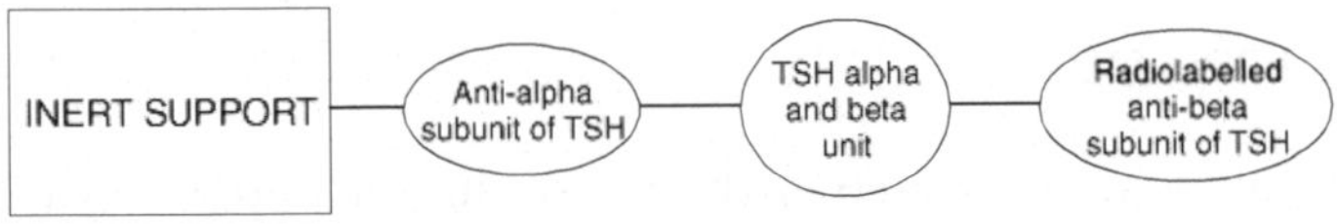

Figure 25.4 Principle of IRMA

RADIOALLERGOSORBENT TECHNIQUE (RAST)

This is a classical technique used to detect allergen-specific IgE in the serum. A specific allergen is bound to a solid-phase support with a microtitre well, glass or magnetized beads or some other inert surface. Patient serum containing IgE, both specific and non-specific for the testing allergen, is incubated with the solid phase material, allowing reaction of the specific IgE in the patient sample. Excess serum and non-allergen specific IgE are then washed away. Radiolabelled anti-IgE antibody conjugate is added (Figure 25.5). During this second incubation period a sandwich complex of "allergen–patient IgE allergen-specific antibody–labelled anti-IgE" is formed. A subsequent wash removes unbound labelled antibody. Measurement of the remaining radiolabelled anti-IgE is directly proportional to the patient's allergen-specific IgE.

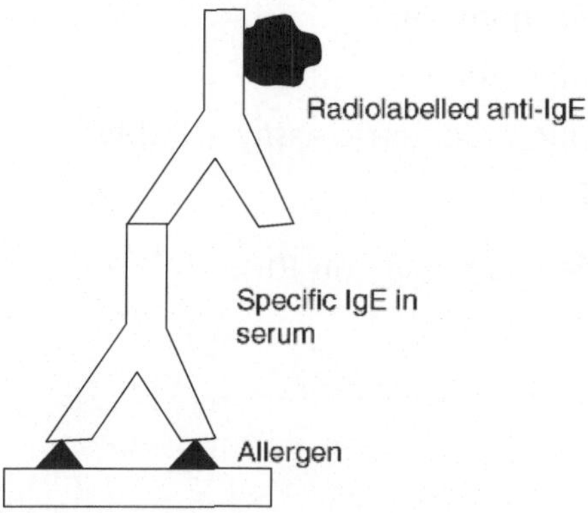

Figure 25.5 Principle of RAST

POINTS TO REMEMBER

- Radioimmunoassay (RIA) is one of the important labelled immunoassay technique used in the detection of hormones, steroids, drugs and microbial antigens.
- It was developed by Rosalyn Yalow and Solomon Aaron Berson in the 1950s.
- The technique is based on the competition between the known fixed amount of labelled antigen and unlabelled antigen for the limited number of antigen binding site of the fixed amount of antibody.

- A calibration curve is made by adding different concentration of unlabelled antigen.
- Two radioisotopes are normally used for radiolabelling of antigens in RIA. They are ^{125}I or ^{3}H.
- Radioimmunoassay is widely used because of its great sensitivity.

REVIEW QUESTIONS

1. Write short notes on:
   i. Radiolabelling
   ii. Immunoradiometric assay (IRMA)
   iii. RAST
2. Write a detailed essay on RIA.

26

ENZYME IMMUNOASSAY

INTRODUCTION

Due to the radioactive hazards involved in radioimmunoassay, immunologists explored an alternative label to perform the immunoassay techniques. The best label that was identified was the enzyme. Many enzymes can be attached to both the antigen and antibody. The enzyme label never requires any sophisticated instruments like gamma counter or liquid scintillation counter like RIA. Since enzymes are highly active and can be detected at low concentrations, they are effective labels. Enzyme-labelled antibodies can be detected when they are exposed to a substrate that enzymes can change. Normally, a substrate that changes colour as a result of enzyme action is used. The amount or rate of colour change can then be used to measure the amount of antibody present. Enzyme labels provide sensitivity similar to that of radioactive labels and have several important advantages—they are stable, inexpensive and safe to use, and they can be used successfully without sophisticated equipment.

Enzyme immunoassay (EIA) is of two types.

1. *Homogenous EIA* It is a homogeneous assay in the sense that it never requires any washing steps. All the reagents are added at a time and the results are read, e.g. enzyme-multiplied immunoassay technique (EMIT).

2. *Heterogeneous EIA* In this assay, the reagents are added in different steps interrupted by the washing procedure, e.g. enzyme linked immunosorbent assay (ELISA).

ENZYME-MULTIPLIED IMMUNOASSAY TECHNIQUE (EMIT)

Enzyme-multiplied immunoassay technique (EMIT) is a homogeneous single phase EIA procedure in which the antigen being measured competes for a limited number of antibody-binding sites with enzyme-labelled antigen. The reagent antibody has the ability to block enzymatic activity when bound with the reagent enzyme–antigen complex preventing its formation of product in the presence of substrate. The free antigen–enzyme complexes resulting from competition with measured antigen in the sample form colour-change products proportional to the concentration of antigen present in the specimen.

EMIT is a common method for screening urine and blood for drugs, whether legal or illicit.

The assay is based on competition between drug in the sample and drug labelled with the enzyme glucose 6-phosphate dehydrogenase (G6P-DH) for antibody-binding sites. Enzyme activity decreases upon binding to the antibody, so the drug concentration in the sample can be measured in terms of enzyme activity. Active enzyme converts oxidized nicotinamide adenine dinucleotide (NAD) to NADH, resulting in an absorbance change that is measured spectrophotometrically.

The technique is relatively non-specific compared to some other analysis methods, such as mass spectrometry but has the advantage of being fast and inexpensive. It has, however, been shown in some cases to be somewhat inaccurate in its findings. The patent for EMIT technology is owned by the Syva Corporation of Palo Alto, CA.

In March 2005, the patent for EMIT technology was sold for an undisclosed price to the ORAK Corporation (San Francisco).

ENZYME-LINKED IMMUNOSORBENT ASSAY (ELISA)

ELISA or enzyme-linked immunosorbent assay is an immunoassay technique involving the reaction of antigen and antibody *in vitro*. ELISA is a sensitive and specific assay for the detection and quantitation of antigens or antibodies. ELISA tests are usually performed in microwell plates. Both direct and indirect ELISA can be used for antigen (or antibody) detection, but indirect ELISA is more common.

It is a method for detection of small concentrations of substances such as proteins, hormones or metabolites in samples.

Direct ELISA

Direct ELISA uses the method of directly labelling the specific antibody. Microwell plates are coated with a sample containing the target antigen and the binding of labelled antibody is quantitated by a colorimetric method. Since the secondary antibody step is omitted, direct ELISA is relatively quick, and avoids potential problems of cross reactivity of the secondary antibody with components in the antigen sample. However, direct ELISA requires the labelling of every antibody to be used, which can be a time-consuming and expensive method. In addition, certain antibodies may be unsuitable for direct labelling. Direct methods also lack the additional signal amplification that can be achieved with the use of a secondary antibody.

Indirect ELISA

The indirect ELISA utilizes an unlabelled primary antibody (specific antibody) in conjunction with a labelled secondary antibody. Since the labelled secondary antibody is directed against all antibodies of a given species (e.g. anti-mouse), it can be used with a wide variety of primary antibodies (e.g. all mouse monoclonal antibodies). The use of secondary antibody also provides an additional step for signal amplification, increasing the overall sensitivity of the assay.

Competitive ELISA

There are two advantages in this ELISA technique.

1. It utilizes only one specific antibody for the antigen to be tested. Other techniques require at least two specific antibodies raised in different animals for the detection of unknown antigen.
2. Another advantage of the competitive ELISA is that non-purified primary antibodies may be used.

There are several different ways for performing the competitive ELISA.

Briefly, an unlabelled purified primary antibody is coated onto the wells of a 96-well microtitre plate. This primary antibody is then incubated with unlabelled standards and unknowns. After this reaction is allowed to go to equilibrium, labelled immunogen (conjugate) is added (Figure 26.1). This conjugate will bind to the primary antibody wherever its binding sites are not already occupied by unlabelled immunogen. Thus, the more immunogen in the sample or standard, the lower the amount of conjugated immunogen bound. The plate is then developed with substrate, and colour change is measured. More the colour, less is the unknown antigen.

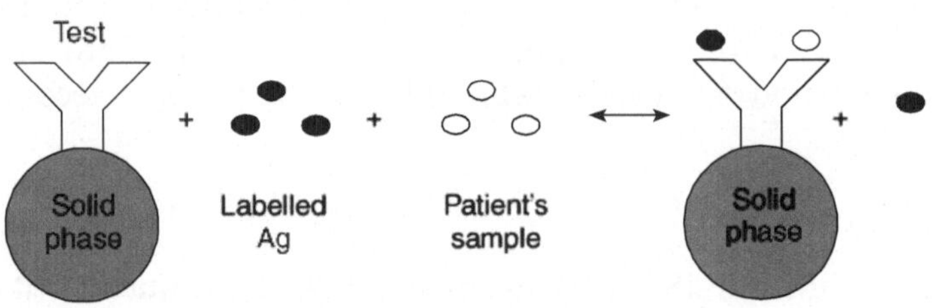

Figure 26.1 Principle of competitive ELISA

Double-antibody sandwich ELISA (DAS ELISA)

This technique is used to detect the unknown antigen. The microtitre plate is coated with specific antibody.

The sample (antigen) is added in the microtitre plate and is incubated. Antigens specific to the bound trapping antibody will attach to it but other proteins remain in solution and are removed by washing (Figure 26.2). The antigen attached to the trapping antibody is detected by adding a labelled antibody specific to the antigen. The label is the enzyme (E) previously conjugated to the antibody. When substrate specific to the enzyme is added in the final step, a colour develops as a result of enzyme action. The amount of colour and its rate of development are correlated to the amount of labelled antibody bound to the antigen which had been trapped by the antibody attached to the plate (Figure 26.3).

This technique can also be referred to as antigen-capture ELISA.

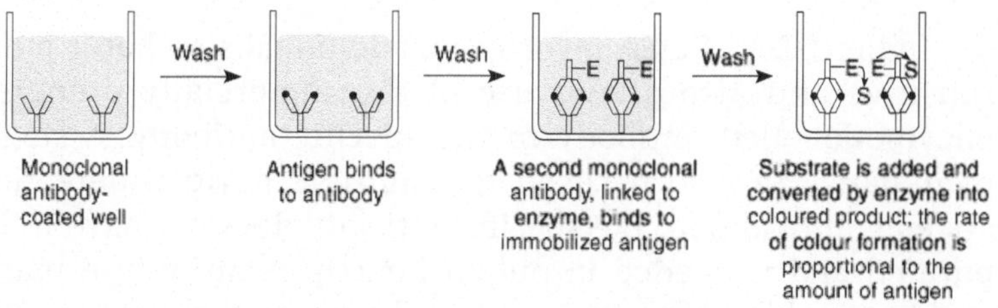

Figure 26.2 Principle of double-antibody sandwich ELISA

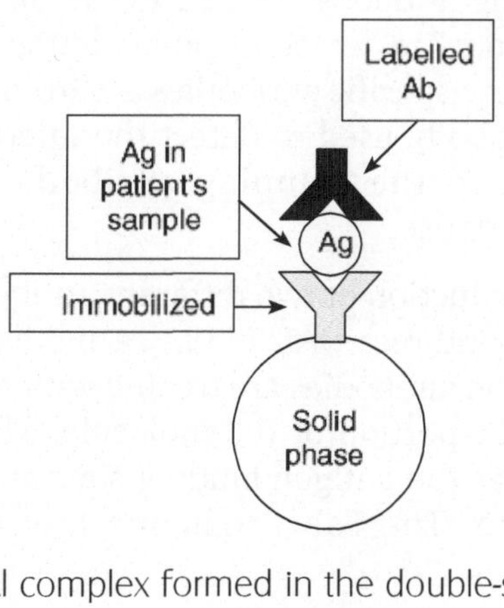

Figure 26.3 Final complex formed in the double-sandwich ELISA

Indirect Double-antibody Sandwich ELISA

It is similar to the DAS ELISA, however, the antigen bound to the trapping antibody (antibody coated in microtitre plate) is detected by an unlabelled specific antibody which is specific to the same antigen but is from an animal species different from the one used to prepare the trapping antibody. For example, if the trapping antibody was prepared in rabbits, the detecting antibody could be from a mouse or a goat. The unlabelled specific antibody which attaches to the antigen is detected by an enzyme-labelled antibody (conjugate) specific to the class of specific antibody. Because the specific antibody is from a species different from the trapping antibody, the conjugate binds only to the specific antibody and no non-specific binding to the trapping antibody occurs. The amount of conjugate is measured by adding substrate and measuring colour change as in DAS ELISA.

Indirect DAS ELISA involves an additional step but is more sensitive and also allows use of a commercially prepared enzyme-labelled antibody to the specific antibody. A single conjugate can also be used for multiple antigen detection systems. In addition, the specific antibody does not have to be purified and is needed in only a limited quantity. The major problem in this technique is that antibodies to the same antigen must be prepared in two different animals (one to be used as "antigen-trapping antibody" coated in the plate and another is a "specific antibody" to detect the bound antigen). If the trapping antibodies and the specific antibodies are from the same species, the labelled antibody used to detect the intermediate antibody will also bind to the trapping antibody and result in a non-specific response.

To avoid production of two different antibodies for the same antigen for the indirect DAS ELISA, a methodology has been standardized. The antibodies are treated with the enzyme pepsin to remove the Fc portion of the molecule. The remaining Fab, fragment still has the antigen binding sites and will bind to the microtitre plate. The Fab fragments are used as trapping antibodies and the whole antibody is used as the specific antibody. An enzyme-conjugated protein A is then used instead of a labelled antibody to detect the specific antibody. It does not react to the trapping antibody because the Fc region has been removed and protein A can attach only to Fc portion and not to Fab′.

Indirect DAS ELISA procedure can be further modified to amplify the reaction. This is commonly done using a biotin–avidin interaction in which the labelled antibody is biotinylated to react with avidin molecules conjugated to multiple enzyme molecules.

The sensitivity of the double-antibody sandwich ELISA is dependent on four factors.

1. The number of molecules of the first antibody that are bound to the solid phase.
2. The avidity of the first antibody for the antigen.
3. The avidity of the second antibody for the antigen.
4. The specific activity of the second antibody.

Solid-phase ELISA or Plate-trapped Antigen ELISA

The solid-phase ELISA or plate trapped antigen ELISA uses antigen-coated microtitre plates, a secondary antibody (for example, goat anti-mouse IgG conjugated with horseradish peroxidase) and a chromogenic substrate. This technique is used to detect the unknown antibody in the patient's serum (Figure 26.4). The output signal (absorbance of the chromogenic substrate following oxidation) depends on the formation of primary antibody–antigen complexes. Under certain conditions, the assay can be linear with respect to antibody concentration and therefore, useful for comparing the relative concentrations of antibodies in different samples. One factor to consider in this assay is that the antigen bound to a surface can be different from the antigen in solution. This is particularly true for small linear peptides (Figure 26.5).

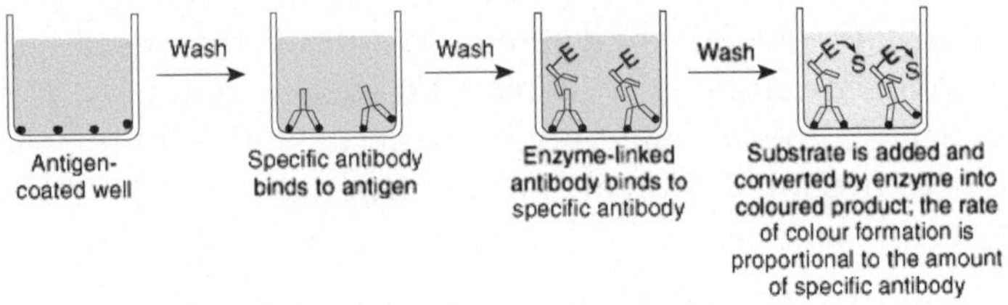

Figure 26.4 Principle of antigen-trapped ELISA

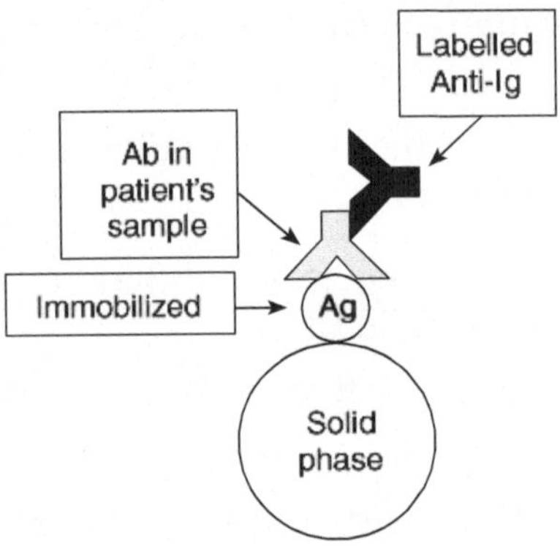

Figure 26.5 Final complex formed in antigen-trapped ELISA

IgM Antibody-capture ELISA (MAC ELISA)

IgM antibody capture ELISA (MAC ELISA) has been developed to detect specific IgM isotype antibodies. Solid-phase support (usually microtitre plate wells) is coated with anti-human IgM antibodies capable of binding all IgM isotype antibodies present in the specimen. The standard antigen is then added, followed by enzyme-labelled antigen-specific antibodies. If IgM antibodies specific for the antigen in question are present, the "sandwich" complex will result in enzymatic colour change proportional to the concentration of IgM-specific antibody present. This technique appears to be the method of choice in many highly specific and more sensitive assays for IgM antibodies for infectious diseases.

MICROPARTICLE ENZYME IMMUNOASSAY (MEIA)

Microparticle enzyme immunoassay (MEIA) is a technique in which the solid-phase support consists of very small microparticles in liquid suspension. Specific antibodies are covalently bound to the microparticles. Antigen, if present, is then "sandwiched" between bound antibodies and antigen-specific, enzyme-labelled antibodies. Antigen–antibody complexes are detected and quantitated by analysis of fluorescence from the enzyme–substrate interaction.

PREPARATION OF ENZYME-LABELLED ANTIBODIES

Conjugated molecules of antibody and enzyme can be prepared in several ways. Alkaline phosphatase is the most widely used enzyme, and the single-step glutaraldehyde method is commonly used to prepare alkaline phosphatase conjugates. Alkaline phosphatase is usually obtained commercially as an enzyme precipitate in a salt solution. The precipitate is recovered from solution by centrifugation and about 5 mg is dissolved in 2 ml of a 1-mg/ml solution of IgG. The mixture is dialysed thoroughly to remove excess salts and then fresh glutaraldehyde is added to a final concentration of 0.05 percent. After 4-hour incubation, the conjugate is dialysed three times to remove the glutaraldehyde and stored at 4°C in phosphate buffered saline (PBS) containing 0.04 per cent sodium azide and 5 mg per ml bovine serum albumin. Good quality conjugates can normally be used at least 1/500 dilution and dilutions as great as 1/5000 or more may be possible.

Conjugates are stable for long periods at 4°C. Freezing or freeze-drying is not recommended unless preliminary testing indicates it is possible. If freezing is necessary, 50 per cent glycerol is added.

Another important enzyme used in EIA is horseradish peroxidase enzyme (HRP). Peroxidase has been found well suited for the preparation of enzyme-conjugated antibodies, due in part to its ability to yield chromogenic products, and in part to its relatively good stability characteristics. Peroxidase-labelled immunoglobulins have been used successfully as immunohistological probes for the demonstration of tissue antigens and in enzyme-amplified immunoassay systems for the quantitative determination of soluble and insoluble antigens.

The HRP labelling of antibodies can be done by the following method.

- Dissolve 2 mg horseradish peroxidase (HRP) in 1 ml water (solution A).
- Dissolve 21.4 mg $NaIO_4$ in 1 ml water (solution B).
- Add 100 microlitre of B into A; colour will change to dark green.
- Wait for 10 minutes at room temperature.
- Put into the dialysis tube.
- Put the tube into 5 mM sodium acetate buffer, pH 4.0 in a 2–3-litre flask.
- Dialyse overnight; colour will change to gold.
- Raise the pH of the HRP solution to pH 9.0 by the addition of 0.2 M sodium carbonate buffer, pH 9.5.
- Mix with the antibody solution (8 mg of IgG in 1 ml), which has been pre-dialysed to 0.01 M sodium carbonate buffer, pH 9.0, overnight.
- Incubate the mixture for 2 hours at room temperature.
- Add freshly prepared 100 microlitre of 0.1 M $NaHBr_4$ in water to the solution.
- Incubate at 4°C for 2 hours.
- Put the mixture into a dialysis tube and dialyse against PBS overnight.
- Now the conjugate solution is ready for use. Add thimerosal to a final concentration of 0.02% for preservation. Add glycerol to a final concentration of 10% (optional).

IMMUNOSORBENTS USED IN ELISA

Another fundamental concept for ELISA is that proteins such as antibodies and antigens will adsorb strongly to the surface of certain plastics such as polystyrene and polyvinyl chlorides. Protein binding also occurs to some forms of cellulose nitrate. These materials are frequently referred to as "immunosorbents" or the "solid phase" in ELISA protocols (Figure 26.6). The protein binding to immunosorbent materials is not specific and is not a serological reaction such as that occurring between antigen and antibody molecules. If a mixture of antibodies is exposed to an immunosorbent plastic, all will bind.

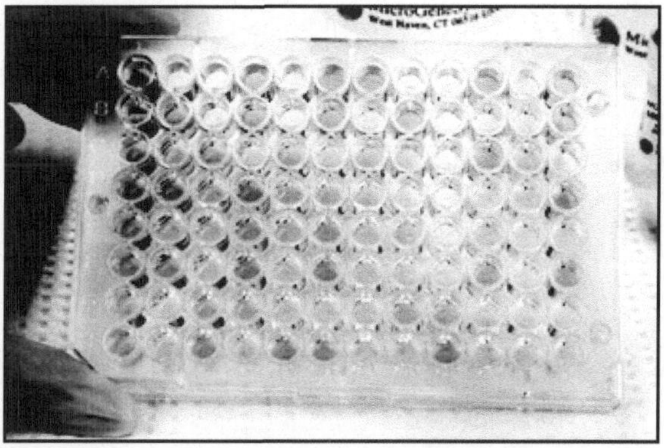

Figure 26.6 Microtitre plate used to perform ELISA technique

Binding either the antibody or the antigen component of a serological system to a solid phase is very useful because the bound component can be subsequently used to probe complex mixtures of potential reactants. Only those that are serologically related will be trapped. All non-reactive components can then be removed by washing and will not interfere with the subsequent steps.

Undesired adsorption of antibody or antigen proteins to the plastic can be avoided by using non-ionic detergents such as Tween 20 by incubating solutions or by adding an excess of a non-specific protein to block all sites not occupied by the desired serological component. For example, the buffer used to coat plates with trapping antibody does not contain Tween 20 but Tween 20 is incorporated in subsequent steps where any non-specific binding of other proteins should be avoided.

POINTS TO REMEMBER

- The EIA is of two types. They are: homogeneous EIA and heterogeneous EIA.
- Homogeneous EIA is a homogeneous assay in the sense that it never requires any washing steps. All the reagents are added at a time and the results are read, e.g. enzyme-multiplied immunoassay technique (EMIT).
- In heterogeneous EIA, the reagents are added in different steps interrupted by washing procedure, e.g. enzyme-linked immunosorbent assay (ELISA).
- EMIT is a common method for screening urine and blood for drugs, whether legal or illicit.
- ELISA or enzyme-linked immunosorbent assay is an immunoassay technique involving the reaction of antigen and antibody *in vitro*.
- Direct ELISA uses the method of directly labelling the specific antibody.
- Indirect ELISA utilizes an unlabelled primary antibody (specific antibody) in conjunction with a labelled secondary antibody.
- IgM antibody-capture ELISA (MAC ELISA) has been developed to detect specific IgM isotype antibodies.

REVIEW QUESTIONS

1. Write short notes on:
   i. EMIT
   ii. Indirect ELISA
   iii. Direct ELISA

iv. Double-sandwich ELISA
v. IgM antibody-capture ELISA
vi. Enzyme labels
vii. Microparticle enzyme immunoassay (MEIA)

27

FLUORESCENCE-ACTIVATED CELL SORTER

INTRODUCTION

The Fluorescence-activated cell sorter (FACS) was invented in the late 1960s by Bonner, Sweet, Hulett, Herzenberg and others to do flow cytometry and cell sorting of viable cells. The fluorescence-activated cell sorter is a machine that can rapidly separate the cells in a suspension on the basis of size and the colour of their fluorescence. Fluorescence-activated cell sorting is a specific type of flow cytometry, which utilizes fluorescent markers (fluorescently labelled monoclonal antibodies) placed on the cells for the purpose of recognizing and sorting the cells.

Fluorescence-activated cell sorters or FACS, possess the ability to perform multiparameter analyses on a single cell. Among the many measurable properties are size, volume, viscosity, the content of DNA, RNA and enzymes and also surface antigens. The applications are very broad for this type of technology.

FLOW CYTOMETER

Cytometry refers to the measurement of physical and/or chemical characteristics of cells or any other biological particle.

Flow cytometry is a process in which such measurements are made while the cells or particles pass, preferably in single file, through the measuring apparatus in a fluid stream. Flow sorting extends flow cytometry by using electrical or mechanical means to divert and collect cells with one or more measured characteristics falling within a range or ranges of values set by the user.

Thus flow cytometry is a method for quantitating components or structural features of cells primarily by optical means. Although it measures on one cell at a time, it can process thousands of cells in a few seconds. Since different cell types can be distinguished by quantitating structural features, flow cytometry can be used to count cells of different types in a mixture.

Instrumentation

Flow cytometers involve sophisticated fluidics, laser optics, electronic detectors, analog to digital converters and computers.

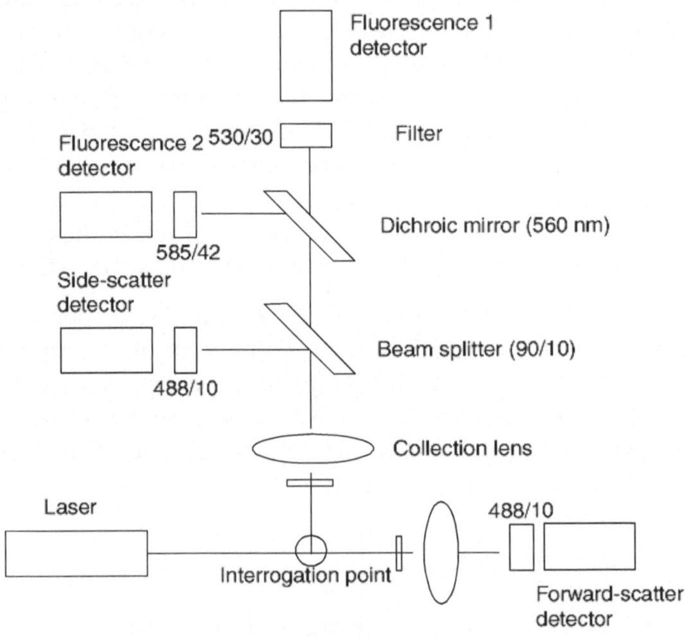

Figure 27.1 Instrument layout of flow cytometer

The optics deliver laser light focused to a beam a few cell diameters across. The fluidics hydrodynamically focus the cell stream within an uncertainty of a small fraction of a cell diameter (Figure 27.1).

In sorters, the stream breaks into uniform-sized droplets to separate individual cells. The electronics quantitate the faint flashes of scattered and fluorescent light. Under computer control, it electrically charges the droplets containing cells of interest so that they can be deflected into a separate test tube or culture wells. The computer records data for thousands of cells per sample and displays the data graphically (Figure 27.2).

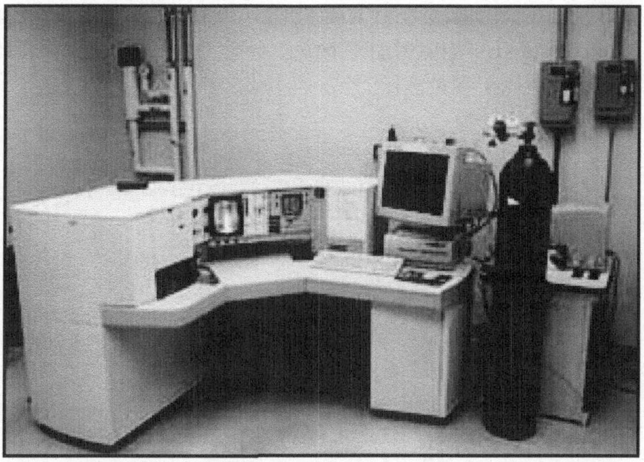

Figure 27.2 Instrumentation of flow cytometer

Working Methodology

The following are the steps involved in FACS:

- Before any procedure is performed, the fluorescence-activated cell sorter must be aligned and calibrated in order to obtain accurate and precise data.
- The desired single cell suspension is isolated from the blood or tissue and labelled with a fluorescent antibody.

- For each individual cellular property being analysed, a different "colour" fluorochrome or monoclonal antibody must be used so that the computer can electronically distinguish the properties of the cell surface.
- These fluorescent markers allow the laser to recognize the cell and record data of that particular cell. These fluorescent markers emit different wavelengths of light that may overlap each other into the band of emissions being collected by the FACS. It is important to electronically compensate for the overlap by subtracting the contaminating signals (Figure 27.3).
- By the application of air pressure, the cells are forced through a nozzle at high speed. As the cells pass through the nozzle, they are met with a liquid jet of saline or water that sheaths the cells for protection.

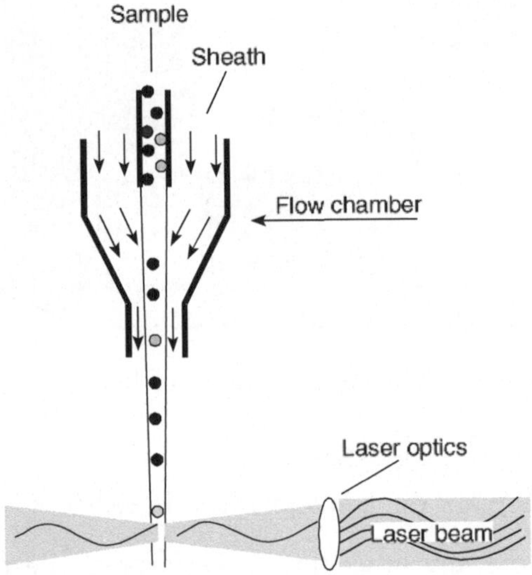

Figure 27.3 Passing of cell suspension through the nozzle

- Vibrations at the tip of the nozzle interrupt the stream in order to break it up into a series of droplets. The droplet size can be regulated in such a way that each droplet contains a single cell (Figure 27.4).

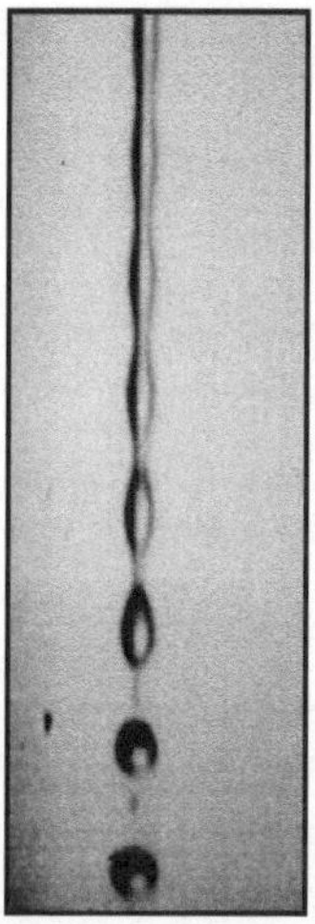

Figure 27.4 Photograph showing the formation of droplets from the nozzle

- A monochromatic laser beam is allowed to illuminate the droplets, which are electronically monitored by fluorescent detectors.
- The droplets that emit the proper fluorescent wavelengths are electrically charged between deflection plates in order to be sorted into collection tubes (Figure 27.5).
- If a heterogeneous suspension of cells need to be analysed, the cytometer will ignore the particles whose light scatter characteristics do not meet the previously defined parameters. This process is known as gating, because it excludes any undesirable particles or cells to provide the desirable single cell suspension.
- Flow cytometric data is primarily displayed as a histogram, or plot. The X-axis of the histogram displays

the fluorescence intensity, which is usually measured on a log scale. The Y-axis displays the number of cells found within each parameter. When measuring three or more parameters, the histogram is usually displayed in a three-dimensional, colour-coordinated display.

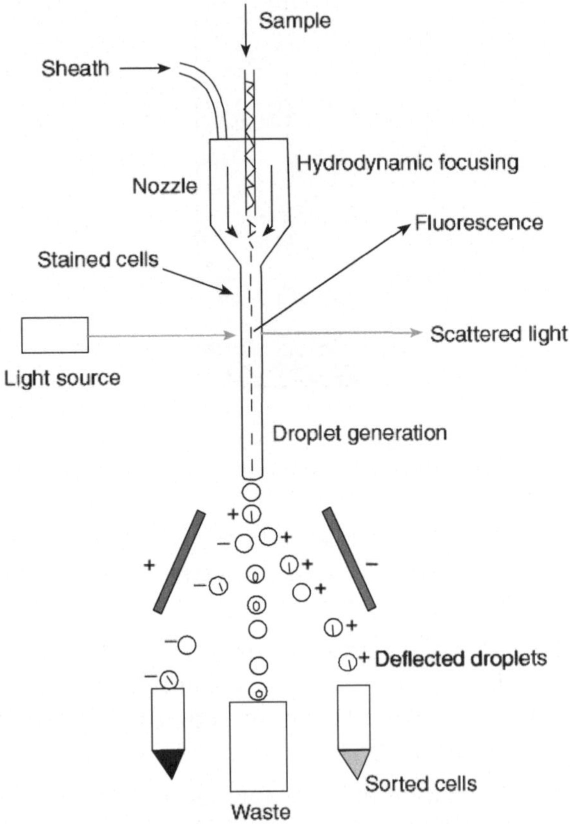

Figure 27.5 Working principle of FACS

APPLICATIONS OF FACS

FACS is a valuable tool in clinical immunology to separate and count various cells involved in immune response (Table 27.1).

Table 27.1 Properties and applications of various fluorescent dyes used in FACS

Fluorochrome	**Excitation (nm)**	**Emission (nm)**	**Laser type**	**Applications**
Fluorescein	495	520	Argon	Phenotypic analysis
Phycoerythrin	495	575	Argon	Phenotypic analysis
Tricolour	488	650	Argon	Phenotypic analysis
Coumarin	357	460	Argon	Phenotypic analysis
Allophyco-cyanin	630	660	Helium–Neon	Phenotypic analysis
Cascade blue	350	480	Argon	Phenotypic analysis
Hoechst 33342	350	470	Argon	DNA analysis/ apoptosis
Hoechst 33258	350	475	Argon	DNA analysis/ chromosome staining
DAPI	372	456	Argon	DNA
Chromomycin A3	457	600	Argon	DNA analysis/ chromosome staining
Propidium iodide	495	637	Argon	DNA analysis
Ethidium bromide	493	620	Argon	DNA analysis
Acridine orange	503	530/640	Argon	DNA, RNA
Fluorescein diacetate	488	530	Argon	Live/dead discrimination

(Contd.)

Table 27.1 (Continued)

Fluorochrome	Excitation (nm)	Emission (nm)	Laser type	Applications
SNARF-1	488	530/640	Argon	pH measurement
Indo-1	349	425/490	Argon	Calcium flux measurement
Fluo-3	488	530	Argon	Calcium flux measurement
Rhodamine 123	515	580	Argon	Mitochondria
To-Pro-3	642	661	Helium–Neon	Live/dead discrimination, DNA staining

Flow cytometry has been the method of choice for monitoring CD4 lymphocyte levels in the blood of AIDS patients. Lymphomas and leukemias are intensively studied for surface markers of diagnostic and prognostic value. In the diagnosis of leukemia, flow cytometry may be used for the immunophenotypic analysis of abnormal cells by focusing on cell lineage. It is helpful when other morphological tests are ambiguous. It is also used in the subclassification of non-Hodgkin's lymphoma.

After the FACS apparatus has sorted the cells, the living cells can be used for cell repopulating in immunodeficient individuals. These cells may also be used to test their ability to co-operate with the immune system or for their response to specific antigens.

In renal, cardiac and bone marrow transplants, flow cytometry is used in discriminating between graft rejection and viral infections in post-operative patients.

POINTS TO REMEMBER

- The fluorescence-activated cell sorter (FACS) was invented in the late 1960s by Bonner, Sweet, Hulett, Herzenberg and others.
- Flow cytometry is a method for quantitating components or structural features of cells primarily by optical means.
- Flow cytometers involve sophisticated fluidics, laser optics, electronic detectors, analog to digital converters and computers.
- FACS is a valuable tool in clinical immunology to separate and count various cells involved in immune response.

REVIEW QUESTIONS

1. Write a detailed account on FACS.

GLOSSARY

Accessory cell A cell required to initiate immune response, often described as antigen-presenting cell.

Accessory molecules Molecules other than the antigen recep-tor and major histo-compatibility complex (MHC) that participate in cognitive, activation, and effector functions of T lymphocyte response.

Acute-phase response An early response stimulated by cytokines which include interleukin-1, interleukin-6, interferons and tumour necrosis factor. As a result, acute phase proteins are formed. These include C-reactive protein, levels of which rise within a few hours of the response. Infection, inflammation, tissue injury, and, very infrequently, neoplasm may be associated with it.

Adaptive immune response The response of antigen-specific lymphocytes to antigen including the development of immunological memory. It is also known as acquired immune response.

Adhesion molecules The molecules which mediate the binding of one cell to other cells or to extracellular matrix proteins. Some of the examples are integrins, selectins, and members of the immunoglobulin gene superfamily. These molecules are important in the operation of the immune system.

Adjuvant A substance, which when given with antigen, enhances the immune response to the injected antigen.

Adoptive transfer The transfer of immunocompetent cells from one individual to another individual.

Affinity A measure of the binding constant of a single antigen combining site with a monovalent antigenic determinant.

Affinity chromatography The purification of a substance by means of its affinity for another substance immobilized on a solid support. For example, an antigen can be purified by affinity chromatography on a column of antigen-specific antibody molecules covalently linked to beads (immunoaffinity chromatography).

Affinity maturation The sustained increase in affinity of antibodies for an antigen with time following immunization. The genes encoding the antibody variable regions undergo somatic hypermutation with the selection of B lymphocytes whose receptors express high affinity for the antigen.

Agammaglobulinemia Absence of immunoglobulin due to lack of mature B cells. Due to defective tyrosine kinase btk, B cell differentiation does not progress beyond pre-B cell. It is a X-linked disease.

Agglutination The aggregation of particulate antigen by antibodies.

Alleles Two or more alternate forms of a gene that occupy the same position or locus on a specific chromosome.

Allelic exclusion The ability of lymphoid cells to produce only one allelic form of antigen-specific receptor (BCR or TCR). The counterpart of the allele's expression in excluded.

Allergen An antigen responsible for producing allergic reactions by inducing IgE synthesis.

Allergic asthma The constriction of the bronchial tree due to an allergic reaction to inhaled allergen. It belongs to type I hypersensitivity.

Allergic rhinitis It is an allergic reaction in the nasal mucosa, also known as hay fever. It causes running nose, sneezing and tears. It belongs to type I hypersensitivity.

Allergy A term used as a synonym for the hypersensitivity reactions.

Allogeneic The term that describes genetic variations or differences among members or strains of the same species. Genetically dissimilar within the same species is described as allogeneic.

Allograft A tissue transplant (graft) between two genetically nonidentical members of a species.

Allotypes Allelic variants of immunoglobulins detected by antibodies raised between members of the same species.

Alpha-foetoprotein A principal plasma protein in the alpha globulin fraction present in the foetus. Patients with liver cancer patients reveal significantly elevated serum levels of α-foetoprotein.

Alternate complement pathway The mechanism of complement activation that does not involve activation of the C1-C4-C2 pathway by antigen–antibody complexes, and begins with the activation of C3.

Alveolar macrophage A macrophage found in the lung alveoli that may be involved in phagocytosis.

Anamnestic It literally, means "does not forget". It is used to describe immunological memory, which leads to a rapid increase in response after re-exposure to antigen.

Anaphylatoxin Substance capable of releasing histamine from mast cells and basophils.

Anaphylaxis Immediate hypersensitivity response to antigenic challenge, mediated by IgE and mast cells. It is a life-threatening allergic reaction, caused by the release of pharmacologically active agents.

Anergy A state of antigen-specific unresponsiveness of T or B cell.

Antibody Serum protein formed in response to antigen stimulation. The antibodies are generally defined in terms of their specific binding to the immunizing antigen.

Antibody-dependent, cell-mediated cytotoxicity (ADCC) A phenomenon in which target cells, coated with antibody, are destroyed by specialized killer cells (NK cells and macrophages), which bear receptors for the Fc portion of the coating antibody (Fc receptors). These receptors allow the killer cells to bind to the antibody-coated target.

Antigen Any foreign material that induces an immune response and is specifically bound by antibody or lymphocytes.

Antigen-binding site The location on an antibody molecule where an antigenic determinant or epitope combines with it. The antigen-binding site is located in a cleft bordered by the N-terminal variable regions of heavy and light chain parts of the Fab region.

Antigenic determinant A single antigenic site or epitope on a complex antigenic molecule or particle which stimulates for the production of a specific antibody.

Antigen presentation The display of antigen as peptide fragments bound to MHC molecules on the surface of a cell to the T cells. The T cells recognize antigen only when it is presented in this way.

Antigen processing The degradation of proteins into peptides that can bind to MHC molecules for presentation to T cells. All protein antigens must be processed into peptides before they can be presented by MHC molecules.

Antigen-presenting cell (APC) A specialized type of cell, bearing cell-surface class II MHC molecules, involved in processing and presentation of antigen to T cells.

Antigen receptor The specific antigen-binding receptor on T or B lymphocytes; these receptors are transcribed and translated from rearrangements and translocation of V, D, and J genes. The antigen binding site in B cell is called as B-cell receptor (BCR) and that of T cell is called as T-cell receptor (TCR).

Anti-immunoglobulin antibodies Antibodies against immunoglobulin constant domains that are useful for detecting bound antibody molecules in immuno-assays and other applications.

Antiserum The serum of an immune individual that contains a heterogeneous collection of antibodies against the molecule used for immunization. Such antibodies bind the antigen used for immunization.

Antitoxin Antibody specific for exotoxins produced by certain microorganisms such as the causative agents of diphtheria and tetanus.

Apoptosis It is the programmed cell death caused by activation of endogenous molecules leading to the fragmentation of DNA.

Appendix A gut-associated lymphoid tissue located at the beginning of the colon.

Arthus reaction A hypersensitivity reaction produced by local formation of antigen–antibody aggregates that activate the complement cascade and cause thrombosis, haemorrhage, and acute inflammation.

Atopy A term used to describe IgE-mediated allergic responses in humans, usually showing a genetic predisposition.

Autochthonous Pertaining to self.

Autograft A tissue transplant from one area to another on a single individual.

Autoimmunity An immune response to self tissues or components. Such an immune response may have pathological consequences leading to autoimmune diseases.

Autologous Derived from the same individual, self.

Avidity The summation of multiple affinities, for example, when a polyvalent antibody binds to a polyvalent antigen.

B7 A co-stimulatory homodimeric immunoglobulin superfamily protein whose expression is restricted to the surface of accessory cells (e.g. B cells and macrophages) that interact with T lymphocytes. The ligand for B7 is CD28.

Basophils White blood cells containing granules those stain with basic dyes. They have a function similar to mast cells.

BCG (Bacillus Calmette–Guerin) A *Mycobacterium bovis* strain that has been used as the vaccine for tuberculosis.

Bence–Jones protein Dimers of immunoglobulin light chains in the urine of patients with multiple myeloma.

B lymphocyte (B cell) The precursor of an antibody-forming plasma cell; expresses immunoglobulin on its surface.

Blocking antibody An antibody molecule capable of blocking the interaction of antigen with other antibodies or with cells.

Bone marrow The site of haematopoiesis, in which stem cells give rise to the cellular elements of blood, including red blood cells, monocytes, polymorphonuclear leucocytes, platelets, and lymphocytes.

Bone marrow transplantation A procedure used to treat both nonneoplastic and neoplastic conditions not amenable to other forms of therapy. It has been especially used in cases of aplastic anaemia, acute lymphocytic leukemia, and acute nonlymphocytic leukemia.

Bronchus-associated lymphoid tissue (BALT) Secondary lymphoid organ connected to the respiratory system.

Bursa of Fabricius An outpushing of the cloaca of the bird considered to be the site of B cell development in birds.

C1 deficiencies A complement deficiency leading to diseases like systemic lupus erythematosus, glomerulonephritis or pyogenic infections as well as an increased incidence of type III (immune complex) hypersensitivity diseases.

C1 esterase inhibitor A serum protein that inhibits the function of activated C1.

C1q An 18-polypeptide chain subcomponent of C1, the first component of complement. It commences the classical complement pathway.

C1q deficiency Deficiency of C1q may be found in association with lupus-like syndromes.

C1r A subcomponent of C1, the first component of complement in the classical activation pathway. It is a serine esterase.

C1s A serine esterase that is a subcomponent of C1. It is the first component of complement in the classical activation pathway. Ca^{2+} binds two C1s molecules to the C1q stalk.

C2 The third complement protein to participate in the classical complement pathway activation. C2 is a single polypeptide chain that unites with C4b molecules on the cell surface in the presence of Mg^{2+}.

C3 The fourth complement component to react in the classical pathway, and also a reactant in the alternative complement pathway. C3 contains α and β polypeptide chains, linked by disulphide bonds and has an internal thiol–ester bond which permits it to link covalently with surfaces of cells and proteins.

C3a A low molecular weight (9 kD) peptide fragment of complement component C3.

C3b The principal fragment produced when complement component C3 is split by either classical or alternative pathway convertases, i.e., C4b2a or C3bBb, respectively. It results from C3 convertase digestion of C3's α chain.

C3 convertase An enzyme that splits C3 into C3b and C3a. There are two types: one in the classical pathway designated C4b2a and one in the alternative pathway of complement activation termed C3bBb.

C3 tickover Alternative pathway C3 convertase perpetually generates C3b. C3 internal thiolester bond hydrolysis is the initiating event.

C4 A complement component that reacts immediately following C1 in the classical pathway of complement activation.

C5 A complement component comprised of α and β polypeptide chains linked by disulphide bonds that react in the complement cascade following C1, C4b, C2a, and C3b fixation to complexes of antibody and antigen.

C5b The principal molecular product that remains after C5a has been split off by the action of C5 convertase on C5. It has a binding site for complement component C6 and complexes with it to begin generation of the membrane attack complex (MAC).

C5 convertase A molecular complex that splits C5 into C5a and C5b in both the classical and the alternative pathway of complement activation.

C5 deficiency A very uncommon genetic disorder that has an autosomal recessive mode of inheritance. Individuals with this deficiency have a defective ability to form the membrane attack complex (MAC), which is necessary for the efficient lysis of invading microorganisms. They have an increased susceptibility to disseminated infections by *Neisseria* microorganisms.

C6 A complement component that participates in the membrane attack complex (MAC).

C6 deficiency A very uncommon genetic defect that has an autosomal recessive mode of inheritance in which affected individuals have only trace amounts of C6 in their plasma. They are defective in the ability to form a membrane attack complex (MAC).

C7 A complement component that forms a complex with C5b67 is formed when C7 binds to C5b and C6. The complex has the appearance of a stalk with a leaf type of structure.

C7 deficiency A very uncommon genetic disorder with an autosomal recessive mode of inheritance in which the serum of affected persons contains only trace amounts of C7 in the plasma. They have a defective ability to form a membrane attack complex (MAC) and show an increased incidence of disseminated infections caused by *Neisseria* microorganisms.

C8 A complement component comprising α and β chains that participates in the membrane attack complex (MAC).

C8 deficiency An uncommon genetic disorder with an autosomal recessive mode of inheritance in which affected individuals are missing one or more C8 chains. This is associated with a defective ability to form a membrane attack complex (MAC).

C9 A complement component comprised of a single chain protein that binds to the C5b678 complex on the cell surface. The interaction of 12 to 15 C9 molecules with one C5b678 complex produces the membrane attack complex (MAC).

C9 deficiency A very uncommon genetic disorder with an autosomal recessive mode of inheritance in which only trace amounts of C9 are present in the plasma of affected persons. There is a defective ability to form the membrane attack complex (MAC).

CAM Cell-surface adhesion molecule.

Carcinoembryonic antigen (CEA) A membrane glycoprotein epitope that is present in the foetal gastrointestinal tract in normal conditions. CEA levels are elevated in cases of carcinoma.

Carrier A large immunogenic molecule or particle to which a hapten or other non immunogenic, epitope-bearing molecule may attach, allowing it to become immunogenic.

Caspases A family of closely related cysteine proteases that cleave proteins at aspartic acid residues. They have important roles in apoptosis.

Cell-mediated cytotoxicity (CMC) Killing (lysis) of a target cell by an effector lymphocyte.

Cell-mediated immunity (CMI) Immune reaction mediated by T cells, in which the cells of immune system directly neutralize the antigen.

Central lymphoid organs The sites of lymphocyte development. In humans, B lymphocytes develop in bone marrow and lymphocytes develop in thymus from bone marrow-derived progenitors.

Chemokines Small cytokines of relatively low molecular weight, released by a variety of cells, and involved in the migration and activation primarily of phagocytic cells and lymphocytes. They have a central role in inflammatory responses.

Chemotaxis It is the phenomenon of migration of cells along a concentration gradient of an attractant.

Chimera An individual containing cellular components derived from another genetically distinct individual. In Greek mythology, an animal possessing the head of a lion, the body of a goat, and the tail of a snake.

Classical complement pathway The mechanism of complement activation initiated by antigen–antibody aggregates and proceeding by way of C1 to C9.

Clonal anergy The interaction of a B or T lymphocyte with an antigen leading to its functional inactivation. Anergy can result when the lymphocyte interacts with antigen in the absence of a second signal (co-stimulation), usually required for cell activation.

Clonal deletion The loss of lymphocytes of a particular specificity due to contact with either self or foreign antigen.

Clonal selection theory The prevalent concept that specificity and diversity of an immune response are the result of selection by antigen of specifically reactive clones from a large repertoire of preformed lymphocytes, each with individual antigenic specificities.

Cluster of differentiation (CD) Cluster of antigens present on the cell surface with which monoclonal antibodies react.

Cold agglutinin An antibody that agglutinates particulate antigen, such as bacteria or red cells, optimally at temperatures less than 37°. In clinical medicine, the term usually refers to antibodies against red blood cell antigens which develop in the "cold agglutinin syndrome."

Colony-stimulating factors (CSF) Glycoproteins that govern the formation, differentiation and function of haematopoietic progenitor cells (stem cells).

Combinatorial joining A phenomenon responsible for the antibody diversity and antigen receptor diversity. It is the joining of segments of DNA to generate new combinations of Ig or TCR genes during the development of B and T cells. Combinatorial joining allows multiple opportunities for two sets of genes to combine in different ways.

Complement A series of serum proteins involved in the mediation of immune reactions. The complement cascade is triggered classically by the interaction of antibody with specific antigen.

Complementarity-determining regions (CDRs) The parts of immunoglobulins and T-cell receptors that determine their specificity and make contact with specific ligands. The CDRs are the most variable part of the molecule and contribute to the diversity and the specificity of these molecules. There are three such regions (CDR1, CDR2, and CDR3) in each V domain.

Complement receptors (CR) Cell-surface proteins on various cells that recognize and bind complement proteins that have bound pathogens or other antigens. CR on phagocytes allows them to identify pathogens coated with complement proteins for uptake and destruction. The different complement receptors include CR1, the receptor for C1q, CR2, CR3, and CR4.

Conformational epitopes Discontinuous epitopes on a protein antigen that are formed from several separate regions in the primary sequence of a protein when brought together by protein folding. Antibodies that bind conformational epitopes bind only native, folded proteins.

Constant region (C region) The invariant carboxyl terminal portion of an immunoglobulin or TCR molecule, as distinct from the variable region at the amino terminus of the chain.

Contact dermatitis Type IV hypersensitivity which develops in response to an allergen applied in the skin.

Co-stimulatory molecules Membrane-bound or secreted products of accessory cells that activate signal transduction events in addition to those induced by MHC/TCR interactions. They are required for full activation of T cells.

C-reactive protein (CRP) It is a serum protein produced by liver cells as part of the acute phase response. It binds to phosphatidylcholine, a component of the surface of many bacteria. As a result of CRP binding to its surface, the bacterium is more easily destroyed by phagocytic cells.

Cross reactivity The ability of an antibody, specific for one antigen, to react with a second antigen. It is a measure of relatedness between two different antigenic substances.

CTLA-4 The high-affinity receptor for B7 molecules expressed on T cells.

Cytokine receptors They are the receptors found on many cells for the attachment of cytokines. Binding of the cytokine to the cytokine receptor stimulates signal transduction resulting in new activities in the cell, such as growth, differentiation, or death.

Cytokines Soluble substances secreted by cells, which mediates a variety of effects on other cells.

Cytophilic antibody An antibody that attaches to a cell surface through its Fc region. It binds to Fc receptors on the cell surface. For example, IgE molecules bind to the surface of mast cells and basophils in this manner. They are homocytophilic antibodies as they bind only to the human mast cells.

Cytotoxic T cells (CTLs) The population of T cells that can kill other cells. Most cytotoxic T cells are MHC class I-restricted $CD8^+$ T cells.

Cytotoxins Proteins made by cytotoxic T cells that participate in the destruction of target cells. Perforins and granzymes are some of the examples of cytotoxins.

Degranulation A mechanism whereby cytoplasmic granules in cells fuse with the cell membrane to discharge the contents from the cell. A classic example is degranulation of the mast cell or basophil in immediate (type I) hypersensitivity.

D gene (diversity gene segments) A small segment of immunoglobulin heavy-chain DNA or T-cell receptor DNA, coding for the third hypervariable region of the receptors.

Delayed-type hypersensitivity (DTH) A cell-mediated immunity elicited by antigen present in the skin. The reaction is mediated by $CD4^+$ T_H1 cells and involves release of cytokines and recruitment of monocytes and macrophages. It is called "delayed-type" because the reaction appears an hour to days (usually 24–48 hours) after antigen is injected.

Dendritic cells Interdigitating reticular cells, derived from bone marrow precursors that are found in T-cell areas of lymphoid tissues. They have a branched or dendritic morphology and are the most potent stimulators of T-cell responses. Dendritic cells present in non lymphoid tissues do not appear to stimulate T-cell responses until they are activated and migrate to lymphoid tissues.

Desensitization A procedure in which allergic individuals are exposed to increasing doses of allergen with the goal of inhibiting their allergic reactions. The mechanism responsible for this effect probably involves shifting the response from $CD4^+$ T_H2 to T_H1 cells and thus changing the antibody produced from IgE to IgG.

Diapedesis The movement of blood cells, particularly leucocytes, from the blood across blood vessel walls into tissues.

DiGeorge syndrome A recessive genetic immunodeficiency disease in which there is a failure to develop thymic epithelium.

Diphtheria toxoid An immunizing preparation generated by formalin inactivation of *Corynebacterium diphtheriae* exotoxins. This toxoid, which is used in the active immunization of children against diphtheria, is usually administered as a triple vaccine, together with pertussis microorganisms and tetanus toxoid (DPT).

DNA vaccine A vaccine prepared by introducing a gene in an expression vector that allows transcription and translation into the protein.

Domain A compact segment in the polypeptide chain of an immunoglobulin or TCR chain, made up of amino acids around S–S bond.

Double-negative thymocytes Immature T cells within the thymus that lack expression of the two co-receptors, CD4 and CD8.

DP, DQ, and DR molecules Human MHC (HLA) class II molecules of humans found on B cells and antigen-presenting cells.

ECAM Endothelial cell adhesion molecule.

Effector cells Lymphocytes that can mediate the removal of pathogens or antigens from the body without the need for further differentiation. They are the end stage of lymphocytes. Effectors are distinct from naive lymphocytes, which must proliferate and differentiate before they can mediate effector functions. They are also distinct from memory cells which must differentiate and sometimes proliferate before they become effector cells.

Elispot assay An adaptation of ELISA in which cells are placed over antibodies or antigens attached to a plastic surface. The antigen or antibody traps the cells' secreted products, which can then be detected by using an enzyme-coupled antibody that cleaves a colourless substrate to make a localized coloured spot.

Endocytosis A mechanism whereby substances are taken into a cell from the extracellular fluid by pinching off plasma membrane vesicles.

Endogenous pyrogens Cytokines like IL-1, TNF-α that can induce a rise in body temperature. They are distinct from exogenous substances such as endotoxin from gram-negative bacteria that induce fever by triggering endogenous pyrogen synthesis.

Enzyme-linked immunosorbent assay (ELISA) An assay in which an enzyme is linked to an antibody and a colour forming substrate is used to measure the activity of bound enzyme. The colour directly will give amount of bound unknown.

Eosinophils Polymorphonuclear leucocytes involved in allergic responses and thought to be important in defence against parasitic infections.

Epitope The site by which the antigen bind to the antibody. It can also be called as determinant site of antigen. The counterpart present in the antibody is called as paratope.

Erythroblastosis foetalis A case of Rh incompatibility. Anti-Rh IgG antibodies present in the mothers' blood will cross the placenta to reach the blood foetus to cause abnormalities in the foetus.

F(ab)′ 2 A fragment of an antibody containing two antigen-binding sites generated by cleavage of the antibody molecule with the enzyme papain, which cuts at the hinge region C-terminally to the inter-heavy-chain disulphide bond.

Fab Fragment of antibody containing one antigen-binding site; generated by cleavage of the antibody with the enzyme papain, which cuts at the hinge region N-terminally to the inter-heavy-chain disulphide bond and generates two Fab fragments from one antibody molecule.

Factor B An alternative complement pathway component which combines with C3b and is cleaved by factor D to produce alternative pathway C3 convertase.

Factor H A regulator of complement in the blood that unites with C3b and facilitates dissociation of alternative complement pathway C3 convertase, designated C3bBb, into C3b and Bb.

Factor I A serine protease that splits the α chain of C3b to produce C3bi and the β chain of C4b to yield C4bi.

Factor P (properdin) A key participant in the alternative pathway of complement activation that combines with C3b and stabilizes alternative pathway C3 convertase (C3bB) to produce C3bBbP.

Farmer's lung A hypersensitivity disease caused by the interaction of IgG antibodies with large amounts of an inhaled allergen in the alveolar wall of the lung, causing alveolar wall inflammation and compromising gas exchange.

Fas A member of the TNF receptor family that is expressed on certain cells and makes them susceptible to killing by cells expressing Fas ligand. Binding of Fas ligand to Fas triggers apoptosis in the Fas-bearing cells.

Fas ligand (FasL) A cell-surface member of the TNF family of proteins. Binding of Fas ligand to Fas triggers apoptosis in the Fas-expressing cell.

Fc The fragment of antibody without antigen-binding sites, generated by cleavage with papain. The Fc fragment contains the C-terminal domains of the domains of the immunoglobulin heavy chains.

Fc receptor (FcR) A receptor on a cell surface with specific binding affinity for the Fc portion of an antibody molecule. Fc receptors are found on many types of cells.

Fcε receptor (FcεR) A mast cell and leucocyte receptor for the Fc region of IgE. When immune complexes bind to Fcε receptors, the cell may respond by releasing the mediators of immediate hypersensitivity, such as histamine and serotonin. There are two varieties of Fcε receptors, designated FcεRI and FcεRII (CD23). FcεRI represents a high-affinity receptor found on mast cells and basophils. FcεRII represents a low-affinity receptor.

Fluorescein isothiocyanate (FITC) A fluorescent dye (a fluorochrome) which emits a yellow-green colour and can be conjugated to antibody or other proteins that are used in immunoassay technique.

Fluorescence microscopy A microscopic method which uses ultraviolet light to illuminate a tissue or cell stained with a fluorochrome-labelled substance such as an antibody against an antigen of interest in the tissue.

Fluorescence-activated cell sorter (FACS) An instrument that uses a laser to differentially deflect cells bound to fluorochrome-linked antibodies thus sorting the cells into fluorescent-positive and fluorescent-negative populations.

Fluorescent antibody An antibody coupled with a fluorescent dye, used to detect antigen on cells, tissues, or microorganisms.

Follicle Circular or oval areas of lymphocytes in lymphoid tissues rich in B cells. They are present in the cortex of lymph nodes and in the splenic white pulp. Primary follicles contain B lymphocytes that are small and medium sized. Antigen stimulation causes development of secondary follicles which contain large B lymphocytes in the germinal centres in which tingible body macrophages (those phagocytizing nuclear particles) and follicular dendritic cells are also present.

Follicular dendritic cells Cells within lymphoid follicles which are crucial in selecting antigen-binding B cells during antibody responses. Their origin is uncertain. They have Fc receptors that are not internalized by receptor-mediated endocytosis and thus hold antigen–antibody complexes on their surface for long periods.

Freund's complete adjuvant An adjuvant, containing oil, killed mycobacteria and an emulsifier, which, when emulsified with an immunogen in aqueous solution, enhances the immune response to that immunogen after injection. Incomplete

Freund's adjuvant will not contain mycobacteria.

Germinal centres Secondary lymphoid structures that are sites of intense B-cell proliferation, selection, maturation, and death during antigen stimulation. They form around follicular dendritic cell networks after migration of B cells into lymphoid follicles.

Glomerulonephritis A group of diseases characterized by glomerular injury. Immune mechanisms are responsible for most cases of primary glomerulonephritis and many of the secondary glomerulonephritis group. Over 70% of glomerulonephritis patients have glomerular deposits of immunoglobulins, frequently with complement components.

Goodpasture's syndrome An autoimmune disease in which autoantibodies against basement membrane or type IV collagen are produced and cause extensive vasculitis. It is rapidly fatal.

Graft-versus-host reaction (GVH) The pathological consequences of a response generally initiated by transplanted tissue immunocompetent T lymphocytes into an allogeneic, immunologically incompetent host. The host is unable to reject the grafted T cells and becomes their target.

Granulocyte-macrophage colony-stimulating factor (GM-CSF) A cytokine involved in the growth and differentiation of myeloid and monocytic lineage cells, including dendritic cells, monocytes and tissue macrophages, and cells of the granulocyte lineage.

Granuloma An organized structure in the form of a mass of mononuclear cells at the site of a persisting inflammation. The cells are mostly macrophages with some T lymphocytes at the periphery. It is a typical delayed hypersensitivity reaction that is persistent due to the continuous presence of a foreign body or infection.

Graves' disease An autoimmune disease in which antibodies against the thyroid-stimulating hormone receptor cause overproduction of thyroid hormone and thus hyperthyroidism.

Gut-associated lymphoid tissue (GALT) Lymphoid tissue situated in the gastrointestinal mucosa and submucosa which constitutes the gastrointestinal immune system. Examples of GALT are Peyer's patches, appendix, and tonsils.

H-2 The major histocompatibility complex of the mouse.

Haemagglutinin Any substance that causes red blood cells to agglutinate. The haemagglutinins in human blood are antibodies that recognize the ABO blood group antigens. Influenza and some other viruses have haemagglutinin molecules that bind to glycoproteins on host cells to initiate the infectious process.

Haematopoiesis The generation of the cellular elements of blood, including the red blood cells, leucocytes, and platelets.

Haematopoietic stem cell A bone marrow cell that is undifferentiated and serves as a precursor for multiple cell lineages. These cells are also demonstrable in the yolk sac and later in the liver of the foetus.

Haemolytic disease of the newborn Erythroblastosis foetalis.

Haplotype A linked set of genes associated with one haploid genome. The term is used mainly in connection with the linked genes of the major histocompatibility complex (MHC), which are usually inherited as one haplotype from each parent. Some MHC haplotypes are over-represented in the population, a phenomenon known as linkage disequilibrium.

Hapten A compound, usually of low molecular weight, that is not itself immunogenic but that, after conjugation to a carrier protein or cells, becomes immunogenic and induces antibody, which can bind the hapten alone in the absence of carrier.

HAT medium Hypoxanthine–aminopterin–thymidine medium used in hybridoma technology.

Heavy chain (H chain) The larger of the two types of chains that comprise a normal immunoglobulin or antibody molecule.

Helper T cells A subset of T cells that helps trigger B cells to make antibody against thymus-dependent antigens. Helper T cells also help in the differentiation of other T cells such as cytotoxic T cells.

Herd immunity The protection of unimmunized members of a community by the resistance of the majority of the population to a particular pathogen generally conferred by vaccinating most of the susceptible individuals.

Heterodimer A molecule comprised of two components that are different, but closely joined structures, such as a protein comprised of two separate chains. Examples include the T-cell receptor comprised of either α and β chains or of γ and δ chains, and class I as well as class II histocompatibility molecules.

Heterophile antigen A cross-reacting antigen that appears in widely ranging species such as humans and bacteria.

High endothelial venules (HEV) Specialized venules found in lymphoid tissues. Lymphocytes migrate from blood into lymphoid tissues by attaching to and migrating across the high endothelial cells of these vessels.

Hinge region A flexible, open segment of an antibody

molecule that allows bending of the molecule. The hinge region is located between Fab and Fc and is susceptible to enzymatic cleavage.

Histamine A vasoactive amine stored in mast cell granules which is released by antigen binding to IgE molecules on mast cells, causing dilation of local blood vessels and smooth muscle contraction. Histamine release produces some of the symptoms of immediate hypersensitivity reactions.

Homodimer A protein comprised of dual peptide chains that are identical.

Human Leucocyte Antigen (HLA) The human major histocompatibility complex that contains the genes coding for the polymorphic MHC class I and II class molecules and many other important genes.

Humanization of antibodies A term used to describe the genetic engineering of mouse hypervariable loops of a desired specificity into otherwise human antibodies. The DNA encoding hypervariable loops of mouse monoclonal antibodies or V regions selected in phage display libraries is inserted into the framework regions of human immunoglobulin genes. This allows the production of antibodies of a desired specificity that do not cause an immune response in humans treated with them.

Humoral immunity Immune responses that involve antibody. This type of immunity can be transferred to another individual using antibody-containing serum.

Hybridoma An immortalized hybrid cell resulting from the *in vitro* fusion of an antibody-secreting B cell with a myeloma. It secretes antibody without stimulation and proliferates continuously, both *in vivo* and *in vitro*. The term is also used for a hybrid T cell resulting from the fusion of a T lymphocyte with a thymoma (a malignant T cell). The T-cell hybridoma proliferates continuously and secretes cytokines upon activation by antigen and APC.

Hyperacute graft rejection An immediate reaction In an allogenic tissue graft caused by natural preformed antibodies that react against antigens on the graft. The antibodies bind to endothelium and trigger the blood-clotting cascade, leading to an engorged, ischemic graft and rapid loss of the organ.

Hypergammaglobulinaemia The condition of elevated serum immunoglobulin levels. A polyclonal increase in immunoglobulins in the serum occurs in any condition where there is continuous stimulation of the immune system, such as chronic infection, autoimmune disease, systemic lupus erythematosus, etc. Hypergammaglobulinaemia may also result from a monoclonal increase in immunoglobulin production, as in multiple myeloma, Waldenstrom's macroglobulinaemia, or other conditions associated with the formation of monoclonal immunoglobulins.

Hypervariable regions Portions of the Ig or TCR chains that are highly variable in amino acid sequence from one molecule to another, and that together constitute the binding site for antigen.

Idiotype The combined antigenic determinants (idiotopes) expressed in the variable region of antibodies of an individual that are directed at a particular antigen.

Immediate-type hypersensitivity Type I hypersensitivity reaction occurring within minutes after the interaction of antigen and IgE antibody.

Immune adherence The adherence of particulate antigen coated with C3b to cells expressing C3b receptors results in enhanced phagocytosis of bacteria by macrophages.

Immune complex Molecules formed by the interaction of a soluble (that is, non-particulate) antigen with antibody molecules. Large immune complexes are cleared rapidly, but smaller complexes formed in antigen excess may deposit in tissues resulting in tissue damage.

Immune modulators They are substances that control the level of the immune response.

Immunodeficiency Decrease in immune response that results from absence of or defect in some component of the immune system.

Immunodiffusion Diffusion of antigen and antibody in the gel to form the antigen–antibody complex leading to the precipitin line.

Immunogen A substance capable of inducing an immune response (as well as reacting with the products of an immune response).

Immunoglobulin (Ig) A general term for all antibody molecules: IgM, IgD, IgG, IgA, IgE; each Ig unit is made up of two heavy chains and two light chains and has two antigen-binding sites.

Immunoglobulin superfamily Proteins involved in cellular recognition and interaction that are structurally and genetically related to immunoglobulins.

Immunoreceptor tyrosine-based activation motif (ITAM) A pattern of amino acids in the cytoplasmic tail of many transmembrane receptor molecules, including Igα and Igβ and CD3 chains, which are phosphorylated and then associate with intracellular molecules as an early consequence of cell activation.

Immunoreceptor tyrosine-based inhibitory motif (ITIM) A pattern of amino acids in the cytoplasmic tail of transmembrane molecules, such as CD32, which negatively regulates cell activation. The ITIM bonds a phosphatase which removes phosphate groups from tyrosine residues in the ITAMs of other membrane molecules.

Inflammation An acute or chronic response to tissue injury or infection involving accumulation of leucocytes, plasma proteins, and fluid.

Innate immunity The antigen-nonspecific mechanisms involved in the early phase of resistance to a pathogen, which include phagocytic cells, cytokines, and complement; not expanded by repeat stimulation with the pathogen.

Integrins A family of two-chain cell-surface adhesion molecules found on leucocytes. It is important in the adhesion of APC and lymphocytes, and in leucocyte migration into tissues.

Intercellular adhesion molecules (ICAMs) 1, 2, and 3 Adhesion molecules on the surface of several cell types, including antigen-presenting cells and T cells that interact with integrins. They are the members of the immunoglobulin superfamily.

Interdigitating dendritic cells (IDC) Thymic bone marrow-derived cells which play a critical role in negative selection of developing thymocytes.

Interferons (IFN) A group of proteins having antiviral activity and capable of enhancing and modifying the immune response.

Interleukins (IL) Glycoproteins secreted by a variety of leucocytes that have effects on other leucocytes.

Intron A segment of DNA that does not code for protein: the intervening sequence of nucleotides between coding sequences or exons.

Isograft Tissue transplanted between two genetically identical individuals (same as syngraft).

Isohaemagglutinins Naturally occurring IgM antibodies specific for the red blood cell antigens of the ABO blood groups; thought to result from immunization by bacteria in the gastrointestinal and respiratory tracts.

Isotype switch The switch which occurs when a B cell stops secreting antibody of one isotype or class and starts producing antibody of a different isotype but with same antigenic specificity; involves joining rearranged VDJ gene unit to a different heavy-chain constant region gene.

Isotypes Also known as antibody classes. Antibodies that differ in the heavy chain constant regions: IgM, IgG, IgD, IgA and IgE. These differences result in distinct biological activities of the antibodies; distinguishable also on the basis of reaction with antisera raised in another species.

J chain (joining chain) A polypeptide involved in the polymerization of IgM and IgA.

J gene A gene segment coding for the J or joining segment in immunoglobulin or T-cell receptor chains.

Janus kinases (JAK) Tyrosine kinases activated by cytokines binding to their cellular receptors.

Killer activating receptors (KAR) Proteins present on NK cells, which when activated cause NK cells to kill a target cell.

Killer inhibitory receptor (KIR) A receptor expressed on NK cells that binds to MHC class I molecules on target cells. The ligation of MHC class I inhibit the signalling that would otherwise lead to target cell killing.

Killer T cell A subset of T cell that kills a target cell expressing foreign antigen bound to MHC molecules on the surface of the target cell.

Langerhans' cell Cell of the monocyte/dendritic cell family that takes up and processes antigens in the epidermal layer of the skin. It migrates through lymphatics to lymph nodes draining the site of exposure to antigen, where it differentiates into a dendritic cell.

Leucocyte common antigen (LCA, CD45) An antigen shared in common by both T and B lymphocytes.

Leucocytes White blood cells; comprise monocytes/macrophages, lymphocytes and polymorphonuclear cells.

Leukemia Uncontrolled proliferation of a malignant leucocyte.

Ligand A molecule or part of a molecule that binds to a receptor.

Ligation The binding of a molecule or a part of a molecule to a receptor.

Light chain (L chain) The chain present in the immunoglobulin molecule. There are two types of light chains. They are kappa and lambda chains.

Linked recognition The requirement for the T helper and B cell involved in the antibody response to a thymus-dependent antigen to interact with different epitopes physically linked in the same antigen.

Lipopolysaccharide (LPS) The component of gram-negative bacterial cell walls. It acts as endotoxin.

Lymph Extracellular fluid that bathes tissues; contains tissue products, antigens, antibodies and cells (predominantly lymphocytes).

Lymph nodes It is the secondary lymphoid organ, in which mature B and T lymphocytes respond to free antigen, or antigen associated with APC, brought in via lymphatic vessels.

Lymphatic system System of vessels through which lymph travels, and which includes organized structures—lymph nodes—at the intersection of vessels. Three major functions: to concentrate antigen from all parts of the body into a few lymphoid organs; to circulate lymphocytes through lymphoid organs so that antigens can interact with rare antigen-specific cells; and to carry products of the immune response (antibody and effector cells) to the bloodstream and tissues.

Lymphocytes Small leucocyte with virtually no cytoplasm, found in blood, tissues, and lymphoid organs such as lymph nodes, spleen, and Peyer's patches. They are responsible in the immune response.

Lymphokine A cytokine secreted by lymphocytes.

Lymphokine-activated killer (LAK) cells The heterogeneous population of lymphocytes, including NK cells, derived from the *in vitro* cytokine-driven activation of peripheral blood lymphocytes from a tumour-bearing patient.

Lymphoma Lymphocyte tumours in lymphoid or other tissues; not generally found in the blood.

Macrophages Large phagocytic leucocytes found in tissues; derived from blood monocytes.

Major histocompatibility complex (MHC) A cluster of genes encoding polymorphic cell-surface molecules (MHC class I and class II) that are involved in antigen presentation to T cells. These molecules also play a major role in transplantation rejection. Several other non-polymorphic proteins are encoded in this region.

Mast cell Bone marrow-derived granule-containing cell found in connective tissues, which releases mediators such as histamine and cytokines following cell activation; plays a major role in allergic responses.

Mature B cell B cells with IgM and IgD on their surface.

Membrane attack complex Terminal components of the complement cascade (C7–C9) which form a pore on the surface of a target cell, resulting in cell damage or death.

Memory A term which in immunology denotes that a second interaction with antigen leads to a more effective and more rapid response than the first interaction (primary response).

Metastasis Spreading of cancer cells from the primary site to distant region(s) in the body.

MHC class I molecule A molecule encoded by genes of the MHC that participates in antigen presentation to $CD8^+$ (cytotoxic) T cells.

MHC class II molecule A molecule encoded by genes of the MHC that participates in antigen presentation to $CD4^+$ T cells.

MHC class III molecules Proteins including complement components C2, C4, and factor B encoded by genes in the major histocompatibility complex (MHC). Although adjacent to MC class I and class II genes that code for molecules critical in cellular interactions in the immune response, molecules coded for by MHC class III genes are not involved in cellular interactions.

MHC restriction The property of T lymphocytes to respond only when they are presented with the appropriate antigen in association with either self MHC class I or class II molecules.

Minor histocompatibility antigens Antigens encoded outside the MHC which stimulate graft rejection, but not as rapidly as MHC molecules.

Mitogen A substance that stimulates the proliferation of many different clones of lymphocytes.

Mixed lymphocyte reaction (MLR) Proliferative response occurring when leucocytes from two individuals are mixed *in vitro*; T cells from one individual (the responder) are activated by MHC antigens expressed by APC of the other individual (the stimulator).

Molecular mimicry Identity or similarity of epitopes expressed by a pathogen and by a self molecule; may explain how autoimmune responses develop.

Monoclonal Means derived from a single clone, the progeny of a single cell. Generally refers to a population of T cells, B cells, or antibody that is homogeneous, and reactive with the same specificity toward an epitope.

Monocyte Phagocytic leucocyte found in the blood. It is the precursor to tissue macrophage.

Motif A pattern of amino acids in the sequence of a molecule critical for the binding of a ligand.

Mucosal-associated lymphoid tissue (MALT) System that connects lymphoid structures found in the gastrointestinal and respiratory tracts and includes tonsils, appendix, and Peyer's patches of the small intestine.

Myeloma A tumour of plasma cells, generally secreting a single monoclonal immunoglobulin.

Natural killer (NK) cells Large granular lymphocyte-like cells that kill various tumour cells *in vitro* and may play a role in resistance to tumours; also participate in ADCC; derived from the lymphoid progenitor but distinct from T and B lymphocytes. They do not exhibit antigenic specificity, and their number does not increase by immunization.

Negative selection A step in development of B and T cells at which cells with potential reactivity to self molecules are functionally inactivated.

Neutralization The ability of an antibody to block or inhibit the effects of a virus.

Opsonization The coating of a particle such as a bacterium with antibody and/or a complement component (an opsonin) that leads to enhanced phagocytosis by phagocytic cells.

Paracortical area (or paracortex) The T-cell area of the lymph node.

Passive cutaneous anaphylaxis (PCA) The passive transfer of anaphylactic sensitivity by intradermal injection of serum from a sensitive donor.

Passive haemagglutination Technique for measuring antibody, in which antigen-coated red blood cells are agglutinated by adding antibody specific for the antigen.

Passive immunization Immunization of an individual by the transfer of antibody synthesized in another individual.

Perforin A molecule synthesized by cytotoxic T cells and NK cells that polymerizes on the surface of a target cell and creates a pore in the membrane, resulting in death of target cell.

Peripheral lymphoid organs organs other than the thymus; include spleen, lymph nodes, and mucosal-associated lymphoid tissue.

Peripheral tolerance Tolerance induced in mature lymphocytes outside the thymus.

Peyer's patches Clusters of lymphocytes distributed in the lining of the small intestine.

Phagocytosis The engulfment of a particle or a microorganism by leucocytes such as macrophages and neutrophils.

Phenotype The physical expression of an individual's genotype.

Phosphatase Enzyme that removes phosphate groups from proteins.

Phospholipase C gamma (PLC-γ) Enzyme involved in T-cell and B-cell activation pathways; splits phosphatidylinositol bisphosphate (PIP_2) into diacylglycerol (DAG) and inositol triphosphate (IP_3) leading to the activation of two major signalling pathways.

Plasma Fluid component of unclotted blood.

Plasma cell The antibody-producing end-stage of B-cell differentiation.

Platelets Bone marrow-derived cells crucial in blood clotting.

Pokeweed mitogen A mitogen that polyclonally activates B cells.

Polyclonal activator A substance that induces activation of many clones of either T or B cells.

Poly-Ig receptor A receptor that binds to IgA at one surface of an epithelial cell, transports it through the cell, and releases it at the opposite lumenal surface. The IgA can then participate in protecting the mucosal system.

Polymorphonuclear leucocytes (PMN) Leucocytes containing cytoplasmic granules with characteristic multilobed nuclei; they are of three major types: neutrophils, eosinophils, and basophils.

Positive selection The process by which developing B and T cells receive signals in the primary lymphoid organ in which they are developing to continue their differentiation; in the absence of these signals, the cells die.

Pre-B cell Cell in the B cell lineage which has rearranged heavy but not light chain genes; expresses surrogate light chains and μ heavy chain at its surface in conjunction with Igα and Igβ all these molecules comprise the pre-B cell receptor (pre-BCR).

Precipitin reaction The mixing of soluble antigen and antibody at different proportions that can result in the precipitation of insoluble antigen–antibody complexes.

Pre-T cell Cell in T-lymphocyte differentiation in the thymus that has rearranged TCR β genes and expresses TCR β polypeptide on the surface with the molecule pTα (gp33), forming the pre-T-cell receptor.

Primary follicle Region of a secondary lymphoid organ containing predominantly unstimulated B lymphocytes. It develops into a germinal centre following antigen stimulation.

Primary lymphoid organs Organs in which the early stages of T- and B-lymphocyte differentiation take place and antigen-specific receptors are first expressed.

Primary response The immune response resulting from first encounter with antigen; generally small, with a long induction phase or lag period, and generates immunological memory. In the primary B-cell response, mainly IgM antibodies are made.

Priming The activation of naive lymphocytes by exposure to antigen.

Pro-B cell Earliest stage of B-cell differentiation in which a heavy chain D-gene segment rearranges to a J-gene segment.

Proteasome Multiprotein cytoplasmic complex, which catabolizes proteins to 8–9 amino acid peptides.

Protein kinase C Enzyme activated by calcium and diacylglycerol during T- and B-lymphocyte activation.

Proto-oncogenes Cellular genes regulating growth control; mutation or aberrant expression can lead to malignant transformation of the cell.

Pyrogen A substance that causes fever.

Radioallergosorbent test (RAST) A solid-phase radioimmunoassay for detecting IgE antibody specific for a particular allergen.

Radioimmunoassay (RIA) A technique for measuring the level of a biological substance in a sample, by measuring the binding of antigen to radioactively labelled antibody or antigen.

***RAG*-1 and *RAG*-2 (recombination activating genes)** Genes and their products which are critically involved in V(D)J recombination in B and T cells.

Receptor Generally a transmembrane molecule that binds to a ligand on the exterior surface of the cell, leading to biochemical changes inside the cell.

Receptor editing The process by which the rearranged genes of a cell in the B-cell lineage may undergo a secondary rearrangement, generating a different antigenic specificity.

Repertoire The complete library of antigenic specificities generated by either B or T lymphocytes to respond to foreign antigen.

Reticuloendothelial system (RES) A general term for the network of phagocytic cells.

Reverse transcriptase Enzyme which transcribes the RNA genome of a retrovirus into DNA; used in molecular biology to convert RNA into complementary DNA (cDNA).

Rheumatoid arthritis Autoimmune, inflammatory disease of the joints.

Rheumatoid factor An autoantibody (usually IgM) that reacts with the individual's own IgG. It is present in rheumatoid arthritis.

Secondary lymphoid organs Organs in which antigen-driven proliferation and differentiation of mature B and T lymphocytes take place following antigen recognition.

Second set rejection Accelerated rejection of an allograft in a primed recipient.

Secretory component Cleaved component of the poly-Ig receptor that attaches to dimeric IgA and protects it from proteolytic cleavage as it is transported through an epithelial cell.

Selectins A family of cell-surface adhesion molecules found on leucocytes and endothelial cells; bind to sugars on glycoproteins.

Serum Residual fluid derived from clotted blood. It contains antibodies.

Serum sickness A type III hypersensitivity reaction resulting from deposition of circulating, soluble, antigen–antibody complexes leading to complement and neutrophil activation in tissues such as the kidney. It is typically induced following therapy with large doses of antibody from a foreign antibodies.

Severe combined immune deficiency (SCID) Disease resulting from early block in differentiation pathways of both B and T lymphocytes.

Signal transducers and activators of transcription (STATs) Intracellular proteins phosphorylated by Janus kinases as a consequence of cytokine–cytokine receptor engagement.

Signal transduction Processes involved in transmitting the signal received on the outer surface of the cell (e.g. by antigen binding to its receptor) into the nucleus of the cell, which lead to altered gene expression.

Slow-reacting substance of anaphylaxis (SRS-A) A group of leukotrienes released by mast cells during anaphylaxis that induces a prolonged contraction of smooth muscle.

Somatic hypermutation Change in the variable region sequence of an antibody produced by a B cell following antigenic stimulation, resulting in increased antibody affinity for antigen.

Spleen Largest of the secondary lymphoid organs; traps and concentrates foreign substances carried in the blood; composed of white pulp, rich in lymphoid cells, and red pulp, which contains many erythrocytes and macrophages.

Superantigen A molecule that activates all T cells with a particular Vβ gene segment, irrespective of their Vα expression.

Suppression A mechanism for producing a specific state of immunological unresponsiveness by which one cell or its products inhibits the function of another.

Surrogate light chains Nonrearranging chains (Vλ5 and V preB) expressed in con junction with μ chain in the pre-B cell; form part of the pre-BCR.

Switch region Region of B cell heavy chain DNA at which recombination occurs in an antigen-stimulated cell. It allows isotype switch (e.g. IgM to IgE).

Systemic lupus erythematosus (SLE) An autoimmune disease which affects many organs of the body, and causes fever and joint pain. Patients produce high levels of antibodies against the components of cell nuclei, particularly DNA, and form circulating soluble antigen–antibody complexes. These complexes deposit in tissues such as the kidney, activate the complement cascade, and result in tissue damage.

T cells The set of lymphocytes whose differentiation requires the thymus.

T_H1 A subset of $CD4^+$ T cells that synthesizes the cytokines IL-2, IFN-γ, and TNF-β. These cytokines activate NK cells, macrophages, and $CD8^+$ T cells.

T_H2 A subset of $CD4^+$ T cells which synthesizes the cytokines IL-4, IL-5, IL-10, and IL-13; these cytokines predominate in the response to allergens and parasites (B-cell class switching to IgE, and eosinophil activation).

TAP-1 and TAP-2 Molecules that selectively transport peptides from the cytoplasm to the endoplasmic reticulum of cells for binding to MHC class I molecules.

T-cell receptor (TCR) A two-chain structure on T cells that binds antigen: $\alpha\beta$ on the major set of T cells, $\gamma\delta$ on the minor set of T cells. The TCR complex comprises the antigen-binding chains associated at the cell surface with the signal transduction molecules CD3 plus ζ or η.

T-dependent antigen An immunogen that requires T helper cells to interact with B cells in order to induce antibody synthesis.

Terminal deoxynucleotidyl transferase (TdT) Enzyme that inserts nontemplated nucleotides at the junctions of V, D, and J gene segments of Ig and TCR locus DNA; these N-nucleotides increase the diversity of antigen-specific receptors.

Thymocytes T cells differentiating in the thymus.

Thymus The primary lymphoid organ for T-cell differentiation, comprising an outer cortex and inner medulla; developing thymocytes interact with epithelial cells and bone marrow derived macrophages and interdigitating dendritic cells in the thymus.

T-independent antigen An immunogen that induces antibody synthesis in the absence of T cells or their products; antibodies synthesized generally only of the IgM isotype, and no memory response.

Titre Used generally as an empirical measure of the avidity of an antibody; it is the reciprocal of the last dilution of a titration giving a measurable effect; e.g. if the last dilution of an antibody giving significant agglutination is 1:128, the titre is 128.

Tolerance Antigen-specific unresponsiveness of B or T cells.

Toxoid A non-toxic derivative of a toxin used as an immunogen for the induction of antibodies capable of cross reacting with the toxin.

Transplantation Grafting solid tissue (such as a kidney or heart) or cells (particularly bone marrow) from one individual to another. *See* allograft and xenograft.

Tuberculin test A clinical test in which antigens derived from the organism causing tuberculosis are injected subcutaneously; individuals who have been exposed to the organism and those who have been previously vaccinated with BCG develop a delayed hypersensitivity response at the injection site 24–48 hours later.

Tumour necrosis factor (TNF) A cytokine with various actions including the selective killing of tumour cells; toxicity may be the result of the production of free radicals following the binding of high-affinity cell surface receptors.

Tumour-specific transplantation antigen (TSTA) Antigens uniquely expressed by certain tumour cells.

Tyrosine kinases A family of enzymes which phosphorylates proteins on tyrosine residues, a critical step in lymphocyte activation. The key tyrosine kinases in T-cell activation are Lck, Fyn, and ZAP-70 and those in B-cell activation are Blk, Fyn, Lyn, and Syk.

Unresponsiveness Inability to respond to antigenic stimulus. Unresponsiveness may be specific for a particular antigen or broadly nonspecific as a result of damage to the entire immune system, for example, after whole-body irradiation.

Vaccination Any protective immunization against a pathogen.

Variable (V) regions The N-terminal portion of an Ig or TCR which contains the antigen-binding region of the molecule; V regions are formed by the recombination of V(D) and J gene segments.

V(D)J recombination Mechanism for generating antigen-specific receptors of T and B cells; it involves the joining of V, D, and J gene segments mediated by the enzyme complex V(D)J recom-binase, and products of the *RAG*-1 and 2 genes.

Western blotting A technique to identify a specific protein in a mixture; proteins separated by gel electrophoresis are blotted onto a nitrocellulose membrane, and the protein of interest is detected by adding radiolabelled or enzyme-labelled antibody specific for the protein.

Wheal and flare Itchy reaction at skin site where antigen is injected into an allergic individual; characterized by erythema (redness due to dilation of blood vessels) and edema (swelling produced by release of serum into tissue).

Xenogeneic Originating from a foreign species.

Xenograft The tissue transplantation between individuals belonging to two different species.

ZAP-70 A T-cell-specific tyrosine kinase involved in T-cell activation.

References

Abbas, A.K., Lichtman, A.H. and Jordan, S.P. 1991. *Cellular and Molecular Immunology.* W. B. Saunders, Philadelphia.

Allen, J.E. and Maizels, R.M. 1997. "T_H1-T_H2: reliable paradigm or dangerous dogma?" *Immunol. Today.* 18: 387.

Azuma, M., Cayabyab, M., Buck, D., Phillips, J.H. and Lanier, L.L. 1992. "CD28 interaction with B7 co-stimulates primary allogeneic proliferative responses and cytotoxicity mediated by small, resting T lymphocytes." *J. Exp. Med.* 175: 353.

Bacci, S., Alard, P., Dai, R., Nakamura, T. and Streilein, J.W. 1997. "High and low doses of haptens dictate whether dermal or epidermal antigen-presenting cells promote contact hypersensitivity." *Eur. J. Immunol.* 27: 442.

Baron. S., Grossberg, S.E., Klimpel, G.R. and Brunell, P.A. 1984. "Mechanisms of action and pharmacology: the immune and interferon systems." In Galasso, G. (ed.) *Antiviral Agents and Viral Diseases of Man.* Raven Press, New York.

Beck, L. Malcolm and Lowell, L. Tilzer. 1996. "Red Cell Compatibility Testing: A Perspective for the Future." *Transfusion Medical Reviews*. p.118.

Beron, W., Alvarez-Dominguez, C., Mayorga, L. and Stahl, P.D. 1995. "Membrane trafficking along the phagocytic pathway." *Trends Cell Biol.* 5: 100.

Benacerraf, B. and Gell, P.G.H. 1959. "Studies on hypersensitivity-I. Delayed and Arthus-type skin reactivity to protein conjugates in Guinea Pigs." *Immunology.* 2: 53.

Biron, C.A. 1994. "Cytokines in the generation of immune responses to, and resolution of, virus infection." *Currrent Opinion in Immunology.* 6: 530.

Boise, L.H., Minn, A.J. and Noel, P.J. 1995. "CD28 co-stimulation can promote T cell survival by enhancing the expression of Bcl-X_L." *Immunity.* 3: 87.

Bokoch, G.M. 1995. "Regulation of the phagocyte respiratory burst by small GTP-binding proteins." *Trends Cell Biol.* 5: 109.

Bouloc, A., Cavani, A. and Katz, S.I. 1998. "Contact hypersensitivity in MHC class II-deficient mice depends on CD8 T lymphocytes primed by immunostimulating Langerhans cells." *J. Invest. Dermatol.* 111: 44.

Brenner, B.G., Grylles, C. and Wainberg, M.A. 1991. "Role of antibody-dependent cellular cytotoxicity and lymphokine-activated killer cells in AIDS and related diseases." *J. Leucocyte Biology.* 50: 628.

Brent, L., Brown, J.B. and Medawar, P.B. 1958. "Skin transplantation immunity in relation to hypersensitivity." *Lancet.* p. 561.

Buchanan, K.L. and Murphy, J. W. 1994. "Regulation of cytokine production during the expression phase of the anticryptococcal delayed-type hypersensitivity response." *Infect. Immun.* 62: 2930.

Bushell, A. and Wood, K.J. 1999. "Permanent survival of organ transplants without immunosuppression: Experimental approaches and possibilities for tolerance induction in clinical transplantation." *Exp. Rev. Mol. Biol.* 29 October.

Chai, J.G. 1999. "Anergic T cells act as suppressor cells *in vitro* and *in vivo*. *Eur. J. Immunol.* 29: 686.

Chase, M.W. 1945. "The cellular transfer of cutaneous hypersensitivity to tuberculin." *Proc. Soc. Exp. Biol. Med.* 59: 134.

Chen, L., Ashe, S. and Brady, W.A. 1992. "Co-stimulation of antitumor immunity by the B7 counterreceptor for the T lymphocyte molecules CD28 and CTLA-4." *Cell.* 71: 1093.

Cher, D.J. and Mosmann, T. R. 1987. "Two types of murine helper T cell clone.II. Delayed-type hypersensitivity is mediated by T_H1 clones." *J. Immunol.* 138: 3688.

Chung, H. T., Samlowski, W.E., Kelsey, D. K. and Daynes, R. A. 1986. "Alterations in lymphocyte recirculation within ultraviolet light-irradiated mice: efferent blockade of lymphocyte egress from peripheral lymph nodes." *Cell Immunol.* 102: 335.

Clark, D.A. 1990. "Murine pregnancy decidua produces a unique immunosuppressive molecule related to transforming growth factor beta-2." *J. Immunol.* 144: 3008.

Coligan, J.E. 1991. *Current Protocols in Immunology*. John Wiley & Sons, New York.

Cooper, K.D., Oberhelman, L. Hamilton, T.A., Baadsgaard, O., Terhune, M., LeVee, G., Anderson, T. and Koren, H. 1992. "UV exposure reduces immunization rates and promotes tolerance to epicutaneous antigens in humans: relationship to dose, CD1a-DR+ epidermal macrophage induction, and Langerhans cell depletion." *Proc. Natl. Acad. Sci., USA.* 89: 8497.

Corry, D.B., Reiner, S.L., Linsley, P.S. and Locksley, R.M. 1994. "Differential effects of blockade of CD28-B7 on the development of T_H1 or T_H2 effector cells in experimental leishmaniasis." *J. Immunol.* 153: 4142.

Croft, M., Bradley, L.M. and Swain, S.L. 1994. "Naive versus memory CD4 T cell response to antigen: memory cells are less dependent on accessory cell co-stimulation and can respond to many antigen-presenting cell types including resting B cells." *J. Immunol.* 152: 310.

Crowther, J.R. 1995. *Methods in Molecular Biology, Volume 42: ELISA, Theory and Practice*. Humana Press, New Jersey.

Cruz, P. D., Jr. 1996. "Basic science answers to questions in clinical contact dermatitis." *Am. J. Contact Derm.* 7: 47.

Dalton, D., Pitts-Meek, S., Keshav, S., Figari, I., Bradley, A. and Stewart, T. 1993. "Multiple defects of immune cell function in mice with disrupted interferon-gamma genes." *Science.* 259: 1739.

Dannenberg, J., A.M. 1991. "Delayed-type hypersensitivity and cell-mediated immunity in the pathogenesis of tuberculosis." *Immunol Today.* 12: 228.

Desjardins, M. and Griffiths, G. 2003. "Phagocytosis: latex leads the way." *Curr. Opin. Cell Biol.* 15: 498.

Desphande, S.S. 1996. *Enzyme Immunoassays: From Concept to Product Development*. Chapman & Hall, New York.

Diamandis, E.P. and Christopoulos, T.K. (eds.) 1996. *Immunoassay*. Academic Press, New York.

Dienes, L. and Schoenheit, E.W. 1929. "The reproduction of tuberculin hypersensitiveness in guinea pigs with various protein substances." *Am. Rev. Tuberc.* 20: 92.

Dummer, W., Rose, C. and Brocker, E. B. 1998. "Expression of CD30 on T helper cells in the inflamatory infiltrate of acute atopic dermatitis but not of allergic contact dermatitis." *Arch. Dematol. Res.* 290: 598.

Ehlers, S., Mielke, M.E. and Hahn, H. 1994. "$CD4^+$ T cell associated cytokine gene expression during experimental infection with *Listeria monocytogenes*: the mRNA phenotype of granuloma formation. *Int. Immunol.* 6: 1727.

Elser, B., Lohoff, M., Kock, S., Giaisi, M., Kirchhoff, S., Krammer, P. H. and Li-Weber, M. 2002. "IFN-γ represses IL-4 expression via IRF-1 and IRF-2." *Immunity*. 17: 703.

Enk, A.H., Angeloni, V. L., Udey, M.C. and Katz, S. I. 1993. "An essential role for Langerhans cell-derived IL-1 beta in the initiation of primary immune responses in skin." *J. Immunol.* 150: 3698.

Ferencik, M. 1993. *Handbook of Immunochemistry*. Chapman & Hall, New York.

Finck, B.K., Linsley, P.S. and Wofsy, D. 1994. Treatment of murine lupus with CTLA-4 Ig. *Science*. 265: 1225.

Fiorentino, D.F., Bond, M.W. and Mosmann, T.R. 1989. "Two types of mouse T helper cell. IV. T_H2 clones secrete a factor that inhibits cytokine production by T_H1 clones." *J. Exp. Med.* 170: 2081.

Fishman, J.A. and Rubin, R.H. 1998. "Infection in organ-transplant recipients." *N. Engl. J. Med.* 338: 1741.

Flores Villanueva, P.O., Harris, T.S., Ricklan, D.E., Stadecker, M.J. 1994. "Macrophages from schistosomal egg granulomas induce unresponsiveness in specific cloned T_H1 lymphocytes *in vitro* and down-regulate granulomatous disease *in vivo*. *J. Immunol.* 152: 1847.

Flores Villanueva, P.O., Reiser, H. and Stadecker, M.J. 1994. "Regulation of T helper cell responses in experimental schistosomiasis by IL-10: effect on expression of B7 and B7.2 co-stimulatory molecules by macrophages." *J. Immunol.* 153: 5190.

Flores Villanueva, P.O., Zheng, X.X., Strom, T.B., Stadecker, M.J. 1996. "Recombinant IL-10 and IL-10/Fc treatment down-regulate egg antigen-specific delayed-hypersensitivity reactions and egg granuloma formation in schistosomiasis." *J. Immunol.* 156: 3315.

Fong, T. and Mosmann, T. 1989. "The role of IFN-gamma in delayed-type hypersensitivity mediated by T_H1 clones." *J. Immunol.* 143: 2887–93.

Gajewski, T.F. 1994. "Anergy of T_H0 helper T lymphocytes induces downregulation of T_H1 characteristics and a transition to a T_H2-like phenotype." *J. Exp. Med.* 179: 481.

Galvin, F., Freeman, G.J., Razi-Wolf, Z., Benacerraf, B., Nadler, L. and Reiser, H. 1993. "Effects of cyclosporin A, FK 506, and mycalamide A on the activation of murine $CD4^+$ T cells by the murine B7 antigen." *Eur. J. Immunol.* 23: 283.

Galvin, F., Freeman, G.J. and Razi-Wolf, Z. 1992. "The murine B7 antigen provides a sufficient co-stimulatory signal for antigen-specific and MHC-restricted T-cell activation." *J. Immunol.* 149: 3802.

Gautam, S., Battisto, J., Major, J. A., Armstrong, D., Stoler, M. and Hamilton, T.A. 1994. "Chemokine expression in trinitrochlorobenzene-mediated contact hypersensitivity." *J. Leukoc. Biol.* 55: 452.

Gell, P.H.G. and Coombs, R.A.A. 1968. "Cinical Aspects in Immunology." Blackwell, Oxford.

Grabbe, S., Bruvers, S. and Granstein, R. D. 1992. "Effects of immunomodulatory cytokines on the presentation of tumor-associated antigens by epidermal Langerhans cells." *J. Invest. Dermatol.* 99: 66.

Gocinski, B.L., Tigelaar, R.E. "Roles of CD4$^+$ and CD8$^+$ T cells in murine contact sensitivity revealed by *in vivo* monoclonal antibody depletion." *J. Immunol.* 1990. 144: 4121.

Gordon, C. and Wofsy, D. 1990. "Effects of recombinant murine tumor necrosis factor-alpha on immune function." *J. Immunol.* 144: 1753.

Guinan, E.C., Gribben, J.G., Boussiotis, V.A., Freeman, G.J. and Nadler, L.M. 1994. "Pivotal role of the B7:CD28 pathway in transplantation tolerance and tumor immunity." *Blood.* 84: 3261.

Haas, W., Pereira, P. and Tonegawa, S. 1993. "Gamma/delta cells." *Annu. Rev. Immunol.* 11: 637.

Hanahan, D. 1998. "Peripheral antigen-expressing cells in thymic medulla: factors in self-tolerance and autoimmunity." *Curr. Opin. Immunol.* 10: 656.

Harding, F.A. and Allison, J.P. 1993. "CD28–B7 interactions allow the induction of CD8$^+$ cytotoxic T lymphocytes in the absence of exogenous help." *J. Exp. Med.* 177: 1791.

Hathcock, K.S., Laszlo, G., Pucillo, C., Linsley, P. and Hodes, R.J. 1994. "Comparative analysis of B7.1 and B7.2 co-stimulatory ligands: expression and function." *J. Exp. Med.* 180: 631.

Hathcock, K.S., Laszlo, G., Dickler, H.B., Bradshaw, J., Linsley, P. and Hodes, R.J. 1993. "Identification of an alternative CTLA-4 ligand co-stimulatory for T-cell activation." *Science.* 262: 905.

Heath, W.R. 1992. "Autoimmune diabetes as a consequence of locally produced interleukin-2." *Nature.* 359: 547.

Heath, V.L. 1998. "Intrathymic expression of genes involved in organ-specific autoimmune disease." *J. Autoimmun.* 11: 309.

Herberman, R.B. and Ortaldo, J.R. 1981. "Natural killer cells: their role in defence against disease." *Science.* 214: 24.

Hernandez, P. R. and Rook, G. A. 1994. "The role of TNF-alpha in T- inflammation depends on the T_H1/T_H2 cytokine balance." *Immunology.* 82: 591.

Higashi, N., Yoshizuka, N. and Kobayashi, Y. 1995. "Phenotypic properties and cytokine production of skin-infiltrating cells obtained from guinea pig delayed-type hypersensitivity reaction sites." *Cell Immunol.* 164: 28.

Hirsch, R.L., Winkelstein, J.A. and Griffin, D.E. 1980. "The role of complement in viral infections. III. Activation of the classical and alternative complement pathways by Sindbis virus." *J. Immunol.* 124: 2507.

Inaba, K., Witmer-Pack, M. and Inaba, M. 1994. The tissue distribution of the B7.2 co-stimulator in mice: abundant expression on dendritic cells *in situ* and during maturation *in vitro*." *J. Exp. Med.* 180: 1849.

Issekutz, T.B., Stoltz, J.M. and Van Der Meide, P.P. 1988. "Lymphocytes recruitment in delayed-type hypersensitivity." *J. Immunol.* 140: 2989.

Janeway, C.A., Jr and Bottomly, K. 1994. "Signals and signs for lymphocyte responses." *Cell.* 76: 275.

Jenkins, M.K., Schwartz, R.H. and Pardoll, D.M. 1988. "Effects of cyclosporine A on T-cell development and clonal deletion." *Science.* 241: 1655.

Jones, L.A. 1990. "Peripheral clonal elimination of functional T cells." *Science.* 250: 1726.

Jones, T. D. and Mote, J. R. 1934. "Phases of foreign protein sensitization in human beings." *N. Engl. J. Med.* 210: 120.

June, C.H., Ledbetter, J.A., Gillespie, M.M., Lindsten, T. and Thompson, C.B. 1987. "T-cell proliferation involving the CD28 pathway is associated with cyclosporine-resistant interleukin-2 gene expression." *Mol. Cell Biol.* 7: 4472.

Kappler, J.W., Roehm, N. and Marrack, P. 1987. "T-cell tolerance by clonal elimination in the thymus." *Cell.* 49: 273.

Kawabe, Y. and Ochi, A. 1991. "Programmed cell death and extrathymic reduction of Vbeta8$^+$ CD4$^+$ T cells in mice tolerant to *Staphylococcus aureus* enterotoxin B. *Nature*. 349: 245.

Kaye, P.M., Rogers, N.J., Curry, A.J. and Scott, J.C. 1994. "Deficient expression of co-stimulatory molecules on Leishmania-infected macrophages. *Eur. J. Immunol.* 24: 2850.

Kitagaki, H., Ono, N., Hayakawa, K., Kitazawa, T., Watanabe, K. and Shiohara, T. 1997. "Repeated elicitation of contact hypersensitivity induces a shift in cutaneous cytokine milieu from a T helper cell type 1 to a T helper cell type 2 profile." *J. Immunol.* 159: 2484.

Kondo, S., Wang, B., Fujisawa, H., Shivji, G. M., Echtenacher, B., Mak, T.W. and Sauder, D.N. 1995. "Effect of gene-targeted mutation in TNF receptor (p55) on contact hypersensitivity and ultraviolet B-induced immunosuppression." *J. Immunol.* 155: 3801.

Kondo, H., Ichikawa, Y. and G. Imokawa. 1998. "Percutaneous sensitization with allergens through barrier-disrupted skin elicits a T_H2-dominant cytokine response." *Eur. J. Immunol.* 28: 769.

Kondo, S., Kooshesh, F., Wang, B., Fujisawa, H. and Sauder, D. N. 1996. "Contribution of the CD28 molecule to allergic and irritant-induced skin reactions in CD28 -/- mice." *J. Immunol.* 157: 4822.

Krasteva, M.,Kehren, J., Horand, F., Akiba, H., Choquet, G., Ducluzeau, M. T., Tedone, R., Garrigue, J. L., Kaiserlian, D. and Nicolas, J.F. 1998. "Dual role of dendritic cells in the induction and down-regulation of antigen-specific cutaneous inflammation." *J. Immunol.* 160: 1181.

Knoerzer, D.B., Karr, R.W., Schwartz, B.D., Mengle-Gaw, L.J. 1995. "Collagen-induced arthritis in the BB rat: prevention of disease by treatment with CTLA-4-Ig." *J. Clin. Invest.* 96: 987.

Kripke, M.L., Munn, C. G. Jeevan, A., Tang, J. M. and Bucana, C. 1990. "Evidence that cutaneous antigen-presenting cells migrate to regional lymph nodes during contact sensitization." *J. Immunol.* 145: 2833.

Lafferty, K.J. and Woolnough, J. 1977. "The origin and mechanism of the allograft reaction." *Immunol. Rev.* 35: 231.

Lamb, J.R. *et al.* 1983. "Induction of tolerance in influenza virus-immune T lymphocyte clones with synthetic peptides of influenza hemagglutinin." *J. Exp. Med.* 157, 1434–1447.

Landsteiner, K. and Chase, M. W. 1942. "Experiments on transfer of cutaneous sensitivity to simple compounds." *Proc. Soc. Exp. Biol. Med.* 32: 688.

Landsteiner, K. 1947. *The Specificity of Serological Reactions.* Harvard Univ. Press, Cambridge.

Lebman, D. and Coffman, R.L.. 1988. "Interleukin-4 causes isotype switching to IgE in T-cell stimulated clonal B cell cultures." *J. Exp. Med.* 168: 853.

Lenschow, D.J., Su GH-T and Zuckerman, L. A. 1993. "Expression and functional significance of an additional ligand for CTLA-4." *Proc. Natl. Acad. Sci. USA.* 90: 11054.

Leonardo, M.R. 1999. "Release of formaldehyde by 4 endodontic sealers." *Oral Surg. Oral Med. Oral Pathol. Oral Radiol. Endod.* 88: 221–225.

Levine, B.L., Ueda, Y., Craighead, N. and Huang, M.L. 1995. June CH. "CD28 ligands CD80 (B7.1) and CD86 (B7.2) induce long-term autocrine growth of $CD4^+$ T cells and induce similar patterns of cytokine secretion *in vitro*." *Int. Immunol.* 7: 891.

Liblau, R.S., Singer, S.M. and McDevitt, H.O. 1995. "T_H1 and T_H2 $CD4^+$ T cells in the pathogenesis of organ-specific autoimmune diseases. *Immunol. Today.* 16: 34.

Linsley, P.S., Greene, J.L., Brady, W., Bajorath, J., Ledbetter, J.A. and Peach, R. 1995. "Human B7.1 (CD80) and B7.2 (CD86) bind with similar avidities but distinct kinetics to CD28 and CTLA-4 receptors. *Immunity.* 2: 203.

Li-Weber, M., Giaisi, M. and Krammer, P.H. 2002. "The anti-inflammatory Sesquiterpene Lactone Parthenolide Suppresses Interleukin-4 Gene Expression in Peripheral Blood T Cells." *European Journal of Immunology.* 32: 3587.

Li-Weber, M., Giaisi, M., Treiber, M. K. and Krammer, P. H. 2002. "Vitamin E Inhibits Interleukin-4 Gene Expression in Peripheral Blood T Cells." *European Journal of Immunology*. 32: 2401.

Li, Y. 1999. "Blocking both signal 1 and signal 2 of T-cell activation prevents apoptosis of alloreactive T cells and induction of peripheral allograft tolerance." *Nat. Med.* 5: 1298.

Liu, Y. and Janeway, C.A. Jr. 1991. "Microbial induction of co-stimulatory activity for CD4 T-cell growth." *Int. Immunol.* 3: 323.

Lucas, P.J., Negishi, I., Nakayama, K., Fields, L.E. and Loh, D.Y. 1995. "Naive CD28-deficient T cells can initiate but not sustain an *in vitro* antigen-specific immune response." *J. Immunol.* 154: 5757.

Lenschow, D.J., Ho, S.C. and Sattar, H. *et al.* 1995. "Differential effects of anti-B7.1 and anti-B7.2 monoclonal antibody treatment on the development of diabetes in the nonobese diabetic mouse." *J. Exp. Med.* 181: 1145.

Levine, B.L., Mosca, J.D. and Riley, J.L. 1996. "Antiviral effect and *ex vivo* $CD4^+$ T-cell proliferation in HIV-positive patients as a result of CD28 costimulation." *Science*. 272: 1939.

Lewis, D.E., Tang, D.S., Adu-Oppong, A., Schober, W. and Rodgers, J.R. 1994. "Anergy and apoptosis in $CD8^+$ T cells from HIV-infected persons." *J. Immunol.* 153: 412.

Mancini, G., Carbonara, A.O. and Heremans, J.F. 1965. "Immunochemical quantitation of antigens by single radial immunodiffusion." *Immunochemistry*. 3: 235.

Matsushima, G.K., Gilmore, W., Casteel, N., Frelinger, J.A. and Stohlman, S.A. 1989. "Evidence for a subpopulation of antigen-presenting cells specific for the induction of the delayed-type hypersensitivity response." *Cell. Immunol.* 119: 171.

Mizutani, H., Ohyanagi, S.Y., Umeda, S., Shimizu, M. and Kupper, T. S. 1997. "Loss of cutaneous delayed hypersensitivity reactions in *Nevus anemicus*. Evidence for close concordance of cutaneous delayed hypersensitivity and endothelial E-selectin expression." *Arch. Dermatol.* 133: 617.

Meade, R., Askenase, P., Geba, K., Neddermann, G. Jacoby, R. and Pasternak, R. 1992. "Transforming growth factor-beta 1 inhibits murine immediate and delayed type hypersensitivity." *J. Immunol.* 149: 521.

Mihara, M., Ikuta, M., Koishihara, Y. and Ohsugi, Y. 1991. "Interleukin-6 inhibits delayed-type hypersensitivity and the development of adjuvant arthritis." *Eur. J. Immunol.* 21: 2327.

Miller, S.D., Vanderlugt, C.L. and Lenschow, D.J. *et al.* 1995. "Blockage of CD28/B7.1 interaction prevents epitope spreading and clinical relapses of murine EAE." *Immunity.* 3: 739–745.

Moretta, A., Bottino, C. and Vitale, M. 1996. "Receptors for HLA class-I molecules in human natural killer cells." *Annu. Rev. Immunol.* 14: 619.

Mosmann, T.R. and Moore, K.W. 1991. "The role of IL-10 in crossregulation of T_H1 and T_H2 responses." *Immunol. Today.* 12: 49.

Müller, I., Kropfe, P., Etges, R.J. and Louis, J. A. 1993. "Gamma interferon response in secondary *Leishmania* major infection: role of $CD8^+$ T cells." *Infect. Immun.* 61: 3730.

Müller, K., Jaunin, F., Masouyé, I., Saurat, J. and Hauser, C. 1993. "T_H2 cells mediate IL-4-dependent local tissue inflammation." *J. Immunol.* 150: 5572.

Nakajima, A., Azuma, M., Kodera, S. *et al.* 1995. "Preferential dependence of autoantibody production in murine lupus on CD86 co-stimulatory molecule." *Eur. J. Immunol.* 25: 3060.

Nandi, D., Gross, J.A. and Allison, J.P. 1994. "CD28-mediated co-stimulation is necessary for optimal proliferation of murine NK cells." *J. Immunol.* 152: 3361.

Natesan, M., Razi-Wolf, Z. and Reiser, H. 1996. "Co-stimulation of IL-4 production by murine B7.1 and B7.2 molecules." *J. Immunol.* 156: 2783.

Ogra, P.L., Leibovitz, E.E. and Zhao, R.G. 1989. "Oral immunization and secretory immunity to viruses." *Curr. Top. Microbiol. Immunol.* 146: 73.

Ohashi, P.S. 1991. "Ablation of 'tolerance' and induction of diabetes by virus infection in viral antigen transgenic mice." *Cell.* 65: 305.

Pape, K.A. 1998. "Direct evidence that functionally impaired $CD4^+$ T cells persist *in vivo* following induction of peripheral tolerance. *J. Immunol.* 160: 4719.

Qin, S. 1993. "'Infectious' transplantation tolerance." *Science.* 259: 974.

Rattis, F. M., Peguet-Navarro, J., Staquet, M.J., Dezutter-Dambuyant, C., Courtellemont, P., Redziniak, G. and Schmitt, D. 1996. "Expression and function of B7.1 (CD80) and B7.2 (CD86) on human epidermal Langerhans cells." *Eur. J. Immunol.* 26: 449.

Rathore, A., Sacristán, C., Ricklan, D.E., Flores Villanueva, P.O. and Stadecker, M.J. 1996. "*In situ* analysis of B7.2 co-stimulatory, major histocompatibility complex class II, and adhesion molecule expression in schistosomal egg granulomas. *Am. J. Pathol.* 149: 187.

Regan, F. and Taylor, C. 2002. "Blood transfusion medicine." *Br. Med. J.* 325: 143.

Riemann, H., Schwarz, A., Grabbe, S., Aragane, Y., Luger, T. A., Wysocka, M., Kubin, M., Trinchieri, G. and Schwarz, T. 1996. "Neutralisation of IL-12 *in vivo* prevents induction of contact hypersensitivity and induces hapten-specific tolerance." *J. Immunol.* 156: 1799.

Röcken, M., Racke, M. and Shevach, E.M. 1996. "IL-4- induced immune deviation as antigen-specific therapy for inflammatory autoimmune disease." *Immunol. Today.* 17: 225.

Saha, B., Das, G., Vohra, H., Ganguly, N.K. and Mishra, G. 1994. "Macrophage–T-cell interaction in experimental mycobacterial infection: selective regulation of co-stimulatory molecules on *Mycobacterium*-infected macrophages and its implication in the suppression of cell-mediated immune response." *Eur. J. Immunol.* 24: 2618.

Santamaria, L.F., Perez Soler, M. T., Hauser, C. and Blaser, K. 1995. "Allergen specificity and endothelial transmigration of

T cells in allergic contact dermatitis and atopic dermatitis are associated with the cutaneous lymphocyte antigen." *Int. Arch. Allergy Immunol.* 107: 359.

Santamaria Babi, L.F., Picker, V.M., Perez Soler, T., Drzimalla, K., Flohr, P., Blaser, K. and Hauser, C. 1995. "Circulating allergen-reactive T cells from patients with atopic dermatitis and allergic contact dermatitis express the skin-selective homing receptor, the cutaneous lymphocyte-associated antigen." *J. Exp. Med.* 181: 1935.

Shelley, W. B. and Juhlin, L. 1977. "Selective uptake of contact allergens by the Langerhans cell." *Arch. Dermatol.* 113: 187.

Schneider, H., Prasad, K.V., Shoelson, S.E. and Rudd, C.E. 1995. "CTLA-4 binding to the lipid kinase phosphatidylinositol 3-kinase in T cells. *J. Exp. Med.* 181: 351.

Schwartz, R.H. 1990. "A cell culture model for T lymphocyte clonal anergy." *Science.* 248: 1349.

Schwarz, A., Grabbe, S., Riemann, H., Aragane, Y., Simon, M., Manon, S., Andrade, S., Luger, T. A., Zlotnik, A. and Schwarz, T. 1994. "*In vivo* effects of interleukin-10 on contact hypersensitivity and delayed-type hypersensitivity reactions." *J. Invest. Dermatol.* 103: 211.

Seder, R.A. and Paul, W.E. 1994. "Acquisition of lymphokine-producing phenotype by $CD4^+$ T cells. *Annu. Rev. Immunol.* 12: 635.

Shahinian, A., Pfeffer, K. and Lee, K.P. 1993. "Differential T cell co-stimulatory requirements in CD28-deficient mice." *Science.* 261: 609.

Shelley, W.B. and Juhlin, L. 1977. "Selective uptake of contact allergens by the Langerhans cell." *Arch. Dermatol.* 113: 187.

Shreedhar, V., Giese, T., Sung, V.W. and Ullrich, S.E. 1998. "A cytokine cascade including prostaglandin E2, IL-4, and IL-10 is responsible for UV-induced systemic immune suppression." *J. Immunol.* 160: 3783.

Sieling R.A. and Modlin, R.L. 1994. "Cytokine patterns at the site of mycobacterial infection." *Immunobiology.* 191: 378.

Silverstein, A. M. 1989. *A History of Immunology*. Academic Press, San Diego.

Singer, G.G. 1994. "Apoptosis, Fas and systemic autoimmunity: the MRL-lpr/lpr model." *Curr. Opin. Immunol.* 6: 913.

Sloop, G.D. and Friedberg, R.C. 1995. "Complications of blood transfusion. How to recognize and respond to non-infectious reactions." *Postgrad. Med.* 98: 159.

Sperling, A.I., Linsley, P.S., Barrett, T.A. and Bluestone, J.A. 1993. "CD28-mediated co-stimulation is necessary for the activation of T cell receptor-gamma delta$^+$ T lymphocytes." *J. Immunol.* 151: 6043.

Steinman, R.M. 2000. "The induction of tolerance by dendritic cells that have captured apoptotic cells." *J. Exp. Med.* 191: 411.

Streilein, J. W., Toews, G. T., Gilliam, J. N. and Bergstresser, P. R. 1980. "Tolerance or hypersensitivity to 2,4-dinitro-1-fluorobenzene: the role of Langerhans cell density within epidermis." *J. Invest. Dermatol.* 74: 319.

Tietz, W. and Hamann, A. 1997. "The migratory behavior of murine CD4$^+$ cells of memory phenotype. *Eur. J. Immunol.* 27: 2225.

Uhr, J.W., Savin, S. B. and Pappenheimer, J. A. M. 1957. "Delayed-type hypersensitivity. II. Induction of hypersensitivity in Guinea pigs by means of antigen–antibody complexes." *J. Exp. Med.* 107: 109.

Ullrich, S.E., Pride, M.W. and Moodycliffe, A.M. 1998. "Antibodies to the co-stimulatory molecule CD86 interfere with ultraviolet radiation-induced immune suppression." *Immunology.* 94: 417.

Vanderlinde, E.S., Heal, J.M. and Blumberg, N. 2002. "Autologous transfusion." *Br. Med. J.* 324: 772.

Viraben, R., Aquilina, C., Cambon, L. and Bazex, J. 1994. "Allergic contact dermatitis in HIV-positive patients." *Contact Dermat.* 31: 326.

Voll, R.E. 1997. "Immunosuppressive effects of apoptotic cells." *Nature.* 390: 350.

Waksman, B.H. 1978. *Clinical Immunology.* Parker, C.W. (ed.) Saunders, Philadelphia.

Waksman, B. H. 1979. "Cellular hypersensitivity and immunity: Conceptual changes in the last decade." *Cell. Immunol.* 42: 155.

Walunas, T.L., Lenschow, D.J. and Bakker, C.Y. 1994. "CTLA-4 can function as a negative regulator of T-cell activation." *Immunity.* 1: 405.

Waterhouse, P., Penninger, J.M. and Timms, E. 1995. "Lymphoproliferative disorders with early lethality in mice deficient in CTLA-4. *Science.* 270: 985.

Webb, S., Morris, C. and Sprent, J. 1990. "Extrathymic tolerance of mature T cells: clonal elimination as a consequence of immunity." *Cell.* 63: 1249.

Wetzler, L.W., Ho, Y. and Reiser, H. 1996. "Neisserial porins induce B lymphocytes to express co-stimulatory B7.2 molecules and to proliferate." *J. Exp. Med.*183: 1151.

Widman Frances. 1997. "Early Observations about the ABO Blood Groups." *Transfusion.* 665.

Winkelstein, A. and Kiss, J.E. 1997. "Immunohematologic diseases." *JAMA.* 278: 1982.

Wu, Y., Guo, Y. and Liu, Y. 1993. "A major co-stimulatory molecule on antigen-presenting cells, CTLA-4 ligand A, is distinct from B7." *J. Exp. Med.* 178:1789

Xu, H., Heeger, P. S. and Fairchild, R.L. 1997. "Distinct roles for B7.1 and B7.2 determinants during priming of effector $CD8^+$ Tc1 and regulatory $CD4^+$ T_H2 cells for contact hypersensitivity." *J. Immunol.* 159: 4217.

Yong, A. J., Grange, J. M., Tee, R.D., Beck, J.S., Bothamley, G.H., Kemeny, D.M. and Kardjito, T. 1989. "Total and anti-mycobacterial IgE levels in serum from patients with tuberculosis and leprosy." *Tubercule.* 70: 273.

Web Sources

www.immunologylink.com/

www.jimmunol.org/

www.nature.com/ni/

immunology.org/

intimm.oxfordjournals.org/

www.aaaai.org/

www.acaai.org/

www.whfreeman.com/kuby/

cvi.asm.org/

www.medimmunol.com/

medind.nic.in/iac/iacm.shtml

www.mi.interhealth.info/

www.immunology.org.au/

www.nii.res.in/

www.annallergy.org/

www.immunology.klimov.tom.ru/

www.bsaci.org/

www.immunology.utoronto.ca/

www.cancerimmunity.org/

immunology.mc.duke.edu/

www.protocol-online.org/prot/Immunology/

Index

C

T